METHODS in MICROBIOLOGY

METHODS in

MICROBIOLOGY

Edited by
T. BERGAN
Department of Microbiology,
Institute of Pharmacy and Department of Microbiology,
Aker Hospital, University of Oslo,
Oslo, Norway

J. R. NORRIS
Agricultural Research Council,
Meat Research Institute,
Bristol, England

Volume 11

 1978

ACADEMIC PRESS
London · New York · San Francisco
A Subsidiary of Harcourt Brace Jovanovich, Publishers

ACADEMIC PRESS INC. (LONDON) LTD
24–28 Oval Road
London NW1

U.S. Edition published by
ACADEMIC PRESS INC.
111 Fifth Avenue
New York, New York 10003

Library of Congress Catalog Card Number: 68–57745
ISBN: 0–12–521511–8

PRINTED IN GREAT BRITAIN BY
ADLARD AND SON LIMITED
DORKING, SURREY

LIST OF CONTRIBUTORS

T. BERGAN, *Department of Microbiology, Institute of Pharmacy, and Department of Microbiology, Aker Hospital, University of Oslo, Oslo, Norway*

H. DIKKEN, *WHO/FAO Leptospirosis Reference Laboratory, Royal Tropical Institute, Amsterdam, The Netherlands*

R. GILLIES, *Department of Microbiology, The Queen's University of Belfast, Belfast, N. Ireland*

P. GUINEE, *National Institute for Public Health, Bilthoven, The Netherlands*

E. KMETY, *Institute of Epidemiology, Medical Faculty of the Komensky University, Bratislava, Czechoslovakia*

H. MILCH, *National Institute of Hygiene, Budapest, Hungary*

F. ØRSKOV, *WHO Collaborative Centre for Reference and Research on Escherichia, Statens Seruminstitut, Copenhagen, Denmark*

I. ØRSKOV, *WHO Collaborative Centre for Reference and Research on Escherichia, Statens Seruminstitut, Copenhagen, Denmark*

S. SLOPEK, *Institute of Immunology and Experimental Therapy, Polish Academy of Science, Warsaw, Poland*

W. TRAUB, *Institute for Klinische Mikrobiology und Infecktionshygiene, Universitat Erlangen-Nurnberg, D-8520 Erlangen, Wasserturmstr. 3. West Germany*

W. VAN LEEVWEN, *National Institute for Public Health, Bilthoven, The Netherlands*

PREFACE

Volume 11 of "Methods in Microbiology" continues the series of four Volumes which describe the methods available for typing the major human pathogens and which started with Volume 10.

Once more we have been fortunate to recruit authors who are international authorities on their subjects. Volume 11 deals with the important Gram-negative rods and various typing methods are described for the different groups of organisms. An opening Chapter discusses the serotyping of the family *Enterobacteriaceae* and this is followed by a similar Chapter on bacteriocin typing. Phage typing of *Escherichia coli*, *Salmonella*, *Klebsiella* and *Proteus* form the topics of separate Chapters. Bacteriocin typing of *Serratia marcescens* is described and finally the serological typing methods used for leptospiras. A chapter on phage typing of *Shigella* will appear in Volume 13.

We hope that Volume 11 will make a valuable contribution to the standardisation and optimisation of methods used for typing this important group of organisms and we would like to take this opportunity of expressing our gratitude to the individual authors who have made it possible for us to put together this compendium of methods.

T. Bergan
J. R. Norris

June, 1978

CONTENTS

CONTENTS OF PUBLISHED VOLUMES

CHAPTER I

Serotyping of *Enterobacteriaceae,* with Special Emphasis on K Antigen Determination

FRITS ØRSKOV AND IDA ØRSKOV

*WHO Collaborative Centre for Reference and Research on Escherichia,
Statens Seruminstitut, Copenhagen, Denmark*

I. INTRODUCTION

The main topic of this Chapter is extensively covered within text-books of bacteriology and in manuals of bacteriological laboratory techniques and has specifically been covered quite recently in several widely distributed excellent books and manuals.

For anyone who wishes to work in this field, "Identification of Enterobacteriaceae" by P. R. Edwards and W. H. Ewing (1972)—two outstanding pioneers in this field—is a must; the 3rd edition is from 1972 and a new edition authored by W. H. Ewing is expected soon. In German a similarly broad and detailed handbook and manual is "Enterobacteriaceae--Infektionen", edited by J. Sedlak and H. Rische, 2nd edition, 1968. Finally, the book by F. Kauffmann, "The Bacteriology of Enterobacteriaceae" 1966, can also be recommended, especially for the historically minded reader who wants a personal account from the man who has contributed most to the foundation of modern serotyping of Gram-negative intestinal bacteria.

The theme of this Chapter is therefore already extremely well covered and a repetition would be redundant if new and hopefully useful information was not included.

However, there are within the K antigens, and especially within the field of polysaccharide capsular and microcapsular antigens, several new developments of importance for serotyping and the main emphasis will therefore be placed on these developments.

A. The family *Enterobacteriaceae*

The *Enterobacteriaceae* is a genetically and physiologically quite homogeneous family of Gram-negative, facultatively anaerobic rods. Several members of the family have their natural habitat in the intestinal tract of animals and man. They constitute only a tiny fraction of the total number of bacteria in the gut, but play important roles, nevertheless, both in intestinal and extraintestinal diseases and for the immunological make up of the host.

Primarily because of its importance in epidemic diseases in animals and man subdivision of the family has been one of the favourite tasks of bacteriologists since early days of bacteriology and thus this group of bacteria can probably be subdivided into more highly refined subunits than any other family.

Definition of the family Enterobacteriaceae

Definition of the family *Enterobacteriaceae* as given in Bergey's Manual, 8th edition (1974): "Small Gram-negative rods; motile by peritrichate flagella or non-motile. Capsulated or non-capsulated. Not spore-forming; not acid-fast. Aerobic and facultatively anaerobic. Grow readily on meat extract media, but some members have special growth requirements. Chemo-organotrophic; metabolism respiratory and fermentative. Acid is produced from the fermentation of glucose, other carbohydrates and alcohols; usually aerogenic but anaerogenic groups and mutants occur. Catalase-positive with the exception of one serotype of *Shigella*; oxidase-negative. Nitrates are reduced to nitrites except by some strains of *Erwinia*. G + C content of DNA 39–59 moles %".

B. Subdivision of the family

Based on a limited number of metabolic traits and G + C% it is possible to subdivide the family into a few groups, in Bergey's Manual, 8th edition, five such tribes (Table I, Bergey 8.1).

These groups can, by inclusion of additional biochemical characteristics and morphological traits, be further subdivided into a number of subgroups or genera (Table II, Bergey 8.3).

Thus, the first step in the subdivision process is mainly based on the physiology of the organisms. However, it was realised at an early date that for epidemiological purposes further subdivision of the genera was necessary. It was also understood that an attempted further subdivision into biotypes based on fermentation or non-fermentation of additional carbohydrates would not bring much help to epidemiological investigations. However, it was at the same early date understood that the immunogenicity

TABLE I

Distinguishing characteristics of the five primary groups (tribes)

	Tribe I *Escherichieae*	Tribe II *Klebsielleae*	Tribe III *Proteeae*	Tribe IV *Yersinieae*	Tribe V *Erwinieae*
Fermentation pattern	Mixed acid	2,3-Butane- diol	Mixed acid	Mixed acid	Mixed acid and 2,3- butane- diol
M.R.	+	D	+	+	
V.P.	–	D	D	–	D
Phenylalanine deamination	–	–	+	–	D
Nitrate reduction	+	+	+	+	D
Urease	–	D	D	D	–
KCN, growth on	D	+	+	–	D
Optimal temp. for growth	37°C	37°C	37°C	30–37°C	27–30°C
G+C %	50–53	52–59	39–42	45–47	50–58

From "Bergey's Manual" (1974).

of the bacterial surface structures would constitute a solid base for a refined subdivision into serotypes. Later it was realised that even this level of refinement in many epidemiological situations did not offer the guidance needed. Therefore, some epidemiologically important serotypes have been further subdivided into biotypes, phage types, colicin types and antibiotic sensitivity pattern types.

In the following we shall primarily treat the principles and procedures for serotyping. In order to make the serotyping principles intelligible short descriptions of the morphology and immunochemistry of those surface structures which are important in serotyping will be given.

II. SURFACE STRUCTURES

A. Morphology

Several review papers have been published recently on the structure of the bacterial cell surface (Freer and Salton, 1971; Shands, 1971). Electron microscopy of enterobacteria shows a rigid double layered cell wall outside the cytoplasmic membrane. The rigidity is determined by the mucopeptide layer. Lipopolysaccharide (LPS) is found probably fixed by non-covalent bonds within this layer and at the same time covering it. The polysaccharide part of LPS contains the serological determinants for the O-antigens, the basis of *Enterobacteriaceae* serotyping. The mucopeptide layer and other deeper antigenic substances do not play any role in ordinary serotyping. Lipoproteins are also found in this region, but they are not involved in the routine serotyping process except as possible causes of O inagglutinability.

While these are basic structures found in all typical *Enterobacteriaceae*, additional structures important for serotyping exist, which are not necessary for the life and multiplication of the organisms and which are not found in all organisms. The structures in question are: (a) A polysaccharide, most often acidic, found as a capsule or microcapsule which probably to some extent covers the lipopolysaccharide (Lüderitz *et al.*, 1968; Jann and Westphal, 1975). The capsular substance can be more or less tightly bound to the surface and, if loosely bound and produced in great amounts, some may be characterised as slime. (b) Motile organisms have flagella. Because of the rich variety in chemical fine structures and because of the fundamental genetic stability (i.e. determination by chromosomal genes), the three structures, the lipopolysaccharide (O antigen), the polysaccharide capsule (K antigen) and the protein flagella (H antigen), make up the three fundamental serotyping antigens.

In addition to the above-mentioned structures many enterobacteria carry fimbriae or pili, thread-like protein structures protruding from the surface as seen in electron micrographs and probably arising in or close to the cell membrane. The word pili is mainly used for fimbriae which are of importance for the recombination process. Both fimbriae and pili are immunogenic and, though several antigenic variants have been described, they are not at present used for ordinary serotyping. They are found in many, probably all *Enterobacteriaceae* (Duguid *et al.*, 1966; Brinton, 1965). However, they do play an important, but poorly defined role during serotyping because of their disturbing interference, either since they can hinder an O, K or H reaction or because they can masquerade as K or H antigens. Finally, some special fimbrial structures which are extensively used for serotyping purposes should be mentioned. These *Escherichia coli* fimbria antigens, K88 (Ørskov *et al.*, 1961) and K99 (Smith and Linggood, 1971; Ørskov *et al.*, 1975a) are genetically determined by extrachromosomal DNA, i.e. plasmids, and play a special role in promoting adhesion to small intestinal epithelial cells in entero-toxic diarrhoeal disease in newborn pigs and calves; lately an as yet unnumbered structure of similar nature has been described in some human diarrhoeal strains (Evans *et al.*, 1975), and another one in some pig strains not having K88 (Isaacson *et al.*, 1977).

B. Chemistry

1. *Lipopolysaccharide (LPS)*

Several brilliant review papers on this theme exist (Lüderitz *et al.*, 1966;

TABLE II

Main biochemical characters of primary groups I to IV

	Group I					Group II				Group III	Group IV
	Escherichia	*Edwardsiella*	*Citrobacter*	*Salmonella*	*Shigella*	*Klebsiella*	*Enterobacter*	*Hafnia*	*Serratia*	*Proteus*	*Yersinia*
Catalase	+	+	+	+	D*	+	+	+	+	+	+
Oxidase	−	−	−	−	d	−	−	−	−	−	−
β-Galactosidase	+	−	+	D	−	+	+	+	+	−	+
Gas from glucose at 37°C	+	+	+	+	−	+	+	+	d	D	−
KCN, growth on	−	−	+	D	−	+	+	+	+	+	−
Mucate (acid)	+	−	+	D	−	d	d	−	−	−	−
Nitrate reduced	+	+	+	+	+	+	+	+	+	+	+
G+C, %	50–51	−	−	50–53	−	52–56	52–59	52–57	53–59	39–42 (1 species = 50)	45–47
Carbohydrates (acid from)											
Adonitol	−	−	−	−	−	d	+	−	d	D	D
Arabinose	+	−	+	+	d	+	+	+	−	−	+
Dulcitol	d	−	d	D	d	d	−	−	−	−	−
Esculin	d	−	d	−	−	+	D	−	−	d	D
Inositol	−	−	−	d	−	D	D	−	d	D	−
Lactose	+ or ×	−	+ or ×	D	D	+	+	−	−	−	−
Maltose	+	+	+	+	D	+	+	+	+	D	+
Mannitol	+	−	+	+	D	+	+	+	+	D	+

Salicin	d	−	−	d	+	+	+	−	d	D
Sorbitol	+	−	+	+	+	+	−	−	D	D
Sucrose	d	+	−	+	+	+	− or ×	+	d	+
Trehalose	+	+	+	+	+	+	+	+	D	D
Xylose	d	−	+	d	+	+	+	d	D	D
Related C sources										
Citrate	−	+	+	−	d	d	+	+	+	−
Gluconate	−	−	−	D	+	+	+	+	−	D
Malonate	−	D	D	−	D	D	+	−	D	−
d-Tartrate	d	+	D	−	d	d	−	−	d	D
Other reactions										
M.R.	+	+	+	+	D	D	−	−	+	+
V.P.	−	−	−	+	D	+	+	D	d	−
Protein reactions										
Arginine	d	−	+	−	−	−	D	−	−	−
Gelatin hydrolysis	−	−	D	+	D	(d)	(+)	−	D	−
H₂S from TSI	+	D	+	−	−	−	−	+	D	D
Indole	+	+	−	D	D	−	−	−	d	D
Lysine decarboxylated	+	−	+	+	d	d	+	+	−	−
Ornithine	d	d	+	d	−	D	+	d	−	D
Urea hydrolysed	−	(+)	−	−	d	+	(d)	−	D	D
Glutamic acid	−	−	−	−	−	−	−	+	+	−
Phenylalanine	−	−	−	−	−	−	−	−	−	−

* D = different reactions given by different species of a genus; d = different reactions given by different strains of a species or serotype; × = late and irregularly positive (mutative). From "Bergey's Manual" (1974).

1968, 1971; Jann and Westphal, 1975) and only a short description of the chemical structure of LPS and of the basis for its antigenicity will be given here.

Figure 1 shows the general stucture of a typical LPS, i.e. that of a *Salmonella* strain belonging to O group B. This complex macromolecule consists of phosphorylated, long chain heteropolysaccharides linked covalently to a glucosamine-containing lipid: lipid A. Lipid A is the determinant for most of the endotoxic capacity of LPS, while the poly-saccharide side chain is the chemical basis for the serological classification of *Enterobacteriaceae* into hundreds of complex O antigen groups, i.e. *Salmonella* into more than 60 and *Escherichia coli* into more than 160 such O groups. The most outlying part of the polysaccharide chain consists of repeating oligosaccharide units joined by glycosidic bonds. This part, the O side chain, is covalently linked to the basal core polysaccharide which in *Salmonella* contains the core oligosaccharide and the heptose-KDO backbone.

All *Salmonella* strains have an identical or very similar basal core, but in other genera like *Escherichia* other basal core structures in addition to that found in *Salmonella* can be found. The inner part of the basal core polysaccharide, the heptose-KDO backbone, is linked to the lipid A. The R antigens (see later) which generally are not used for serotyping purposes have their chemical basis in the core polysaccharide. However, when many independent *Salmonella* R mutants are examined it will be found that they can be grouped chemically (and serologically) into five chemotypes, Ra, Rb, Rc, Rd and Re, each of them corresponding to a mutational step which leads to a simpler core chain. Thus, the LPS of the R forms contains in its side chain only the innermost part of the polsaccharide chain found in the O form. When hidden, as it normally is in the O form, the R side chain is not, or is only weakly, immunogenic and does not react with possible R antibody.

The LPS of smooth forms can in its O specific side chain contain many different sugars, some of which are different from those found in the basal core polysaccharide (e.g. hexoses, 6-deoxyhexoses, 3–6 dideoxy-hexoses, amino sugars and neuramic acids). The sugars are generally arranged as repeating units like those found in the *S. typhi* (O antigen 9, 12) side chain or in a *Salmonella* from the E group (O antigen 3, 10), or in *E. coli* O86 (Fig. 2).

Since the discovery that serological specificity was correlated with small structural units in the side chains, an understanding of the background for the overwhelming diversity among O antigens of *Enterobacteriaceae* has now been achieved. At the same time the many cross-reactions within the genera, between genera and between enterobacteria and other bacteria can be simply explained.

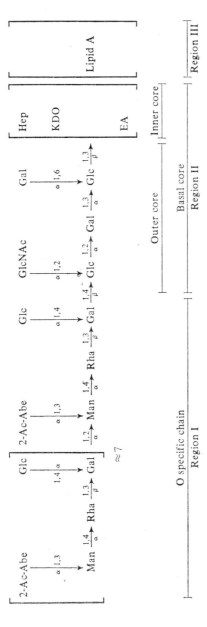

FIG. 1. Structure of lipopolysaccharide of *Salmonella typhimurium* according to Lüderitz (1970) and Lüderitz *et al.* (1973). modified (from Jann and Westphal, 1975). Ac = acetyl, Abe = abequose, Man = mannose, Rha = rhamnose, Gal = galactose, Glc = glucose, GlcNAc = N-acetylglucosamine, Hep = heptose, KDO = 3-deoxy-D-mannooctulonic acid, EA = ethanolamine.

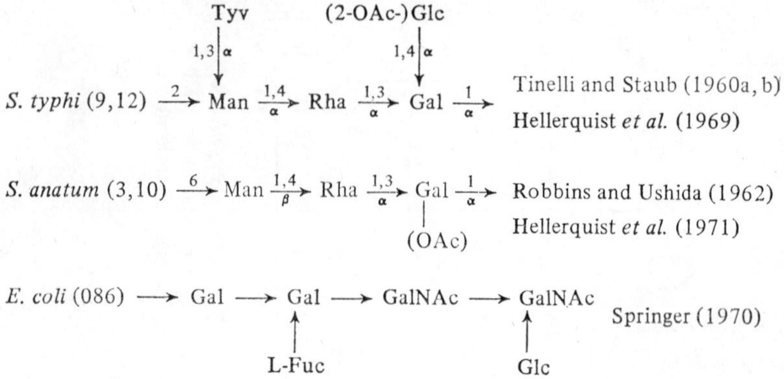

Tyv   (2-OAc-)Glc

$S.\ typhi\ (9,12)$ $\xrightarrow{2}$ Man $\xrightarrow[\alpha]{1,4}$ Rha $\xrightarrow[\alpha]{1,3}$ Gal $\xrightarrow[\alpha]{1}$ Tinelli and Staub (1960a, b)

Hellerquist et al. (1969)

$S.\ anatum\ (3,10)$ $\xrightarrow{6}$ Man $\xrightarrow[\beta]{1,4}$ Rha $\xrightarrow[\alpha]{1,3}$ Gal $\xrightarrow[\alpha]{1}$ Robbins and Ushida (1962)

(OAc)   Hellerquist et al. (1971)

$E.\ coli\ (086)$ $\longrightarrow$ Gal $\longrightarrow$ Gal $\longrightarrow$ GalNAc $\longrightarrow$ GalNAc

Springer (1970)

L-Fuc   Glc

FIG. 2. Structure of O specific polysaccharide (repeating unit) from three O antigenic lipopolysaccharides (from Jann and Westphal, 1975). Tyv = tyvelose, OAa = O-acetyl, Glc = glucose, Man = mannose, Rha = rhamnose, Gal = galactose, GalNAc = N-acetylgalactosamine, L-Fuc = L-fucose.

Chemotypes. Most O specific polysaccharides isolated from *Entero-bacteriaceae* strains contain five or more sugars. Based on sugar patterns it is possible to establish chemotypes of which more than 50 different types are known from examination of *Salmonella* and *Escherichia* strains (for a discussion of the current concepts of chemotypes, see Jann and Westphal, 1975). It is characteristic that strains from one O serogroup contain the same sugars; however, many examples can be found where O antigens belong to the same chemotype but do not cross-react serologically. In such cases the single sugars must be linked differently, undetected constituents or conformational differences must exist. Complete understanding of the background for serological specificity is therefore only possible after chemical structural analysis. More and more such analyses have been carried out, a few examples of which have been given above. With modern chemical techniques it will be possible to explain any of the O antigen factors. Lüderitz, Staub and Westphal (1966) summarised their review with the following statements: (1) The serological specificities demonstrable on the surface of the bacterial cell by agglutination, i.e. the O factors of the Kauffmann–White scheme, are carried by specific polysaccharides. (2) The different O factors of a given *Salmonella* serotype are not part of different molecules, but they are carried by the same polysaccharide molecule.

2. Capsular polysaccharides (K antigen)

Capsular and microcapsular polysaccharides are found in many genera of *Enterobacteriaceae* (for reviews, see Lüderitz *et al.*, 1968; Jann and Westphal, 1975). In one genus, viz. *Klebsiella*, they are so well developed and found so constantly that serological typing is based on the capsular antigen. Klebsiellae were also the first capsulated *Enterobacteriaceae* described and the carbohydrate nature of the capsule was also described first in this genus by Toenniessen (1920). However, very similar capsules can be found in other genera, e.g. *Escherichia*, *Enterobacter* and *Serratia*. The term capsule is most often used for an external substance which covers the complete surface of the bacterium and which is bound so strongly to other surface structures that it is not solubilised completely in fluid medium —this last qualification to keep slime substances out of the definition. Many text-books would also include in the definition that the capsule can be made visible in the microscope. These two requirements are difficult to maintain as immunochemically identical polysaccharides can be found in otherwise identical serotypes as (1) capsule, (2) slime, (3) mixtures of capsule and slime, or (4) microcapsules which can hardly be detected in the microscope but only by serological or chemical means. In fact, all capsular substances are solubilised to some extent from the living bacteria and probably both the magnitude of the natural solubilisation of this substance and the quantity produced are genetically determined. We would therefore use the term capsule polysaccharide when a surface polysaccharide is detected which is independent of the lipopolysaccharide. We would not include the proteinaceous fimbria or fimbria-like antigens (K88 or the like) among capsule antigens, though they may constitute a veritable fur around the bacterium such that they concur with the definition of K antigens. One important characteristic of polysaccharide capsules, already mentioned above, is the quantitative variations found within the same capsular type. In some instances from one and the same *E. coli* or *Klebsiella* strain, capsule type variants (mutants) may be isolated concomitantly. In one variant the capsule is easily detected in the light microscope, whereas microcapsules (less developed capsular K antigens) in the other variants are only detectable by serological or chemical techniques, in some instances even by serology alone.

Most capsular polysaccharides isolated from *Enterobacteriaceae* are linear polysaccharides of high molecular weight containing acidic groups such as hexuronic acids, neuraminic acids or phosphate. For example, the Vi antigen of *Salmonella*, first described by Felix and Pitt (1934), is a homopolysaccharide consisting of D-galactosaminuronic acid (Clark *et al.*, 1958). Other examples are the *E. coli* K42 antigen which consists of galacturonic acid, galactose and fucose in molar ratio 1:1:1 (Jann *et al.*, 1965) and the *E. coli* K1 antigen which is a polyneuraminic acid (Barry and Goebel, 1957; Kasper *et al.*, 1973).

The considerations about the relationship between chemical structure and serological factors, the immunodeterminant group, discussed above under the O antigens, are in principle also true for the capsular polysaccharides. For details, the reader is referred to recent reviews mentioned above.

3. Flagellar protein

By mild acid treatment *Salmonella* flagella yield a homogeneous, soluble protein, named flagellin, of about 40,000 molecular weight (Weibull, 1950).

4. Proteinaceous surface structure

Fimbriae, fimbria-like antigens and sex pili have already been mentioned above (page 5).

Other surface components interfering with serotyping undoubtedly exist. Some of these are also immunogenic, but as they are not used for general serotyping purposes, and as their chemistry is unknown, they will not be discussed here (see page 30).

C. Genetics

It is impossible in this Chapter to cover the extensive literature concerning the genetics of *Enterobacteriaceae* antigens. The interested reader should consult a number of review papers (Mäkelä and Stocker, 1969; Stocker and Mäkelä, 1971; Iino and Lederberg, 1964; Iino, 1969; Ørskov et al., 1977).

III. SEROLOGICAL TECHNIQUES

A. Bacterial agglutination

1. General considerations

Bacterial agglutination is defined as clumping of bacteria caused by a specific antiserum. The method is simple and has been used since the early days of bacteriology for identification within many bacterial families. It is still the fundamental method for refined antigenic subdivision of the *Enterobacteriaceae*. The Kauffmann–White scheme covering the serological types of *Salmonella* is the most brilliant example of how successfully this agglutination and agglutination absorption technique can be applied.

The agglutination test is carried out either in tubes, in plastic trays or on glass slides and consists of mixing bacteria suspended in saline (0.15M sodium chloride) with appropriate dilutions of antiserum. At neutral pH, most *Enterobacteriaceae* carry a negative surface charge which, together with a number of other physical factors, is responsible for their ability to remain evenly suspended for long periods. Higher salt concentrations may modify this negative charge, thus causing agglutination without the presence of antibody. Contrarily, lower salt concentrations or distilled water can be used to counteract spontaneous agglutination as seen with rough mutants (no O polysaccharide side chains in LPS), which at the same time lack acidic polysaccharide capsules. If the salt concentration in the mixture is too low, the agglutinability of smooth cultures in their specific antisera may also be impaired.

It should be emphasised that bacterial agglutination can be used quantitatively only with difficulty and without great precision. It is a qualitative or, at the most, semi-quantitative procedure. Where the primary antigen–antibody reaction occurs on the surface of the comparatively large bacterial body and that the reaction is the secondary clumping of these corpuscles.

It should also be stressed that agglutination is only as simple as it sounds in cases where one is certain that not more than one antigen–antibody system is involved. One such very important case is the routine diagnosis of *Salmonella* O antigens. The sera used for this purpose are O sera, i.e. rabbit sera generally produced by immunisation with culture heated at 100°C for 2½ h. From a practical point of view such antisera contain only O antibodies. Agglutination reactions found in these sera could be considered as O antigen reactions as long as no other thermostable, highly immunogenic antigen is left on the bacterial surface after heating at 100°C for 2½ h. This is true for *Salmonella*, *Shigella* and many *Escherichia coli* strains, but not for heavily capsulated strains like *Klebsiella* and certain *E. coli*. In these cases the O sera may contain detectable amounts of antibody against the polysaccharide K antigens and varying amounts of O antibody at the same time.

The situation is quite a bit more complex with so-called OK sera, i.e. sera produced with a live or formalinised culture. Such sera, e.g. typical *E. coli* OK sera, contain antibodies against most of the known and unknown structures found on the surface. If examined with homologous live culture, the agglutination observed is the combined consequence of several independent antigen-antibody systems. If the same antiserum is used for the examination of an unknown culture, and if an agglutination is observed, it will not be simple to demonstrate whether one or more possible surface antigens of the unknown culture are involved. With the aid of appropriately absorbed sera (page 38) and/or by using several OK sera produced with different known combinations of the antigens involved it will, however, in many cases be possible to reach a reliable typing result. It is possible by certain cultivation procedures to inhibit the development of a special surface structure (*vide infra*) and thereby eliminate its interference with agglutination, e.g. to inhibit the development of the *E. coli* fimbria-like proteinaceous K88 antigen by culturing at 18°C (page 32).

2. *Slide agglutination*

Slide agglutination is an extremely simple, sensitive and for many purposes useful method. Most simply a small amount of culture from an agar plate is transferred and suspended by a straight platinum wire into a droplet of diluted serum on a microscopic slide. The slide is rocked by hand and after a short time, most often 10 (or 20) sec, the agglutination is read by the naked eye supported by the aid of a hand lens. Preferably it should be read against a black background using some sort of indirect light. Instead of direct suspension from an agar plate culture, heavy suspensions of bacteria can also be used, transferred either by loop or pipette. When not working with highly pathogenic bacteria wooden tooth picks can, with great saving of time, replace the straight wire, also making the use of a Bunsen burner superfluous. Tooth picks can be autoclaved and used many times without previous cleaning.

Slide agglutination is primarily used to screen many colonies in a limited number of sera, e.g. for common *Salmonella* and *Shigella* serotypes. Frequently the slide agglutination test needs confirmation by an agglutination titration under more standardised conditions in tubes or trays.

Slide agglutination can be much influenced by the ratio between number of bacteria and antibody content and, especially in cases where only low titre sera are available, negative reactions may be replaced by clear-cut positive reactions by a reduction in the amount of bacteria used; this is very likely the result of a prozone phenomenon. Another pitfall hidden in this technique which, however, could also be looked upon as a special asset is caused by the variability in the reaction between single colonies: Variation within a culture which cannot be detected by examination of a fluid culture or of confluent growth from an agar plate can be revealed by slide agglutination of several single colonies. A negative slide agglutination based on the examination of one single colony or of the mass culture may also be misleading because of this variation phenomenon, e.g. from variation of some of the *Salmonella* O antigens and the similar variation found in the *E. coli* K1 antigen (page 33).

3. *Tube agglutination*

Tube agglutination or agglutination in plastic trays can be used both as a primary screening method, e.g. when heated cultures of e.g. *E. coli* strains are examined in pooled O antisera (see below), and for titration. Most frequently similar amounts (e.g. 0·2 ml) of bacterial suspensions and serum dilutions are mixed followed by incubation at the recommended temperature and time.

(*a*) *Titration*. The agglutination technique is widely used as a titration assay, generally in tubes, but nowadays quite often in plastic trays. To a series of tubes each containing e.g. 0·2 ml antiserum diluted in two-fold dilution steps the same amount of a bacterial suspension of a certain density is added; most often 10^8 to 10^9 organisms per ml is suitable. It is advisable to use a standard density of bacteria. Depending on the antigens examined, the mixture is incubated at 37°C or 50°C for a defined period of time (pages 20, 25, 27). The titre which indicates the relative strength of the serum is expressed as the reciprocal of the highest dilution causing agglutination which can be seen with the naked eye. Due to slight variation in dilutions, agglutination titres are not precise ($\pm$ one titre step). A comparison of the titres of an unknown and the homologous strain in a test serum with known potency is a rough guide to the relationship between the two antigens. However, it should be borne in mind that this is only true in rather well defined systems like the O system. In agglutination systems involving unheated bacteria and OK sera it is extremely difficult to draw any conclusions from the titres about the closeness of a possible antigenic relationship of two strains. It is for example a well-known fact already demonstrated by Felix for the Vi antigen (1938) and by Vahlne for *E. coli* (1945) that capsulated bacteria will give a lower titre than mutants of the same strain with less capsular substance in one and the same homologous OK serum. Likewise, an actively motile culture may give a higher titre than a less motile variant of the same strain when examined in one and the same H serum.

(*b*) *Prozone phenomenon*. From a theoretical point of view, no agglutination would be expected to occur when antibody molecules are in great excess relative to antigenic determinants on the bacterial surface because the simultaneous binding of the two binding sites of antibody molecules to different bacteria would be unlikely. In this situation a tube titration may show no agglutination in the tubes with low serum dilutions but increased agglutination in tubes with higher dilutions. This phenomenon, called the prozone effect, is not a great problem in routine *Enterobacteriaceae* serotyping, but, it is not at all uncommon when one is working with live (or formalinised) bacteria and OK sera, as in K antigen typing of *E. coli*.

(*c*) *Macroscopic appearance of agglutinates*. A few words should be said about the macroscopic appearance of agglutinates. They may look very different from one another and text-books often state that O agglutinates are "granular", H5 agglutinates are "floccular" and capsular antigen agglutinates are "disc-like". These descriptions are only true with strong qualifications. O agglutinates are granular and difficult to disperse by shaking when boiled suspensions are examined in O sera. H agglutinates are more fluffy looking, like very loose cotton balls when highly motile cultures are examined in specific sera that contain no antibodies against other surface structures of the agglutinated organism. But, as mentioned above, if live or formalinised cultures—and in fact also boiled cultures—are assayed in OKH sera it is difficult to tell anything about the antigens and antibodies involved from the appearance of the agglutinates. Fimbria-like antigens, e.g. K88, give distinct floccular agglutinates which, however, can easily be disrupted to fine granules. Special agglutinoscopes have been developed. The authors have no personal experience with such agglutinoscopes which presumably are of special value when using a microtitre system in plastic trays.

4. *Automated typing methods*

The typing procedure, especially of O and H antigens, can be automated, thus saving labour. Naturally, such automation is most advantageous in genera where many single antigenic factors have been found. The interested reader is referred to papers by Bettelheim *et al.* (1975) and Guinée (1972) who have both developed such systems.

B. Indirect haemagglutination

Indirect or passive haemagglutination has not been much used for serotyping of *Enterobacteriaceae*. However, it has been widely used for detection of antibodies against such bacteria and also in immunochemistry as a quantitative measure for antigens. The principle is that certain antigenic macro-molecules will bind to the surface of mammalian erythrocytes, most frequently human blood group O cells or sheep erythrocytes. Either extracts of bacteria treated in different ways or purified substances are used. In the classical technique described by Neter *et al.* (1952) for *E. coli*, boiled extracts of bacteria are employed. After heat treatment of the culture or after treatment of the purified lipopolysaccharide with alkali (Neter *et al.*, 1956) it will bind to the erythrocyte surface. When the strain—and the extract—also contains a capsular polysaccharide this too will bind to the surface. In experiments with OK sera, the agglutination reaction observed will be a mixed O and capsular K antigen reaction. If a non-boiled extract is used, the capsular K antigen will usually fix to the erythrocyte surface. Ordinarily, proteinaceous antigens in the unheated extracts will not coat the red cell surface. However, a tannic acid treatment of erythrocytes developed by Boyden (1951) has made it possible to coat red cells with protein antigens.

Passive haemagglutination is most frequently carried out in plastic trays and the settling pattern of the red cells at the bottom of the cups is used as an indication of the degree of agglutination obtained. Latex particles have been used in the same way as red cells (Kabat and Mayer, 1961).

C. Immunofluorescence

This technique has not been much used for general serotyping of *Enterobacteriaceae*. However, it can be useful as a screening method for a few serotypes, such as the special *E. coli* O groups (O111, O55, etc.) which are associated with infantile diarrhoea in children's institutions (Le Minor *et al.*, 1962), and has also been used for detection of *Salmonella* (Thomason *et al.*, 1959). The basic principle is that antibodies coupled to fluorescent dykes like fluorescein or rhodamine will keep their specificity and thus after binding to the bacterial cell can be visualised in the fluorescence microscope. For serotyping purposes the test can be carried out in two ways:

(1) Direct test: Antibacterial rabbit antibody is directly conjugated with the fluorochrome and applied directly to e.g. a smear from faeces.

(2) Indirect test: Antibacterial rabbit serum is applied to the smear and the antibody specifically bound to bacteria is demonstrated by the application of fluorochrome conjugated antirabbit immunoglobulin.

The reader is referred to special articles about details of these techniques (Beutner, 1961; Le Minor *et al.*, 1962; Walker *et al.*, this Series, Vol. 5A).

D. Capsular quellung reaction

The capsular quellung reaction is another old technique, originally developed by Neufeld (1902) for type determination of pneumococci. It is a rather time-consuming, but useful and precise technique for serotyping of heavily capsulated bacteria, e.g. many *Klebsiella* strains (for details, see later under *Klebsiella*).

E. Gel precipitation

Until recently, precipitation techniques have not been used for serotyping of *Enterobacteriaceae*. With the development of gel precipitation techniques and with the increased interest in distinguishing between different antigens on the bacterial surface such techniques have acquired increasing importance. For serotyping bacterial extracts are more useful than purified antigens. Simple extracts, produced by suspension of overnight 37°C growth from broth agar plates in saline ($10^{10}-10^{11}$ bacteria per ml), can be recommended. The suspension is heated at 60°C for 20 min in a water-bath followed by centrifugation at 20,000 g for 20 min. Half of the supernatant is used as "60°C extract", while the other

2

half is heated at 100°C for 1 h and labelled "60°/100°C extract". The 60°C extract will contain lipopolysaccharide, a possible acidic polysaccharide K antigen, and different protein antigens, while the 60°/100°C extract from a serotyping point of view will only contain the lipopolysaccharide and acidic polysaccharide antigens.

1. Gel precipitation in two dimensions

Gel precipitation in two dimensions can be carried out according to the original description by Ouchterlony (1958), but for many purposes a simple method with filter paper discs (Ørskov and Ørskov, 1970) can be used: Filter paper discs, 6 mm in dia., are dipped into antigen or serum solution until soaked with fluid. The soaked discs are placed in any chosen pattern drawn on a piece of paper placed underneath a 14 cm dia. plastic Petri dish containing ca. 2 mm layer of agar. A high quality agar, or agarose recommended for gel precipitation, should be used and 0·065% sodium azide added. The Petri dishes are placed in a plastic bag and incubated at 37°C overnight followed by room temperature for 3–4 days. The plates will not dry out for several weeks. The precipitation results are recorded by drawing after 24 h and repeatedly during the following days. The plates are not easy to stain, but since the method is primarily a qualitative one, we feel that drawings give as much, or perhaps more information than classically stained preparations. A still simpler, and at the same time more gentle way of preparing an antigenic extract is to place the unused filter paper discs directly on confluent growth on an agar plate. With forceps the discs are moved to another unused location of the confluent growth. This procedure may be repeated two or three times. Finally, the bacteria-soaked filter paper discs are transferred to the gel precipitation plate. In order to increase the amount of antigens released from the bacteria the broth agar plate with bacterial growth and filter paper discs may be placed in an oven, e.g. for 1 h at 60°C (Ørskov et al., 1976).

2. Immunoelectrophoresis

Immunoelectrophoresis, as described by Grabar and Williams (1953), is a most useful technique when the problem is to distinguish between antigens which move differently in an electric field. It has been extremely useful for differentiation between lipopolysaccharide and other polysaccharide antigens in Enterobacteriaceae. In our laboratory the micromethod developed by Scheidegger (1955) is used. In this Chapter we discuss on several occasions the problems involved in the classical agglutination technique when applied to non-heated bacteria agglutinated by OK(H) sera. Using immunoelectrophoresis and the 60°C and 60°/100°C extracts described above, one will often obtain a clearcut separation and thus an easily interpretable result. For a definite diagnosis of capsular acidic polysaccharide antigens in e.g. *E. coli* we believe at present that immuno-electrophoresis is indispensable.

3. Crossed immunoelectrophoresis

A further development of immunoelectrophoresis was conceived by Laurell (1966) who added a second electrophoretic step to the first separation of differently charged molecules. The second step is carried out perpendicularly to the first step and into an antibody containing gel. This technique, which makes quantitative studies possible, has many variations by which it is possible to make direct examination for identity or, elegantly, to make absorption directly on the electrophoresis plate. For very precise and easy-to-follow accounts of the many possibilities of this technique the reader is referred to two recent books by Axelsen *et al.* (1973) and Axelsen (1975).

Crossed immunoelectrophoresis will probably find its main role in more refined serotyping of acidic polysaccharide K antigens.

4. Counter-current immunoelectrophoresis

Another technique which has been developed for examination of capsular antigens such as those of *Neisseria meningitidis* and *Haemophilus influenzae* has also been adapted for the determination of acidic polysaccharide K antigens of *Enterobacteriaceae*. The principle is that in a gel subject to an electric potential difference the negatively charged acidic polysaccharide will move towards the anode, while the precipitating antibodies will move towards the cathode. The antigen and the antibody are therefore filled into two wells close to each other in the direction of the current, in such a position that they will meet in high concentrations in a narrow area between the wells during the electrophoretic run. This method, which has only very recently been adapted to *E. coli* polysaccharide K antigens, seems to offer a fast and reliable way of typing for this group of antigens. The first step in the procedure is to produce suitable extracts. For *E. coli* polysaccharide K antigens, the 60°/100°C extracts described above will be suitable; however, a suspension heated to 100°C can be used directly without centrifugation (Semjen *et al.*, 1977). The use of such boiled extracts or suspensions may be difficult in the very few cases where the polysaccharide K antigens are partly destroyed or changed by heating, e.g. the K1 antigen. To simplify the procedure, the unabsorbed OK(H) sera are pooled with 4 sera in each pool, each serum thus being diluted 1: 4. The extract is first used at a dilution of 1: 200 or higher; if this test is negative, lower dilutions of 1:100 or 1: 50 may be tried. If 1: 50 or less diluted antigens are used, possible precipitations will often occur very close to or in the serum well. The results may be read within 1 h. In crossed immunoelectrophoresis and counter-current electrophoresis, a high quality agarose (e.g. Litex agarose, Litex, Glostrup, Denmark) is used instead of agar.

IV. THE ANTIGENS

A. The O antigens

The lipopolysaccharide (LPS) O antigens are heat-stable so that boiled bacteria are suitable for immunisation, agglutination, and agglutinin absorption. For serological purposes, smooth colonies are selected that are not spontaneously agglutinable in saline. Cultures are heated at 100°C for $2\frac{1}{2}$ h for O antiserum production and similar cultures, formalinised or heated at 100°C, are used for tube agglutination. Result is read after incubation at 50°C overnight. However, in routine bacteriology of *Salmonella* and *Shigella* slide agglutination with O sera and unheated agar plate cultures are often used. With non-heated cultures, whether in tube or slide agglutination, it is important to remember that many surface structures may inhibit O agglutination. For examination of O antigens of heavily encapsulated cultures such as *E. coli* A forms or most *Klebsiella* strains, even prolonged boiling will not destroy the O inhibitory capacity of the capsule. In such cases it is necessary either (1) to isolate K^- mutants or (2) to heat the culture at 120°C for 2 h. The simplest and often most efficient way to isolate K^- mutants from mucoid cultures is to let the agar plate, which has grown overnight at 37°C, stand for one or more days at room temperature (ca. 21°C). At this time non-mucoid sectors will often appear at the edge of colonies still growing slowly at this low temperature; such K^- sectors will be directly agglutinable in O serum and give typical O agglutinations after heating at 100°C for 1 h. Occasionally, non-mucoid sectors will consist of mutants which produce less capsule substance, and which are only agglutinable in O serum after heating at 100°C for 1 h. The agglutination obtained with a mucoid culture heated at 120°C for 2 h will often be weak, but generally sufficient for O grouping purposes.

B. The K antigens

In order to understand the nature of what we would describe today as the K antigens of *Enterobacteriaceae* it will be necessary to give a short historical introduction.

1. *Discussion of presently used terminology (L, A and B antigens)*

The currently used definitions of K antigens in *Enterobacteriaceae* have been based to a large extent on principles laid down by Kauffmann in the forties (1966). Many years before the term K antigen was introduced, antigens belonging to this category had been described. Probably the *Klebsiella* capsular antigens (Toenniessen, 1915; Julianelle, 1926), the Vi antigen of *Salmonella* (Felix and Pitt, 1934) and some mucoid capsular antigens of *E. coli* (Smith, T. and Bryant, 1927; Smith, T., 1927; Smith, D., 1927) were the first antigens described belonging to this group. They were not however described as K antigens, a term introduced later by Kauffmann and Vahlne (1945) whose definition of the K antigens was based exclusively on bacterial agglutination experiments. When Kauffman (1943) began his serological examination of *Escherichia* strains he very soon found that many strains, especially from pathological processes, were "O inagglutinable", i.e. they were not agglutinable in the living state in the homologous O serum—a situation well known from Vi antigen-equipped *Salmonella typhi* strains. He showed that this O inagglutinability could be overcome by heating at 100°C for 1–2 h. Kauffmann further demonstrated that the agglutinin binding capacity of these new *E. coli* surface antigens was destroyed by heating at 100°C and, since this was in contrast to the Vi antigen, he coined the new term L antigen (L for labile) for such antigens. Another subgroup of K antigens was found in O inagglutinable strains in which agglutinin binding capacity of the antigen was not destroyed by heating; these were called B antigens (Knipschildt, 1946). *E. coli* strains which were O inagglutinable even after heating at 100°C for 2½ h were designated A forms by Kauffmann (1944). Knipschildt (1945a) showed that such strains possessed an antigen which he called A antigen. This antigen was characterised by a greater heat stability; 120°C for 2 h was required to overcome the inagglutinability in homologous O serum. This type of K antigen was similar to the capsular antigens of carbohydrate nature found in mucoid *E. coli*, which were thoroughly described by Theobald Smith and his co-workers in 1927 (see above). In Table III we have collected the fundamental experimental data, all based on bacterial agglutination experiments on which Kauffmann and his collaborators defined the three K antigen varieties, L, B and A (Kauffmann, 1966).

The table is necessarily a simplified one. Unheated culture, for example, will often show some agglutination, say a titre of 80 in O serum in case of both L and B antigens. But the important trait is that the unheated culture is what is called O inagglutinable, i.e. the live culture gives a significantly lower titre than the boiled culture in the homologous O serum; live culture is agglutinable in OK serum. It is seen that the fundamental difference between L and B antigens is that the absorption of an OL serum with boiled bacteria will leave some agglutinins for the live culture, in contrast to the B antigens where nothing is left in an OB serum after absorption with boiled

TABLE III

Schematic presentation of the agglutination results on which hitherto used
definitions of K antigens (L, B, A) are based

		OK serum		
Antigen	O-serum	Absorbed by culture heated at 100°C 1 h	Unabsorbed	
Live (or formalinised)	" – "	+	+	L
Boiled (100°C 1 h)	+	—	+	
Live	" – "	—	+	B
Boiled	+	—	+	
Live	—	—	+	A
Boiled/autoclaved	—/+	—	+	

" – " = negative or significantly lower than that of the boiled culture.
— = no agglutination.
+ = agglutination.

culture. The difference between A and B is that A antigen culture requires
heating at 120°C for 2 h before it agglutinates in homologous O serum.

The above-described definition of *E. coli* L, B and A antigens and the
establishment of the first more than 50 K antigens was published in the
forties and the establishment of further K antigens according to these
principles continued until recently. However, it is striking that very few
reports exist where general K determination using all K typing sera has
been carried out. Based on our earlier experience, one explanation could be
that general K determination according to the above principles is often
very difficult to carry through to a reliable result. As mentioned previously,
OK sera, i.e. antisera produced by immunisation with unheated cultures,
usually contain antibodies against several different surface structures.
Therefore, the fundamental L agglutination, i.e. that of live culture in
the absorbed OK serum, will often, just as the agglutination in the unabsorbed OK serum, be a consequence of the sum of several serologically
different structures.

The typical L carrying strain has, in addition to the LPS, an acidic
polysaccharide envelope antigen and furthermore flagella and most likely
fimbriae and other undefined antigens. For the absorption procedure the

culture is heated at 100°C and much of the acidic polysaccharide and the LPS will be solubilised. When gel precipitation or immunoelectrophoresis were introduced to reach a better and more precise understanding of K antigens, three non-motile strains containing L antigen were examined (Ørskov and Ørskov, 1970). It was impossible to absorb undiluted OK antiserum (use of undiluted serum is necessary for gel diffusion techniques) with boiled homologous culture without removing all precipitating antibody to the acidic polysaccharide. By bacterial agglutination and indirect haemagglutination it could be demonstrated that the agglutinating antibodies to the acidic polysaccharide K antigens were also removed. This was particularly the case when the absorbed serum was undiluted, while the same agglutinins were only incompletely removed when an antiserum 1: 10 diluted was absorbed. These findings have been confirmed later by examination of additional strains. Thus, L antigen is solubilised but is chemically stable at 100°C. Sufficient L antigen is usually left on the cell surface to preserve the agglutinin binding capacity. It thus seems possible to "change" a classical L to a B antigen by introduction of additional methods for demonstration of the antigen.

It remains to explain the antigenic background for the L antigen as defined by Kauffmann (1966). According to our present knowledge, all strains described by Kauffmann as L strains contain acidic polysaccharide antigens independent of the LPS O antigens. When L-carrying cultures are heated at 100°C for 2 h, and used for absorption according to Kauffmann (1966), i.e. with OK serum diluted 1: 10, enough antibody is often left to give a reduced agglutination titre. OK sera absorbed in this way, however, will also contain antibodies to other structures like flagella and fimbriae and, possibly, fimbria-like antigens. Such a true thermolabile antigen (not H) was found for example in the test strain of antigen K12 in addition to the polysaccharide K12 antigen (Ørskov and Ørskov, 1970). These thermolabile structures may be developed to a varying extent depending on cultural conditions, and thus the antibody content against them may also differ, even in antisera produced against the same strain.

As a probable effect of the existence of these "uncontrolled" additional antigens, it is not unusual that the same strain examined in two laboratories is called L in one laboratory and B in the other, or that the same strain examined at different times in the same laboratory may be labelled either L or B. We therefore feel that the time has come to redefine these K antigens.

The *Escherichia* A antigens belong to the acidic polysaccharide K antigens. Most frequently they are found in heavily capsulated strains and they are in many respects similar to the *Klebsiella* K antigens. Because of their close association with a few O groups (O8, O9 and O101) and because of genetic characteristics (Ørskov and Ørskov, 1962; I. Ørskov et al., 1976) and chemical structure (Jann and Westphal, 1975) they have a special position within this group of antigens. Vahlne (1945) described the possibility of isolating less heavily capsulated variants from the typical A forms. Such mutants were inagglutinable in the living state in O serum, but heating at 100°C for 1 h abolished this inagglutinability. The boiled bacteria would remove all anti-A antibodies from an OK serum and such mutants thus behaved exactly like B antigens (see Table III).

The B antigens, B1, B2 and B3, described by Knipschildt (1946) are the only officially numbered B antigens against which it has been possible to produce a pure K serum. This was obtained by absorption of OK anti-serum with a heterologous strain with the same O antigen and a different K antigen. The three test strains have the following formula: O2: K56 (B1): H7 (H17B), O8: K25 (B2): H9 (Bi7575/41) and O9: K57 (B3): H32 (H509d).

Recent examinations (partly unpublished) make it reasonable to class K56(B1), K25(B2) and K57(B3) with the K antigens as defined in this Chapter (see below). Like the K antigen strains described above, these three strains thus have an acidic polysaccharide K antigen and the differences described by Knipschildt do not warrant the establishment of a special B category. The serological differences between L and B antigens (see Table III) were not always present and were probably caused by minor differences in their tendency to solubilise during heating. As already mentioned, it has been a common experience that a strain could be labelled either L or B, dependent on time and/or laboratory. Likewise, some test strains originally designated A antigens to-day should be named B antigens because they now are O agglutinable after boiling, if the old definition were followed. The higher B numbers like K58(B4), K59(B5) and K60(B6) found in O111: K58(B4), O55: K59(B5) and O26: K60(B6) and in fact most of the other K antigens hitherto described as B, nearly all of which are strains from diarrhoea in animals and man, should not be considered as K antigens to-day. Most of these B antigens are always found in the same OK combination and it has proved impossible to produce a pure K58 serum because O antigen 111 was never found in other K combinations. These B antigens were solely designated on the single criterion that the living cultures were totally or partly inagglutinable in the homologous O sera and at the same time gave agglutination in an OK serum. It has never been possible in immunoelectrophoretic examinations to detect any poly-saccharide K antigen in most of the E. coli test strains for K antigens of the B type (Ørskov et al., 1971; Ørskov and Ørskov, 1972). Chemical examination of such strains has also supported the idea that they do not contain a special acidic polysaccharide (Lüderitz et al., 1968). Similar

K antigens of the B type were described within several *Enterobacteriaceae* genera. They have never been closely examined genetically or chemically and, until further confirmation is produced, we will advance the hypothesis that most such B antigens are not morphological, genetical or chemical entities which can be separated from the LPS O antigen.

2. *Proposals for new K antigen terminology*

Based on the discussion above we propose that all acidic polysaccharide capsular (microcapsular or envelope) antigens which are independent of the O specific LPS are named K antigens or polysaccharide K antigens, a terminology in accordance also with that used for several other bacterial groups. If necessary, the designation K(A) may be used to give the information that this strain has a capsular antigen possessing certain special physical and other characteristics.

A certain group of established K antigens are of protein nature. Only few of these exist, viz. *E. coli* K88 and K99, which are similar in many respects. Both are genetically determined by plasmids and both play a special role causing bacterial adhesion to the epithelial lining of the small intestine of young animals, K88 in piglets and K99 in calves and lambs. According to the rules laid down by Kauffmann, such K antigens are typical L antigens (see Table III) which are actually heatlabile and, when necessary, could have the notation K88(L) and K99(L), where the (L) indicates the special labile (protein) nature of these antigens. Especially in cases where both an acidic polysaccharide and a protein K antigen are found in the same strain, e.g. O8: K87, K88(L): H19, a common serotype from piglet diarrhoea, such a notation, which is already widely adopted, will be useful; another possibility would be to call these protein antigens (P) e.g. K88(P). If it will be deemed advisable in the future to number further labile surface antigen, e.g. fimbrial antigens, they should probably also have a similar notation.

Although most of the examples pertain to the *Escherichia* group, most of the above statements are also applicable to the other *Enterobacteriaceae* groups due to close similarities for the whole family. In Table IV some fundamental characters of established K antigens are summarised.

3. *K antigen determination*

The polysaccharide K antigens were originally determined by slide or tube agglutination of live or formalinised cultures in OK sera (see pages 12–16, 20–25). Tube agglutination times are read after incubation for 2 h at 37°C followed by room temperature overnight. In well controlled systems, e.g. when a few, well examined, strains and sera are involved, slide agglu-

TABLE IV
Established *Enterobacteriaceae* K antigens

K antigen	Genus	Designation	Genetic basis	Morphology	Serological technique suggested (see text)
	Escherichia	K1 to K103*			Slide agglutination,‡ gel precipitation
Polysaccharide (acidic)	*Salmonella*	Vi	Chromosomal	Capsules (including microcapsules)	Slide agglutination‡
	Klebsiella	K1 to K82†			Slide agglutination,‡ quellung reaction
	All genera	M			Slide agglutination‡
Protein	*Escherichia*	K88(L) and K99(L)	Plasmid	Fimbria-like protein threads	Slide agglutination‡

"K1" (= α antigen) and "K2", both from Providence group (*Proteus inconstans*), "K" antigens from *Serratia*, Vi antigen from *Citrobacter freundii*, "K" antigen of *Shigella flexneri* 6 = *Sh. newcastle* and other K antigens under consideration have not been included.

* Several K numbers deleted (see Table V).

† Some K antigens doubtful (see page 63).

‡ See text about the possible errors inherent in the slide agglutination technique when used with non-heated culture and OKH sera.

tination can be a useful and simple procedure. Furthermore, slide and tube agglutination can be used occasionally with success for more general K typing purposes, but the result should be confirmed by a gel precipitation technique such as immunoelectrophoresis or counter-current immunoelectrophoresis (see pages 18, 19). Heavily capsulated *Enterobacteriaceae* like *Klebsiella* can also be examined with advantage by the capsular swelling technique (see page 17). The heat-labile K antigens of the *E. coli* K88 and K99 type may be examined by slide agglutination in OK sera produced preferably with non-motile strains, unabsorbed or absorbed by boiled culture. When unabsorbed OK(H) sera are employed, use of two different antisera produced with two strains both containing K88 (or K99) but differing in O, polysaccharide K and H antigens is recommended. As these antigens are not developed at 18°C, cultures grown at this temperature are useful as controls.

Hindrance of flagellar development on agar plates by different chemicals may be important during K agglutination procedures (see pages 34, 35).

C. The H antigens

The H antigens are the serological determinants of the protein flagellar organs found in motile organisms. Usually they are completely destroyed by heating at 100°C for 1 h. Slide or tube agglutination is used for their determination (see page 14). It is highly important for a successful determination that all the organisms are actively motile. In some *Enterobacteriaceae*, e.g. *Salmonella*, it is generally no problem to obtain such actively motile cultures, but in others like *Escherichia* many strains are only sluggishly motile. *Salmonella* cultures will often have well developed flagella when grown on ordinary agar plates, while e.g. only some *E. coli* strains will be actively motile under such conditions. However, it is usually possible to get well flagellated bacteria by passages of the organisms through semisolid agar media. We do not use U-tubes (Kauffmann, 1966) but instead the following, more simple procedure: With a straight wire a 3–4 cm long stab is made in an ordinary tube containing 10 cm semisolid agar (Ørskov and Ørskov, 1975). After incubation at 37°C, or probably better at 30°C (Chandler and Bettelheim, 1974; Aleksic *et al.*, 1973) for 20 h, most motile cultures will have spread down close to the bottom of the tube. With a Pasteur pipette a 2–3 cm long agar cylinder is withdrawn from an area close to the front edge of the advancing culture. The pipette is withdrawn and one or two drops of its contents delivered to a tube with broth (5–7 ml), care being taken not to dip the pipette into the broth, as its outside is covered by culture from the surface growth in the semisolid agar tube. The broth culture is incubated at 37°C or 30°C for 5–6 h, at which time motility may be checked under the microscope. If all organisms are actively motile, the culture will be suitable after the addition of 0·5% formaldehyde. Generally, one passage will be sufficient. When this is not the case, two or more passages in semisolid agar may be helpful. Other methods for obtaining motile cultures can be found in Edwards and Ewing (1972). H agglutination occurs rapidly and is therefore read after incubation in a water-bath at 50°C for 2 h thereby avoiding interference by the late developing O agglutination. The typical agglutinates are loose and fluffy but, like agglutinates of other surface antigens, may be strongly influenced in shape and structure by other antigens present. Phase variation occurs in strains of the *Salmonella* group (see page 57, and for details Edwards and Ewing, 1972, and Kauffmann, 1966).

D. Other surface antigens

For serotyping of *Enterobacteriaceae* usually only the O, K and H antigens are used. We have already mentioned several times that other surface antigens exist. Most frequently these antigens have been detected because they interfered in some way or other with routine antigen determination. Some of the best described of these antigens are listed below. Generally, little is known about their chemical nature, but most of them are probably protein in nature.

Fimbriae (pili) are thread-like structures projecting from the bacterial surface, first described by Anderson (1949) and thoroughly examined by Duguid and his co-workers who have also written several papers on the serology of fimbrial antigens and their ability to cause direct haemagglutination (Duguid, 1959; Duguid and Campbell, 1969; Duguid et al., 1955, 1966; Gillies and Duguid, 1958; Brinton, 1965). Fimbriae are to be found in all Enterobacteriaceae genera. Their development is dependent on growth conditions. They are best developed in fluid cultures and fimbriation is here often associated with pellicle formation; however, they are often well developed on ordinary broth agar plate media. It is our experience that most Salmonella serotypes show direct haemagglutination (type 1 fimbriae) from ordinary broth agar plates, also after several passages on such plates (unpublished observation). This is in contrast to general descriptions in the literature. The capacity to cause direct haemagglutination of washed guinea pig erythrocytes is so closely associated with fimbriation that it can be used for detection at least of type 1 fimbriae.

Several different types of fimbriae have been described based on morphology and other characteristic traits (Duguid et al., 1966). Duguid and his co-workers (see Duguid and Campbell, 1969) also examined the antigenic properties of type 1 fimbriae. A common fimbrial antigen was found in all Salmonella serotypes examined and in certain serotypes there were probably also other fimbrial antigens present. In Shigella flexneri, E. coli and Klebsiella aerogenes a common fimbrial antigen, different from that in Salmonella, was found. The Shigella flexneri strains contained in addition a "flexneri-specific antigen".

A more systematic examination of antigenic properties of fimbriae of Enterobacteriaceae has not been carried out, but it is apparent that there are several different fimbrial antigens within Enterobacteriaceae and many cross-reactions between fimbriae from different genera. The same strain may possess several fimbrial antigens simultaneously. At present much interest in serotyping is centred on the possible interference of fimbriae with accepted agglutination procedures for the determination of H and K antigens. As fimbrial antigens are detached though not totally denatured at 100°C, even O antisera may contain fimbrial antibodies. Gillies and Duguid (1958) mention that the typical fimbrial agglutinate develops rapidly and is loosely floccular. Anti-fimbrial sera are high-titred. Fimbriae tend to mask the O antigens and confer relative O inagglutinability. If examined according to the test hitherto used for subdivision of E. coli

K antigens into L, B and A antigens (Table III), fimbrial antigens would be labelled L antigens. In our experience it is difficult, by agglutination procedures, to differentiate between agglutinations caused by fimbriae alone and those due to collaboration with other surface antigens (see discussion above under K antigens of the L type). Gillies and Duguid (1958) found that fimbrial antigens bear important resemblances to several previously described cross-reacting antigens, in particular to the X antigen (Topley and Ayrton, 1924) but also to the α antigen of Stamp and Stone (1944) and the β antigen of Mushin (1949, 1955) (see below).

E. "Common" antigens

Although some of these antigens have been mentioned previously, they are listed again under this heading because they belong to a group of antigens which serologically, due to their abiquity, have one important character in common, viz. they interfere with the normal serotyping process.

1. *Polysaccharide antigens*

(a) *Kunin antigen.* The Kunin antigen, often called the common antigen or CA, was detected by the indirect haemagglutination technique by Kunin *et al.* (1962) and is only one of several common antigens. In fact, it rarely interferes with ordinary typing procedures, since it is not immunogenic in strains having an O specific polysaccharide. The *E. coli* strain generally used for production of CA antiserum, i.e. the test strain for *E. coli* antigen O14, is, when examined today, devoid of an O specific polysaccharide chain in its LPS. The Kunin antigen, which is found in most, if not all *Enterobacteriaceae* strains, is normally hidden by the O polysaccharide chain (for details, see Edwards and Ewing, 1966).

(b) *M antigen.* This polysaccharide antigen has been described under many different names: M antigen (Kauffmann, 1935, 1936), capsular antigen (Markovitz, 1964), colanic acid (Goebel, 1963), slime wall antigen, mucous antigen and several others.

This group of closely related antigens has been examined serologically by Henriksen (1949, 1950) and Kauffmann (1966). The genetics has been examined in great detail by Markovitz (1964). Variational phenomena, especially the influence of media and growth temperature, were investigated by Anderson and Rogers (1963) (see page 33). For information on chemical composition, see Lüderitz *et al.* (1968) and Garegg *et al.* (1971). The M antigen is found in many genera of *Enterobacteriaceae*.

2. Thermolabile antigens (proteins)

From our own experience we know that non-immunised rabbits may often contain gel precipitating antibodies to one or more of at least four different thermolabile antigenic determinants found in most *E. coli* strains examined. The same antigens are found in several other *Enterobacteriaceae* genera. After immunisation with any unheated *Escherichia* strain the precipitation lines will generally be stronger from sera which were positive before immunisation and they will also be found in previously negative rabbits. It is likely that some of these common thermolabile precipitating antigens are similar to some of those mentioned below (unpublished observations).

(*a*) α *antigen* (Stamp and Stone, 1944). α agglutinins can be found in unimmunised rabbits but mostly after immunisation. The agglutinating and agglutinin binding capacity is destroyed by heating at 100°C for 15 min. α titres are often high (1 : 20,000 or more). For details, see Edwards and Ewing (1972).

(*b*) β *antigen* (Mushin, 1949, 1955). This antigen is similar in some respects to the α antigen of Stamp and Stone. The agglutinating and agglutinin binding properties are destroyed by heating at 100°C for 1 h. Both α and β antigens are widespread among *Enterobacteriaceae*, see Edwards and Ewing (1972).

(*c*) *Fimbrial antigens* (see pages 5, 28). Several fimbrial antigens within the *Enterobacteriaceae* group have been described, but the number of different antigens is probably rather limited and their occurrence so frequent that they should be listed among the common antigens. Fimbrial antigens may cause O inagglutinability.

(*d*) *ATA antigen* (anodic thermolabile antigen). Seltmann (1971), Seltmann and Reissbrodt (1975) and Larsson *et al.* (1973) have described a thermolabile antigen which in immunoelectrophoresis moves towards the anode. It is probably a glycoprotein and has been detected in all *Enterobacteriaceae* strains examined and furthermore in *Pseudomonas* (Seltmann and Reissbrodt, 1975).

V. VARIATIONAL PHENOMENA OF IMPORTANCE FOR SEROTYPING

It is not possible to write about serotyping without stressing the great influence different variational phenomena can have on the outcome of such examinations. A thorough description is given by Edwards and Ewing (1972) and Kauffmann (1966) and only an annotated list of the most important phenomena will be given here. The long history of *Entero-bacteriaceae* bacteriology has had its influence—sometimes more confusing than illuminating—on the terminology used to cover some of these observations.

A. Lipopolysaccharide O antigens

1. Smooth (S) to rough (R) variation

Smooth (S) to rough (R) variation might be called O+ to O- variation. This type occurs in all genera and is caused by mutational events which block different steps in the synthesis of the O specific polysaccharide chain or of the basal core (see page 8). The typical rough strain is spontaneously agglutinable in physiological saline, but may be stable in lower salt concentrations. It is devoid of the typical O antigen. However, many strains exist that are only partly rough, either because they are a mixture of smooth and rough variants or because the enzymatic block is incomplete. Therefore, it is of the utmost importance for O antigen determination, and indeed for all serotyping procedures, to select smooth looking colonies whenever subcultures are prepared. It should be kept in mind that mutants exist which have lost their O specificity, but have retained their capsular antigen, frequently look like smooth colonies and are stable in saline suspension when the cells are not boiled, but autoagglutinable after boiling.

2. S-T-R variation

S-T-R variation, first described by Kauffmann (1956, 1957) concerns a mutational phenomenon affecting the LPS synthesis of certain *Salmonella* serotypes (see page 55).

3. Form variation

Form variation, first described by Kauffmann (1940) in *Salmonella*, is a *non-mutational* variation involving certain O antigenic factors manifested by different degrees of agglutinability (from ± to + +) of single colonies from the same strain in homologous O antiserum; the changes are reversible. For details, see the description by Kauffmann (1966) and Edwards and Ewing (1972) and also page 55. A probably related phenomenon has been described in *E. coli* antigen O141 by Ørskov *et al.* (1961) (at the time when this form variation phenomenon was described it was considered to be a variation in the K85 antigen, an antigen which today is regarded as part of the O antigen (see Table V)). In *Shigella flexneri*, similar phenomena have been described, see Edwards and Ewing (1972).

4. Phage induced changes in O antigens, antigenic (lysogenic) conversion

Genetic material may be transferred from one strain to another by transduction, lysogenisation and conjugation. These processes influence analysis of the antigens. Antigenic (lysogenic) conversion, i.e. modifications of the O antigens caused by infection with a lysogenic phage is particularly important for the understanding of differences between certain *Salmonella* O antigens (see page 55, Edwards and Ewing, 1972; Sedlak and Rische, 1968; Le Minor, 1968a). Similar changes in the O antigens of *Shigella flexneri* have been described by Ketyi *et al.* (1971) and Gemski *et al.* (1975) and they probably occur in all *Enterobacteriaceae* genera (see also Stocker and Mäkelä, 1971).

5. *Influence of composition of growth media*

Schlecht and Westphal (1966) showed that an increase in glucose content will be followed to a certain extent (0·3 to 0·4%) by a relative increase in the amount of LPS produced by *Salmonella*. It is our experience that the O inagglutinability of some *E. coli* will be decreased, i.e. living bacteria will be more agglutinable in O serum when grown on media with such high glucose content (unpublished observation).

6. *Influence of growth temperature*

As far as we know, no systematic studies on the influence of growth temperature on the development of the O antigen characters have been carried out. Ørskov *et al.* (1961) demonstrated that two types of O antibodies were elicited in rabbits with 100°C treated cultures of *E. coli* strain G7 (O8: K87: H19) after growth at 18°C, but only one type after conventional growth at 37°C; for a discussion of this special phenomenon see Ørskov *et al.* (1977).

B. K antigens

1. *Polysaccharide K antigens*

(*a*) K^+ *to* K^- *variation.* K^+ to K^- variation, earlier sometimes called KO to O variation, is determined by mutational blocks in the synthesis of the polysaccharide K antigen. This type of mutational event occurs in all genera where strains with polysaccharide capsules or microcapsules are found, e.g. *Salmonella* (Vi antigen), *Klebsiella* and *Escherichia*. It is characteristic that intermediate forms containing decreasing amounts of K antigen can often be isolated. It is also characteristic that reversion of K^- to K^+ is a rare event, normally not observed. The K antigen is one of the factors that can make a strain inagglutinable in O serum and therefore, when the K antigen is lost (K^+ to K^- mutation), the non-heated strain will often show stronger agglutination in O serum. It should be remembered that several other surface components such as flagella and fimbria may also be responsible for O inagglutinability. K+ organisms may often be selected by their inagglutinability in O serum, but one may also take advantage of the colonial morphology since the K+ colonies are generally more opaque and dome-shaped than the corresponding K⁻ forms. When it is important to select K⁻ forms, it is advantageous to let agar plates with well separated colonies stand for several days at 21°C (room temperature). The colonies will continue to grow and often produce acapsular mutant sectors which are easy to isolate.

(b) *Form variation.* A variational phenomenon of the *Salmonella* Vi antigen has been described by Kauffmann (1966). A phenomenon similar to that found in O antigens was detected in *E. coli* K1 antigen by Ørskov *et al.* (1971, 1978b).

(c) *Influence of composition of growth media.* No systematic studies have been carried out, but it is a common experience that capsular development may vary on different media and also that the capsule is best developed on rich media with a high sugar (glucose) content. For the special polysaccharide K antigen, the M antigen (see page 29), Anderson and Rogers (1963) have shown that solid media with concentrations of solutes giving a high osmotic pressure (either single or different mixed simple salts or sucrose will do) will cause many, normally non-mucoid, *Enterobacteriaceae* to produce abundant amounts of mucous substance.

(d) *Influence of growth temperature.* For a long time it has been known that growth temperature influences the formation of *Salmonella* polysaccharide Vi antigen. This antigen is developed optimally at 37°C in *S. typhi*, but only poorly or not at all at 20°C (Felix *et al.*, 1934). Jude and Nicolle (1952) made more extensive studies and confirmed that the Vi antigen in *S. typhi* and *S. paratyphi* C was best developed at 37°C, whereas the similar Vi antigen in Ballerup strains (now *Citrobacter freundii*) and *E. coli* had its optimal development at 18°C. Ørskov *et al.* (in preparation) investigated the development of polysaccharide K antigens of *E. coli* and found that many of these are less well, or not at all developed at 18°C, while others are produced at both 18° and 37°C. The polysaccharide M antigen found in several *Enterobacteriaceae* can be developed at 37°C, but usually better so at room temperature.

2. *Plasmid determined K antigens*

(a) *K+ to K⁻.* The plasmid determined antigens, such as *E. coli* K88 antigen, can be lost when the plasmids are lost and can only be acquired again after reinfection with the plasmids.

(b) *Influence of growth media.* The *E. coli* K99 antigen, often associated with strains from calf and lamb diseases, can be difficult to detect because of the richly developed polysaccharide capsules often present in such strains. Experience has shown that some media are better than others for demonstration of this antigen (Ørskov et al., 1975; Guinee et al., 1976, 1977).

(c) *Influence of growth temperature.* The *E. coli* K88 and K99 antigens are not developed at 18°C, a characteristic which can be very useful for serotyping (Ørskov et al., 1961, 1975). This fact is also important to bear in mind when agar plates have been kept at room temperature for some time following incubation; under such circumstances separated colonies will continue to grow and the temperature sensitive K antigens will not be formed. However, it will frequently be possible on the same plate to detect these antigens in the areas of confluent growth, where little subsequent growth has occurred.

C. H Antigens

1. H^+ to H^- variation

H^+ to H^- variation has traditionally been called H to O variation. The historical explanation for the designation of these antigen changes is that some motile *Proteus* cultures growing as thin films on agar plates were named by the German word "Hauch" meaning "breath". The designation O comes from the German word "ohne" meaning "without" (ohne Hauch). This term was first applied to non-swarming variants of *Proteus*. H^+ to H^- variation is a mutational event which changes a motile strain to a non-motile one, most often caused by loss of the capacity to form flagella. Unfortunately, no simple routine technique exists for the selection of H^- mutants.

2. *Phase variation* (see page 57)

3. *Influence of growth media*

Serotyping of flagellated strains involves two particular problems. One is the growth of sufficiently motile strains so that abundant flagellar antigens are present for dependable H antigen determination and good production of anti-H-antibodies. This may be solved by procedures such as passage through semisolid agar or growth on swarm agar (Gard, 1938) (see page 27 and Edwards and Ewing, 1972). The other problem involves the prevention of developing flagella which will disturb the serological examination of other surface antigens when live or formalinised cultures and OK(H) antisera are employed. Different chemicals have been used as additives to agar media to stop swarming and the development of flagella, e.g. 0·1% phenol or short ethanol treatment of the agar surface. We have been satisfied with the addition to the agar medium of 0·1% of the anion active detergent Pril or 0·01% of a similar substance (Maranil). Motile *Escherichia* cultures grown on such plates generally show no H agglutination.

D. Other antigens

1. *Fimbrial antigens*

According to the literature, development of fimbriae depends on growth conditions (see e.g. Duguid and Campbell, 1969). However, it is our experience that it is not easy to obtain non-fimbriate cultures, even if the directions given in the literature are followed. For further details and references, see Edwards and Ewing (1972).

2. *M antigen*

Influence of media and growth temperature, see page 33.

VI. ANTISERA

A. Production of antisera

1. *General procedures*

In general, the production of *Enterobacteriaceae* antisera in rabbits is simple. The methods used are similar for all genera, though they may vary somewhat from author to author (Kauffmann, 1966; Sedlak and Rische, 1968; Edwards and Ewing, 1972). On the whole, the sera obtained against O and polysaccharide K antigens are equally useful for bacterial agglutination and gel precipitation.

We shall here give the main guidelines for production of O, K and H antisera with whole-cell vaccines. Where special procedures are used, mention will be made in the sections dealing with the single genera.

An important first step in antiserum production is to select suitable strains, culture media and incubation conditions in which the relevant antigen is well developed. Two or more rabbits are injected intravenously in the marginal ear vein with the antigen preparation, usually five or six injections with a 5-day interval. The rabbits are given increasing doses (e.g. 0·25, 0·5, 1·0, 1·5, 2·0 and 2·5 ml) of the antigen preparation containing 5×10^8 to 5×10^{10} bacteria/ml and exsanguinated a week after the last injection. If the antibody titre is not satisfactory after the sixth injection, no further injection is recommended. To preserve the antisera, we add both merthiolate and chloroform at a final concentration of 0·01% and 1% respectively. Some workers use equal volumes of glycerol—a preservation method which cannot be recommended if the antiserum is to be employed for gel precipitation tests. If it is important to avoid additives, the sera should be kept at −20°C or lower temperatures.

2. O antisera

To produce O antisera, it is important that the antigen culture is smooth (S) and not rough (R) or on the verge of becoming R, because presence of R agglutinins in the antisera may lead to confusing cross-reactions. Care must be taken to inspect and examine single colonies for smooth appearance and for stability in suspension after heating at 100°C. A rough culture may look smooth on an agar surface if it has retained a K antigen and this fact may only be disclosed by autoagglutinability of the culture after boiling.

O antigens are generally immunogenic after heating at 100°C, in contrast to H and K antigens, except e.g. *Klebsiella* K antigens and the *E. coli* K(A) antigens. Therefore, O antisera are prepared against cultures heated at 100°C for $2\frac{1}{2}$ h, either as a suspension from an agar surface or as a broth culture incubated at 37°C for 18–20 h. Higher doses than usual can be given if the LPS (endotoxin) solubilised by heating is removed by centrifugation. Cultures with K antigens of heat-stable immunogenicity should be autoclaved at 121°C for 2 h to prevent the production of K antibodies. When possible, however, K⁻ forms should be selected for O antiserum production. The boiled O antigen preparation may be preserved with formalin, as it is to be used for all vaccinations.

O antisera may also be produced according to the method of Roschka (1950). In this method the sediment of a heated bacterial suspension is treated with alcohol and acetone. For details, see Edwards and Ewing (1972).

Aleksic *et al.* (1973) have recently described a method which, according to these authors, gives pure *Salmonella* O antisera. Some O antigen factors, such as O2 and O5 (see page 55), are partly destroyed by heating at 100°C and it is therefore highly desirable, the authors say, to produce pure O antisera with non-heated whole bacteria. Aleksic *et al.*, grow a culture at 43°C for 20 h, wash it repeatedly and treat it in a mixer in order to remove all remnants of flagella. After repeated washings, 1% formalin is added. We have no personal experience with this method which gives remarkably high titres. However, we will warn against the use of unheated cultures for O antiserum production. Also when so-called O forms (no H antigen) are used for immunisation because such sera are likely to contain antibodies directed against other surface structures in addition to O antibodies. More comparisons between the conventional method using heated culture and the one proposed with unheated culture are required.

Finally, we might direct the interested reader to a systematic study of the influence of various parameters on the production of *Salmonella* O antibodies (agglutinating, haemagglutinating and precipitating), the results of which are very likely also valid for other *Enterobacteriaceae* O antigens (Schlecht and Westphal, 1967, 1968a, b).

3. *OK antisera*

For OK antiserum production, rabbits are immunised with unheated cultures, either a suspension from an agar plate or a young broth culture. Prior to immunisation the culture should be examined for presence of K antigen, if its identity is known. If only the O antigen is known, colonies, with the least possible agglutination in O antiserum should be selected; although such O inagglutinability is no sure proof of K antigen presence, it is a useful aid in selecting K+ forms. If available, non-motile cultures are preferable.

Many authors recommend that formalin-killed antigen preparations be injected at the beginning of the immunisation course and live suspensions thereafter. The antisera thus produced may have higher titres than those obtained using formalinised suspensions alone. However, it should be emphasised that live vaccine may be difficult to administer in a completely aseptic manner and, unless special precautions are taken, live bacteria may spread in the animal house.

Since OK antisera are produced with non-heated cultures, they contain antibodies additional to those required. They always contain O antibodies and, in case of motile strains, varying amounts of H antibodies, even if Pril supplemented plates are employed for cultivation (see page 35); therefore, the designation OK(H) is frequently used. H antibodies may be disturbing, but no certain method is known which can prevent some formation of H-specific immunoglobulins. However, the antibodies causing most trouble are those elicited by inadequately defined heat labile bacterial constituents, e.g. different types of fimbriae (see page 28).

4. *H antisera*

For production of H antisera preparations of actively motile cells inactivated by formalin should be used as antigen. Insufficient motility may be improved by passage through semisolid agar. The preparation may be a young broth culture or a suspension from an agar plate which must be freshly prepared and moist before use. Aleksic *et al.* (1973) recommend suspensions prepared from semi-solid agar plate cultures grown for not more than 16 h at 30°C. By this change from the usual 20 h at 37°C, they obtained H sera with four times higher titres. Fey and Wetzstein (1975) have also modified the methods in order to obtain high titre *Salmonella* H sera with only small amounts of O antibody. For details of this method, which uses the detached flagella for immunisation, the reader is referred to the original paper.

In general, H antisera also contain antibodies against constituents other than the flagella, but two factors make this less troublesome than expected. H agglutination is characterised by a loose, floccular appearance which is easily distinguished from other reactions and the H antibody titre is usually much higher than those for O and K antibodies. Antibodies against fimbriae may be difficult to distinguish from H antibodies. In doubtful cases, an immobilisation test carried out by means of the microscope may be helpful (see Ørskov, 1954).

5. Pooled and polyvalent antisera

When it is desirable to use several antisera simultaneously, a pool consisting of a mixture of single antisera can be used. As the titre of each serum is decreased by this pooling, it may be preferable to produce polyvalent antisera. The antigens of the mixture are prepared as for monovalent antisera, but the injection volume is increased, at least after the first few injections, and the course of the immunisation is prolonged to obtain higher amounts of antibodies. If a trial bleeding shows a low titre for a particular component, the amount of that component is increased. If no improvement is obtained, a monovalent antiserum of this specificity should be added to the polyvalent antiserum.

B. Absorbed antisera

Absorption of antisera is carried out if: (1) unwanted antibody specificities are present. For instance O antibodies may be removed from OK or H antisera, (2) a factor serum is required, e.g. sera with specificity against a particular determinant or (3) if a closer identification is required.

The antiserum to be absorbed will generally be used at a dilution of 1:5 or more, or undiluted if it is to be used in gel precipitation. Most authors employ suspensions of agar plate cultures for all kinds of absorptions. O antibodies are usually removed by heated cultures (boiled for 1–2 h). Unheated cultures may be employed, but other antigens in such cultures may prevent the binding of O antibodies.

For absorption of K or H antibodies, unheated live or formalinised cultures are used. Antibodies against polysaccharide K antigens are most often removed by a boiled absorbing culture, while this is not the case with H antibodies or protein K antibodies.

For O antibody absorption we use growth from evenly inoculated plates which are suspended in saline, heated (2 h for *E. coli* and 1 h for *Salmonella*) and centrifuged. Antiserum is mixed thoroughly with the cells. For antiserum dilutions phenolised (0·25%) physiological saline is used. The mixture is kept at 37°C for 1–2 h and usually stored in the refrigerator overnight. Then the bacteria are removed by centrifugation. Absorptions of K and H antibodies are performed in the same way, except that the culture suspensions are not boiled but killed by addition of formalin (0·5%), or the bacteria are suspended directly in the antiserum. Plates for the H absorbing culture must be inoculated with an actively motile culture and, the plates not too well dried. When unheated cultures are used for absorption, the absorbed antisera should be preserved e.g. by addition of a small amount of chloroform, particularly if they are not diluted in phenolised saline.

For all absorptions, a two-step procedure is recommended, as this is more effective than a single absorption. The same amount of a cell suspension as for a single absorption is divided into two tubes before centrifugation. Antiserum is added to the sediment of one tube, centrifuged the next day and mixed with the sediment of the other tube which has been kept in the cold after the first centrifugation.

The amount of antigen used for each absorption must be sufficient to remove all homologous antibodies. This amount varies with different antisera and different cultures. In our laboratory, we use the growth from ten large (14 cm dia.), thickly poured broth agar plates as a standard amount to absorb the content of O agglutinins in 10 ml of a 1 : 10 dilution of antiserum. This process generally removes all homologous O agglutinins, however in many cases less antigen would suffice. For H antibody absorption, the same amount of growth is added to 10 ml of H antiserum, but this time the antiserum is diluted 1:100. For absorption of K antibodies, no rule is given regarding the required amount of absorbing culture, as this varies considerably with different K antigens. However, smaller amounts than the growth from ten plates (14 cm dia.)—often from one or two plates —will generally suffice to remove the antibodies against polysaccharide K antigen in 1 ml serum.

VII. SEROTYPES, DEFINITION

Before turning to serotyping of the separate genera of the family *Enterobacteriaceae*, a few general remarks about the labelling of serotypes and the construction of antigenic schemes should be given. Some typical examples of serotypes would be:

Salmonella: 1, 4, 5, 12: b: 1, 2 (*S. paratyphi* B), where 1, 4, 5, 12 is the O antigen and b: 1, 2 covers the two phases of the H antigen (see page 57).

Klebsiella: O1: K1, or just K1 because the O determination is usually not carried out.

Escherichia coli: O1: K1: H5 or 1: 1: 5.

Shigella: II: 1, 3, 4 (*Sh. flexneri* 2a), where both II and 1, 3, 4 represent O antigen determinants, II being the so-called type specific O factor and 1, 3, 4 the group factors.

We shall not dwell on the historical background for these unfortunate differences in labelling, but just point out that most likely they do not cover any inherent basic differences in the immunochemical structure of the single antigens.

Simplified serological labels can be found in the literature, quite often in relation to *E. coli*. One may find *E. coli* O111 or O75 described as serotypes, although they are O serogroups. The serotype of strains belonging to one of these O groups could be e.g. O111: H2 or O75: K100: H5. The term O: H serotype, e.g. O75: H5, can also be found and is often used because K antigen determination has not been carried out. Naturally it is important in a given situation to remember for what purpose serotyping is carried out and not to carry it further than necessary.

VIII. ANTIGENIC SCHEMES

The single antigens and the combinations of these have traditionally been collected in antigenic schemes, the most famous of which is the Kauffmann–White scheme of *Salmonella*. As was the case with the serotype formulas, some basic differences exist between the structure and principles of some of the antigenic schemes.

The *E. coli* O groups, e.g. O group 1, consist of a series of interrelated complex O antigens usually characterised by the fact that they give an O titre in the O test serum not less than two steps from that of the test strain. The exact relation of these many different sub O antigens has not been worked out or, expressed in the *Salmonella* terminology (see below), their composition in O factors have not been worked out. The *coli* O1 group corresponds to the *Salmonella* O groups like A, B, C etc., as each of these groups consists of interrelated O antigens precisely described by their O antigen factors. A total of 66 different *Salmonella* O antigens and O factors have been described to date.

Some *Salmonella* groups, mainly those with high numbers, contain one antigen (antigenic determinant), while many of the lower O groups like A, B, C, D, and E consist of groups of several antigenic factors (antigenic determinants), each group defined by one or two main factors.

The systems according to which the complex H antigens are described in *Salmonella* and *Escherichia* follow the same two principles as outlined above for O antigens. It should be remembered that the numbering of an O antigen (O factor) does not imply that no more than one antigenic determinant is present on the polysaccharide side chain, but only that the possible antigenic determinants have hitherto always been found together in the same strains. It is thus possible that further immunochemical analyses will show that some *Salmonella* O antigens, which to-day have only one antigen number, contain more than one antigenic determinant. It should therefore be kept in mind that the serological typing system, based as it is on the reaction of the rabbit to the basic chemical structures on the bacterial surface, is a symbolic picture. For the other genera of the *Enterobacteriaceae*, the principles used for serological analysis and for possible antigenic schemes have been either those for the *Escherichia* or those for *Salmonella*. The *Shigella* antigenic scheme, the fundamental details of which were described many years ago, deviates to some extent from the above two principles (for details, see Edwards and Ewing, 1972). As an appropriate illustration to these two principles of antigenic analysis and antigenic scheme construction, we would like to refer to two recent papers, published simultaneously by French and British authors, both establishing antigenic schemes for the same *Enterobacteriaceae* group, viz. genus *Levinea malonatica* = species *Citrobacter koseri*. The French group (Popoff and Richard, 1975) follows the *Salmonella* way, while the British group (Gross and Rowe, 1975) uses principles from *Escherichia* serology (see page 42).

IX. THE SINGLE GENERA

The remaining part of this Chapter will cover serotyping procedures recommended for each separate group or genus within the family *Enterobacteriaceae*. As already emphasised in the introduction, several manuals exist which fulfil this purpose brilliantly in practically every aspect. Therefore, we shall only point out the special traits that characterise serotyping of the genus in question, and thus stress the possible differences from the general description of the family given above. We shall also elaborate further when newer developments make it relevant.

The historical background for the delimitation of the single genera and species and details about biochemical reactions will be found extensively covered in Edwards and Ewing (1972) and Sedlak and Rische (1968) where numerous further references are given.

A. *Escherichia coli*

This genus consists of one species, *E. coli*, the main biochemical characters of which can be found in Table II (page 6).

1. O antigens

Antiserum production and antigen determination follow the general procedures described above (pages 12–16, 36). Kauffmann (1944) described the first 20 O antigens and Knipschildt (1945b) added five more. A further 38 O antigens were examined by Kauffmann and 47 by Knipschildt. These 85 O groups were not included in the first antigenic scheme (see below). In 1944, 110 O groups, O1 to O110, were established, but two of these, O31 and O47, were later cancelled and O groups 67, 72 and 94 were removed because they turned out to be *Citrobacter freundii*. Since 1945, new O groups brought the number up to 148 at the time of publishing the 2nd edition of "Identification of *Enterobacteriaceae*" by Edwards and Ewing (1972). However, new O groups are constantly being established, the most recent ones, O163, being published by Ørskov *et al.* (1975b), and O164 by Rowe *et al.* (1977).

With the presently established O antigens and the corresponding O sera, it is possible to group around 90% of smooth *E. coli* strains received in our laboratory. However, most of these strains arrive from geographical areas in the temperate zone and mainly from man and domestic animals. We know that the O group distribution is not the same in all animals and that it may differ according to geographical area; therefore it is possible that geographical areas and milieus exist where less than 90% are groupable. Undoubtedly it would be possible to continue the establishment of new O groups at a much increased scale, as it is not difficult to detect new *Escherichia* O groups, but, in agreement with other *Escherichia* typing centres, it has been our policy not to establish new official O groups, unless: (1) the strains have been through a complete examination, i.e. the new strain must be examined in all known O test sera and all O antigenic test strains have to be examined in an O antiserum produced with the strain in question. Furthermore, we have tried to stick to rule, (2) that only strains of a certain general interest, either from an epidemiological or a scientific point of view, should be established as new O group test antigens.

For the actual O antigen determination, the many O sera are mixed in pools of cross-reacting O antisera. We carry out the primary steps of the O antigen determination in perspex plates, while the final titration is made in glass tubes (Ørskov and Ørskov, 1975). Many cross-reactions exist between the single O groups. A record of cross-reactions between O1 to O148, is given by Edwards and Ewing (1972). A number of cross-absorbed O factor sera are therefore required when new strains are to be grouped within the O system. At present we employ more than 300 such O factor sera in conjunction with the 164 O standard sera; for details, see Ørskov and Ørskov (1975) and Edwards and Ewing (1972). Even such a determination does not decide if the O antigens of the unknown strain and that of the

O group test strain in question are identical. For a definite determination, O antiserum production with the unknown strain and mutual O absorptions is necessary.

2. *K antigen*

In the general part of this Chapter, an attempt has been made to describe the present state within this controversial field (pages 20–25). In the combined list of O and H antigens given below (Table V), K antigens of polysaccharide nature, marked K or K(A), and the plasmid determined protein K antigens K88 and K99 have been included. The notation (A) denotes the polysaccharide K antigens, found in those O8, O9 and O101 strains which have to be autoclaved to become agglutinable in sera. K determination is preferably carried out by counter-current immunoelectrophoresis (Semjen *et al.*, 1977) but may be performed by slide agglutination in well controlled systems, slide agglutination being especially applicable for strains with K(A) antigens.

Primary demonstration of existence of polysaccharide K antigen is carried out by agarose electrophoresis combined with second dimensional Cetavlon precipitation (Ørskov, 1976).

A few words should be said about typing of the so-called enteropathogenic serotypes, such as O111: H2, O55: H6 etc. Traditionally, these strains are detected by slide agglutination in antisera produced by immunisation with unheated organisms (OK sera). In our experience, such strains do not contain special polysaccharide K antigens (earlier called e.g. O111:B4, O111:K58(B4) or O111:K58), see pages 24–25. In spite of this fact, O antisera will most often be of no use in the primary slide agglutination test because of O-inagglutinability. What is determined by slide agglutination in OK sera is the O antigen as found in the living unheated strains. A positive slide agglutination in the OK serum should always be confirmed by agglutination titration of the culture boiled at 100°C for 1 h. This last step being introduced in order to prevent the many possible false positive reactions caused by antiflagellar, antifimbrial and other antibodies found in the OK serum.

The K numbers at present established run from 1 to 103 but, since several of the K(B) antigens have been deleted and some K antigens have been removed for other reasons, the actual number is 74. Two of these are of protein nature and the rest polysaccharide.

3. *H antigens*

H antiserum production and determination follow general procedures. It is most often necessary to passage the strains through semisolid agar in order to obtain well flagellated organisms. All strains are monophasic.

TABLE V

Escherichia coli antigenic scheme comprising all O, K
and H antigenic test strains (December 1977).
Arranged according to O-antigen numbers

O	K	H	Culture no.
1*	1	7	U5-41
1	51	—	A183a
2	1	4	U9-41
2	56	7	H17b
2	1	6	A20a
2	2	1	Su1242
2	n.e.‡	8	Ap320c
3	2ab	2	U14-41
3	n.e.	31	K15 (=HW33)
3	n.e.	44	781-55
4	3	5	U4-41
4	6	5	Bi7457-41
4	12	—	Su65-42
4	52	—	A103
5	4	4	U1-41
6	2ac	1	Bi7458-41
6	13	1	Su4344-41
6	15	16	F8316-41
6	53	—	PA236
6	54	10	A12b
6	13	49	2147-59
7	1	—	Bi7509-41
7	7	4	Pus 3432-41
8	8	4	G3404-41
8	25	9	Bi7575-41
8	27	—	E56b
8	40	9	A51d
8	41	11	A433a
8	42	—	A295b
8	43	11	A195a
8	44	—	A168a
8	45	9	A169a
8	46	30	A236a
8	47	2	A282a
8	48	9	A290a
8	49	21	A180a
8	50	—	PA80c
8	84	—	H308b§
8	87	19	D227(=G7, K88⁻)
8	102	—	6CB10-1
8	n.e.	20	H330b
8	n.e.	21	U11a-44

TABLE V (*continued*)

O	K	H	Culture no.
8, 60	n.e.	*51*	C218-70
9	*9*	*12*	Bi316-42
9a	*26*	—	Bi449-42
9ab	*28*	—	K14a
9	*29*	—	Bi161-42
9	*30*	12	E69
9	*31*	—	Su3973-41
9	*32*	19	H36
9	*33*	—	Ap289
9	*34*	—	E75
9	*35*	—	A140a
9	*36*	19	A198a
9	*37*	—	A84a
9	*38*	—	A262a
9	*39*	9	A121a
9	*55*	—	N24c
9	*57*	32	H509d
9	n.e.	*19*	A18d
10	*5*	*4*	Bi8337-41
11	*10*	*10*	Bi623-42
11	n.e.	*33*	K181 (=HW35)
11	n.e.	*52*	C2187-69
12	5	—	Bi626-42
13	*11*	*11*	Su4321-41
14	7	—	Su4411-41
15	*14*	*4*	F7902-41
15	n.e.	*17*	P12b
15	n.e.	*25*	N234 (=HW26)
15	n.e.	*27*	K50 (=HW28)
16	*1*	—	F11119-41
16	n.e.	*48*	P4
17	*16*	*18*	K12a
18ab	(*76*)†	*14*	F10018-41
18ac	(77)	7	D-M3219-54
19ab	n.e.	7	F8188-41
20	*17*	—	P7a
20	*83*	26	CDC134-51
20	*84*	26	CDC2292-55
20	*101*	—	1473
21	*20*	—	E19a
22	13	1	E14a
23	*18*	*15*	E39a
23	*21*	15	H38
23	*22*	15	H67
24	+	—	E41a
25	*19*	*12*	E47a

TABLE V (*continued*)

O	K	H	Culture no.
25	23	1	H54
26	(60)	—	H311b
26	(60)	—	F41
26	(60)	46	5306-56
27	—	—	F9884-41
28	—	—	K1a
28	(73)	—	Kattwijk
29	—	10	Su4338-41
30	—	—	P2a
32	—	19	P6a
33	—	—	E40
34	—	10	H304
35	—	10	E77a
36	—	9	H502a
37	—	10	H510c
38	—	26	F11621-41
38	n.e.	30	N157 (=HW32)
39	—	—	H7
40	—	4	H316
41	—	40	H710c
42	—	37	P11a
43	—	2	Bi7455-41
44	74	18	H702c
45	+	10	H61
45	n.e.	23	K42 (=HW23)
46	—	16	P1c
48	—	—	U8-41
49	+	12	U12-41
50	—	4	U18-41
51	—	24	U19-41
51	n.e.	24	K72 (=HW25)
52	—	10	U20-41
52	n.e.	45	4106-54
53	—	3	Bi7327-41
54	—	2	Bi3972-41
55	(59)	—	Su3912-41
55	(59)	6	Aberdeen 1064
56	+	—	Su3684-41
57	—	—	F8198-41
58	—	27	F8962-41
59	—	19	F9095-41
60	—	33	F10167a-41
61	—	19	F10167b-41
62	—	30	F10524-41
63	—	—	F10598-41
64	—	—	K6b

TABLE V (*continued*)

O	K	H	Culture No.
65	—	—	K11a
66	—	25	P1a
68	—	4	P7d
69	—	38	P9b
70	n.e.	42	P9c
71	—	12	P10a
73	—	31	P12a
73	92	34	6181-66
74	—	39	E3a
75	95	5	E3b
75	100	5	F147
76	—	8	E5d
77	96	—	E10
78	(80)	—	E38
79	—	40	E49
80	—	26	E71
81	97	—	H5
82	—	—	H14
83	—	31	H17a
83	24	31*	H45
84	—	21	H19
85	—	1	H23
86	—	25	H35
86	(61)	—	E990
86	62	2	F1961
86	(64)	36	5017-53
86	(61)	34	BP12665
86	n.e.	47	1755-58
87	—	12	H40
88	—	25	H53
89	—	16	H68
90	—	—	H77
91	—	—	H307b
92	—	33	H308a
95	+	33	H311a
96	—	19	H319
97	—	—	H320a
98	—	8	H501d
99	—	33	H504c
100	—	2	H509a
101	—	33	H510a
101	99(L)	—	B41
101	103	—	8CE275-6
102	—	8	H511
103	+	8	H515b
104	—	12	H519

TABLE V (*continued*)

O	K	H	Culture no.
105	—	8	H520b
106	—	33	H521a
107	*98*	27	H705
108	—	10	H708b
109	—	19	H709c
110	—	39	H711c
111	*(58)*	—	Stoke W
112ab	*(68)*	18	1411-50
112ac	*(66)*	—	Guanabara (1685)
113	*(75)*	21	6182-50
114	*(90)*	32	26w (=K10=HW34)
115	—	18	27w
116	+	10	28w
117	98	4	30w
118	—	—	31w
119	*(69)*	27	34w
120	+	6	35w
121	—	10	39w
123	—	16	43w
124	*(72)*	30	Ew227
125ab	*(70)*	19	Canioni
125ac	*(70)*	6	Ew2129-54
126	*(71)*	2	E611
127a	*(63)*	—	4932-53
127ab	*(65)*	4	2160-53
128	*(67)*	2	Cigleris
129	—	11	Seeliger 178-54
130	—	9	Ew4866-53
131	—	26	S239 (=HW27)
132	+	28	N87 (=HW30)
133	—	29	N282 (=HW31)
134	—	35	4370-53
135	—	—	Coli Pecs
136	*(78)*	—	1111-55
137	*(79)*	41	RVC1787
138	*(81)*	—	CDC62-57
139	82	1	CDC63-57
139	—	56	SN3N/1
140	—	43	CDC149-51
141	*(85)*	4	RVC2907
141	*85ab*(L)	4	E68
142	*(86)*	6	C771
143	—	—	4608-58
144	—	—	1624-56
145	—	—	E1385(3)
146	—	21	CDC2950-54

TABLE V (*continued*)

O	K	H	Culture no.
147	(*89*)	19	D357 (=G1253 K88⁻)
147	(89), *88ac*(L)	19	G1253
148	—	28	E519-66
148a	n.e.	*53*	E480-68
149	(*91*)	10	D616 (=CS1483 K88⁻)
150	*93*	6	1935
151	—	10	880-67
152	—	—	1184-68
153	—	7	14097
154	*94*	4	E1541-68
155	—	9	E1529-68
156	—	47	E1585-68
157	88ac(L)	19	A2
158	—	23	E1020-72
159	—	20	E2476-72
160	—	34	E110-69
161	—	*54*	E223-69
162	—	10	10B1/1
163	—	19	SN3B/1
164	—	—	145/46

* Italicised numbers = test (reference) antigens.
† Italicised numbers in parentheses = former reference antigens, now deleted. These numbers will not be used in the future (see page 24).
‡ Presence of K antigen not examined.
§ Former test strain of 093 (Ørskov *et al.*, 1977).
— in K column = no K antigen detected.
+ in K column = not numbered polysaccharide K antigen.
— in H column = non-motile.
The following polysaccharide K antigens have been found to be closely related or identical (Semjen *et al.*, 1977): K*2ab*~K2ac~K*62*, K7=K*56*, K*12*~K*82*, K*13*~K*23*, K*18*~K*22*, K*16*~K*37*~K*97*, K*53*~K*93*, K*54*~K*96*.
This scheme only contains the serotype formulae of the reference strains used at the WHO Collaborative Centre for Reference and Research on *Escherichia*. It therefore contains all hitherto officially established *Escherichia* antigens.
It is evident that nothing about prevalence of single antigens can be deduced from the data found in this scheme. The strains have been collected over more than 30 years and many different considerations have determined the selection.

56 H antigens have been described, but two, H13 and H22, have been removed as being *C. freundii* and H50 has been withdrawn (Ørskov *et al.*, 1975b). The H sera are combined in pools for primary H determination. A recent paper describes the pools and further procedures as used in our laboratory (Ørskov and Ørskov, 1975); see also Edwards and Ewing (1972).

Cross-reactions between *E. coli* antigens H1 to H51 can be found in Edwards and Ewing (1972).

4. Antigenic schemes

Kauffmann (1944) published the first antigenic scheme which comprised 20 O groups. Later an extended scheme according to Kauffmann, Knipschildt and Vahlne, comprising 25 O groups, 55 K antigens and 21 H antigens, was presented (see Kauffmann, 1966).

As explained in the general Section (page 22), a common K determination has only been carried out to a limited extent and a complete antigenic scheme comparable to that of *Salmonella* cannot be presented to-day. Because of the many single antigens now described in the *Escherichia* group and because most of these can be found in different combinations leading to several different O: K: H combinations an antigenic scheme like that of Kauffmann–White would be unwieldy. Detailed serological analysis of *Escherichia* strains has only been carried out on a limited number of strains in comparison with the high number of *Salmonella* strains examined through the years. However, even this rather low number of *E. coli* strains shows a rich variety of antigenic combinations. As an illustration, it can be mentioned that O group 111 was found in combination with 14 and O8 together with 29 different H antigens; O8 is found in many different O:K combinations, and the number of permutations accordingly is high. If we compare the situation in *Salmonella* and *Escherichia*, it should also be remembered that each *E. coli* O group consists of several non-identical O antigens with a common main factor and some partial factors, i.e. *E. coli* O groups 1, 2, 3 etc. could probably be compared to *Salmonella* O groups A, B, C etc.

We have decided to publish here a combined, restricted antigenic scheme containing all the *E. coli* antigenic test strains (Table V).

5. Cross-reactions

Many cross-reactions between *E. coli* and other *Enterobacteriaceae* have been reported. Best documented are probably the crosses to the genetically closely related *Shigella* group (see the *Shigella* chapter in Edwards and Ewing, 1972). In that book, a description of cross-reactions to most of the other genera is also found. Similar records are given by Lachowicz (1968), Kauffmann (1966) and Ørskov et al. (1977). A recent paper by Refai and Rohde (1975) reports the re-examination of O antigenic relations between *E. coli*, *Salmonella*, and *C. freundii*.

B. Edwardsiella

The main biochemical characteristics can be found in Table II. Only one species, *Edwardsiella tarda*, is recognised. This was first named the Asakura group by Sakazaki (1967) who also published the first antigenic scheme.

1. O antigens

Antiserum production and antigen determination follow the general rules (pages 36, 12–16). Sakazaki (1967) described 17 O antigens and McWhorter *et al.* (1967) increased the number to 48. Few cross-reactions were described.

2. K antigens

K antigens have not been reported. Live cultures were agglutinable in O antiserum.

3. H antigens

Sakazaki (1967) described 11 H antigens and McWhorter *et al.* (1967) increased the number to 37. Sakazaki emphasises that, in order to obtain well developed H antigens for serum production and agglutination, it is advantageous to passage the culture through semisolid agar at pH 6·6 and to use merthiolate 0·1% instead of formalin for preservation.

4. Antigenic scheme

Sakazaki (1967) published an antigenic scheme containing 17 O antigens and 11 H antigens and McWhorter *et al.* (1967) presented a larger, provisional scheme including Sakazaki's strains. The latter scheme has since been enlarged and according to Edwards and Ewing (1972) it now comprises 49 O antigens and 37 H antigens, but its details have not been published.

5. Cross-reactions

Sakazaki (1967) examined his O and H antigenic test strains in the majority of O and H test antisera from *Salmonella, Escherichia, Citrobacter, Proteus, Enterobacter cloacae* and *Hafnia*. However, no significant agglutination reactions were detected. See also Edwards and Ewing (1972).

C. Citrobacter

Details of biochemical characters are given in Table II. "Bergey's Manual", 8th edition (1974), describes two species, *C. freundii* and *C. intermedius*, but states that the genus *Citrobacter* is in the melting pot and that different authors have suggested subdividing it into several genera and/or species. All serotyping within this group has hitherto been centred around *C. freundii*. However, the above-mentioned new development, which was started by the proposal for a new species, *C. koseri*, has been followed by an antigenic analysis of such strains (see below). For a discussion of the taxonomic and nomenclatural problems connected with strains of the provisional *C. koseri* (syn. *C. diversus* and *Levinea malonatica*), the reader is referred to the papers by Frederiksen (1970), Ewing and Davis (1971), Gross and Rowe (1974), and Richard (1975). In the present Chapter the new group will be described provisionally as *C. koseri*.

1. *C. freundii*

(a) *O antigens.* Antiserum production and antigen determination follow the normal procedures. 42 O antigens, many of them complex, have been examined. Many cross-reactions are found, particularly among the low O antigen numbers. Slide agglutination is recommended. Recommendations for production of diagnostic polyvalent antisera have been given by Edwards and Ewing (1972).

(b) *K antigens.* An antigenic determinant similar to that of the *Salmonella* Vi antigen has been found in *C. freundii* of O groups O5 and O29. No description of other K antigens in this group exists.

(c) *H antigens.* Antiserum production and antigenic determination are according to general principles. The H antigens are complex and a total of 90 H antigen factors have been described, many of which occur in several different H antigens (Edwards and Ewing, 1972; Sedlak and Rauss, 1968a). Tube agglutination is recommended by Edwards and Ewing. Only monophasic cultures have been detected.

(d) *Antigenic scheme.* The first analysis of strains which to-day are called *C. freundii* was carried out on a group of strains which fermented lactose slowly or not at all and were assumed to be closely related to *Salmonella*. They were called the Bethesda–Ballerup group. An antigenic scheme based on *Salmonella* principles containing 32 O groups and 76 complex H antigens was established by West and Edwards (1954). A number of lactose-positive strains were also included, and it was concluded that the *C. freundii* group comprised both lactose-positive and lactose-negative strains and that this trait was not correlated with the antigenic characters. Sedlak and Slajsova (1966) extended this antigenic scheme to include 42 O groups and 90 complex H antigens. With these sera nearly 80% of 6000 strains could be typed (Sedlak and Rauss, 1968a).

(e) *Cross reactions.* Antigenic relationships between the O antigens of *C. freundii* and O antigens of other genera have been studied, particularly regarding the cross-reaction with *Salmonella* and *Escherichia* (Sedlak and Rauss, 1968; Edwards and Ewing, 1972).

2. *C. koseri* (syn. *C. diversus, Levinea malonatica*)

(*a*) *O antigens.* Two serological studies have been published almost simultaneously. Gross and Rowe (1974) examined strains received from different authors and showed that Frederiksen's *C. koseri*, Young's *Levinea malonatica*, the new *Citrobacter* species of Booth and McDonald, and *C. diversus* of Ewing and Davis formed a biochemically homogenous group. An antigenic scheme comprising 7 O antigens was proposed. This scheme was extended to 14 antigens by Gross and Rowe (1974). Popoff and Richard (1975) described 14 O antigens (factors) in their publication. O serum production and determination followed the general rules in Gross and Rowe's examination (agglutination in perspex trays). Popoff and Richard incubated the agglutinations 2 h at 37°C followed by 20 h at room temperature, as opposed to the usual incubation in a water-bath at 50°C overnight. Tube agglutination was used.

(*b*) *K antigens.* K antigens have not been described.

(*c*) *H antigens.* H serum is produced according to standard procedures, but again Popoff and Richard (1975) incubated the agglutination for 2 h at 37°C instead of using a 50°C water-bath.

(*d*) *Antigenic schemes.* Gross and Rowe's antigenic scheme (1974) comprises 14 O antigen groups and these are established according to the principles used for *Escherichia*. The scheme of Popoff and Richard (1975) is structured like that of *Salmonella* and contains O groups A to F, each consisting of several O factors, some of which can be found in several groups. Only few cross-reactions exist between the 7 H antigens detected. Popoff and Richard write that their scheme is difficult to compare with that of Gross and Rowe, partly because two somewhat different approaches for designation of O antigens have been used.

(*e*) *Cross-reactions.* Only few and weak O cross-reactions to *E. coli* O antigens have been detected (Gross and Rowe, 1974).

D. *Salmonella*

The main biochemical properties are shown in Table II from "Bergey's Manual, 8th edition (1974)". No subdivision into species is described in the manual and it is stated that none of the present methods of nomenclature of *Salmonella* is satisfactory. However, support is given to divide *Salmonella* into four subgroups as proposed by Kauffmann (1966). This is shown in Table VI, where subgenus I contains biochemically typical *Salmonella* strains, subgenera II and IV atypical strains, while genus *Arizona* is incorporated as subgenus III. This incorporation is not favoured by e.g. Edwards and Ewing (1972).

TABLE VI

Distinguishing characters of the subgenera
of *Salmonella*

	I	II	III	IV
Dulcitol	+	+	—	—
Lactose	—	—	+ or ×*	—
β-Galactosidase	—	×	+	—
d-Tartrate	+	—	—	—
Mucate	+	+	d	—
Malonate	—	+	+	—
Gelatin	—	(+)	(+)	(+)
KCN	—	—	—	+

* × = late and irregularly positive; (+) = late, but always
positive.
From "Bergey's Manual" (1974).

Since an extensive literature exists about *Salmonella*, its antigens and
serological technique (Kauffmann, 1966; Edwards and Ewing, 1972;
Buczowski, 1968), we shall only call attention to some facts which we
consider of particular importance.

1. *O antigens*

Most *Salmonella* strains are agglutinable in O antisera without being
heated. Consequently, the O antigens may be determined by means of a
slide agglutination.

With the *Arizona* strains included, the number of O antigens in the
Salmonella group is 66, but some of the numbers have been deleted for
different reasons. This concerns numbers 29, 31, 32, 33 and 49. The O
antigens are contained in 42 O groups of the Kauffmann–White scheme
named by letters A, B, etc. to Z—comprising O antigens 1 to 50—followed
by O groups named by Arabic numerals 51 to 66. O groups C, D, E, and G
are subdivided further into C_1, C_2, C_3, C_4; D_1, D_2, D_3; E_1, E_2, E_3, E_4 and
G_1, G_2.

Strains of O groups A to H—with the exception of F—each contains
more than one antigen or O antigen factor. This means there are more than
one antigenic determinant on the O antigen chain of the LPS molecule,
e.g. O factors 1, 2, 12 in *S. paratyphi A* or O group A.

If a *Salmonella* strain fails to agglutinate in any O antiserum, this is
to-day unlikely to be due to the presence of a new O antigen but, rather to
the presence of an outer surface component rendering the strain O inagglutinable. This may be the Vi antigen, the M antigen, flagella or fimbriae.

In such cases, the O antigen examination must be repeated, but with boiled culture.

As regards the production of O antisera, the general procedures can be followed, meaning that boiled culture is used for immunisation. Since some antigenic determinants are partly destroyed by heating at 100°C, e.g. O factors 2 and 5, antibodies against these should be raised by injection of formalin-killed bacteria (see also page 36).

2. Form variation and lysogenic conversion

Two phenomena should be remembered when *Salmonella* O antigen determination is carried out. One is form variation and the other lysogenic conversion. The term O form variation, first defined in O1 by Kauffmann (1940), covers the observation that some colonies (1 in 10–20) from a certain strain contain large amounts of an antigen, while others contain slight amounts, and that each form gives rise to the other on subculture. Many O antigens are subject to form variation, viz. 1, 6_1, 8, 12_2, 19, 20, 22, 23, 24, 25, 27, 36, 37 and 50_3 (Kauffmann, 1966; see also page 31).

Lysogenisation by converting bacteriophages brings about modification of the O antigens, in some cases causing a change not only in the O antigenic formula, but also in the name. Iseki and Sakai (1953a, b) discovered that the presence of bacteriophage ϵ15 changes antigen 3, 10 to 3, 15, which changes the designation *S. anatum* to *S. newington*. In O groups A, B, D and G the presence of O antigen 1 is associated with lysogenisation without causing any change in the name of the organism. Several O factors are connected with lysogenisation; for further details, see Le Minor (1968a, b).

Many of the O antigen modifications determined by bacteriophage genes are subject to form variation; no matter whether the form variation is associated with presence of phage, the mechanism of it is as yet obscure (Stocker and Mäkelä, 1971).

3. T antigens

Kauffmann (1956, 1957) has described two T antigens in the *Salmonella* group. The T forms are regarded as transient (T) serological forms between S (smooth) and R (rough) forms. They are smooth but become rough with a high frequency. As a result of mutation the O antigen specificity is lost and the T specific determinant is produced. However, the T_1 side chain can be present simultaneously with the O side chain (Sarvas and Mäkelä, 1965), but it is uncertain whether both can be present on the same molecule (Lüderitz et al., 1971).

4. K antigens

Two polysaccharide K antigens, the Vi antigen and the M antigen, can be found in *Salmonella* strains and of these only the Vi antigen is used for serotyping. In his more recent publications Kauffmann also lists the 5 antigen as belonging to this group. This antigen, which is determined chemically by acetyl groups added to the polysaccharide side chain of the 4, 12 O antigen, is thus an O antigen factor and should not be listed among K antigens just because its sensitivity to different chemical and physical treatment differs slightly from most other O antigen factors.

(*a*) *The Vi antigen* is extensively treated by Kauffmann (1966) and Edwards and Ewing (1972). It is an acidic polysaccharide (Baker *et al.*, 1959) which causes O inagglutinability (Felix and Pitt, 1934) and is subject to what Kauffmann has called V–W or Vi variation, meaning that different colonies give strong, weak or no agglutination in Vi antiserum simultaneously with no, weak or strong agglutination in O antiserum. Many authors will determine the presence of *Salmonella* Vi antigen by use of a Vi antiserum elicited against a *Citrobacter* strain containing a closely related Vi antigen.

(*b*) *M antigen.* The antigenicity of mucus formed by many *Salmonella* strains was first demonstrated in *S. schottmuelleri* (*paratyphi-B*) by Kauffmann (1935, 1936). The mucus, which was denoted the M antigen, is structured as a capsule in many strains, but not in all. It is an acidic polysaccharide (Anderson and Rogers, 1963). Capsulated forms give capsular swelling with M antiserum, but it should be emphasised that the immunogenicity of the M antigen may be poor—very poor. The M antigens produced by mucoid *Salmonella* strains are mutually strongly related and also related to the ubiquitous M antigen found in other *Enterobacteriaceae* genera (see page 29). However, *Klebsiella* or *E. coli* strains with K antigens of the A type are not mucoid due to possession of M antigen, but some of the K antigens of these strains cross-react with the M antigen. Antisera against e.g. *E. coli* K30 (Hungerer *et al.*, 1967), *Klebsiella* K13 (Perch, 1950) or *Klebsiella* K8, 11, 21, 26 and 35 (Henriksen, 1954) may contain cross-reacting antibodies against the M antigen. Only some of the immunised rabbits will develop these. Thus, antisera against these types cannot always be used for the demonstration of M antigen. The presence of M antigen is disturbing as it inhibits O agglutination when carried out with non-heated culture. Strains with M antigen should be heated for O antigen determination or a search made for non-mucoid colonies, e.g. by selection of highly motile bacteria, as suggested by Edwards and Ewing (1972).

5. *Other surface structures*

Salmonella strains possess fimbriae (Duguid *et al.*, 1966; Old and Payne, 1971) which ordinarily do not confer O inagglutinability. Recently,

however, a new type of fimbria has been described in *S. typhimurium* and
S. enteritidis which strongly inhibits O agglutination, even by primary
isolation from nutrient agar plates (Rohde *et al.*, 1975). This fimbrial
antigen represents a special type of *Salmonella* fimbriae, morphologically
different from type 1 and type 2 (Duguid *et al.*, 1966; Old and Payne, 1971)
and not causing positive haemagglutination. As revealed by electron microscopy, these filaments form a fringe of closely packed threads at the border
of the cell wall and are thinner and longer than ordinary fimbriae. Rohde
et al. (1975) discuss whether this heat labile antigen should be regarded as a
K antigen or a fimbrial antigen and decide to call it "fimbrial".

6. *H antigens*

H antigen determination of *Salmonella* can be carried out by tube or
slide agglutination; for a discussion of the two possibilities, see Edwards
and Ewing (1972).

In order to H antigen type *Salmonella* and to prepare H antisera, it is
important to be familiar with the variation phenomenon termed phase
variation (Andrewes, 1922) which is a unique phenomenon only described
in *Salmonella* (and *Arizona*) strains. Many *Salmonella* cultures manifest
two alternative types of H antigen. One is called the "specific phase" or
"phase 1" and the other the "non-specific phase" or "phase 2". Each such
Salmonella strain has its own specific type of phase 1 and phase 2 H
antigens. When a mass culture of the strain is plated it dissociates into
colonies of phase 1 and phase 2. During subcultures of the two types, each
population gives rise to cells of the alternative phase. The frequency of
phase variation differs in different strains, but is always low.

In order to make the phenomenon of phase variation better understood,
we shall briefly mention the genetics of H antigen determinants as described by Iino and Lederberg (1964). As a result of transduction experiments it has been shown that the antigenic specificities of phase 1 and
phase 2 are determined by genes at independent loci on the bacterial
chromosome which are symbolised by H1 and H2. In the expression of the
antigenic phases, the H2 gene plays a decisive role. H2 exists in two
different states, active and inactive. Active H2 inhibits the production of
the phase 1 antigen, while it determines the production of phase 2 antigen.
When H2 changes to the inactive state corresponding to the change from
phase 2 to phase 1, the production of phase 2 antigen stops and, alternatively
the production of phase 1 antigen, specified by H1, proceeds. *Salmonella*
strains expressing phase 1 and phase 2 alternatively are called diphasic;
those which express only one antigenic phase, either phase 1 or phase 2, are
called monophasic-1 or monophasic-2 types, respectively. Occasionally
three or four-phasic strains appear, these are probably duplications of H1

and/or H2 genes on the chromosome. For further references, see Iino (1969).

Many H antigen types appear only in phase 1 as a, b, c, d, etc. and others only in phase 2 as 1, 2; 1, 5; 1, 6 etc. However, there are several antigen types like e, l and w which appear both in phase 1 and phase 2 indicating that these antigen types are not specific to a particular antigen phase. The same antigen l, w is thus determined by H2 gene in *S. wien* b: l, w, and by H1 in *S. dar-es-salaam* l, w: e, n, z18. There are occasions where the same type of antigenic subunit appears in both phase 1 and phase 2 of the same strain as in *S. salinatis* d, e, h: d, e, n, z15.

Since the frequency of variation from one phase to another is low, as mentioned above, it is often necessary after having determined the H antigens of one phase to select for the other phase. This is done by incorporation into the medium of an H antiserum against the phase already formed. For more details about the historical development and the serology of *Salmonella* H phases, see Edwards and Ewing (1972) and Kauffmann (1966).

7. *Antigenic relationships*

For antigenic relationships between *Salmonella* and other *Enterobacteriaceae*, see Kauffmann (1954), Kampelmacher (1959), Refai and Rohde (1975).

8. *Antigenic scheme*

In the Kauffmann–White scheme the different serotypes are arranged primarily according to O antigen in O groups as described above. All strains of the same O group are subdivided according to their H antigen type. To-day the scheme contains more than 1500 different serotypes, as seen in "Bergey's Manual (1974)" supplemented by Le Minor *et al.* (1974, 1975).

E. Shigella

The main biochemical characteristics can be found in Table II. It is apparent that *Shigella* is closely related to the *Escherichia* group and they could be considered as one group were it not for historical reasons. At the time when these groups were given taxonomic status *Shigella* strains were primarily isolated as the cause of serious dysentery disease in man, while *Escherichia* was looked upon as rather harmless in the intestine. The finding that completely typical *Escherichia* strains may also cause dysentery-like disease (Sakazaki *et al.*, 1967; Novgorodskaya, 1968) further underlines the similarities between these two groups. Investigations of similarities between DNA base sequences have also shown high concordance between the two groups (Brenner *et al.*, 1972). However, the serological descriptions of *Escherichia* and *Shigella* follow different lines, and again history offers the best explanation for this divergence.

The subdivision of *Shigella* into four species or subgroups, *Shigella dysenteriae* (subgroup A), *Sh. flexneri* (subgroup B), *Sh. boydii* (subgroup C) and *Sh. sonnei* (subgroup D) is largely, but not solely based on biochemical characteristics, as e.g. *Sh. flexneri* strains are closely related biochemically to those of *Sh. boydii*, but are kept separate because they constitute a serologically quite uniform group.

Because of the early recognition of *Shigella* strains as a cause of serious and epidemic diarrhoea, a rich literature has accumulated. It is impossible in this book to give a satisfactory description of this important group. However, the interested reader will find a brilliant account of both its intricate historical development and its present status in the book of Edwards and Ewing (1972) which also contains all required technical details for serological typing.

1. *O antigens*

The O group determination is the dominant serotyping procedure used since no motile strains exist and no special K antigens have yet been officially designated.

Within *Sh. dysenteriae* there are ten serotypes, showing only a few cross-reactions. Serotype 1 produces the so-called Shiga-neurotoxin which until recently was regarded as the only exotoxin produced by *Enterobacteriaceae*.

Sh. flexneri comprises more than ten different, but related serotypes. Both the biochemical background and the genetic basis for this diversity in the LPS side chains of these strains have been elegantly elucidated during the last decade (Simmons, 1971; Seltmann, 1972; Seltmann and Beer, 1974).

Sh. boydii consists of 15 serotypes which do share some minor antigens, but generally show few cross-reactions.

Sh. sonnei, finally, comprises late lactose-fermenting strains. These belong only to one serotype which is described as existing in two phases.

Because of the very refined analysis required to sort out the many interrelated strains of the *Sh. flexneri* group, absorbed O factor sera have to be applied for this purpose. For details about the production of such sera and the actual typing techniques—mostly slide agglutination—the reader is referred to Edwards and Ewing (1972).

We would like to point out here that the historical development has caused a curious difference in the terminology in cases where the main chemical structures, as we know them today, are basically similar. In the *Shigella* literature one will find differentiation between group and type antigens within the O antigens, where group antigens mean those which are common to more strains within a group (viz. *Sh. dysenteriae* or *Sh. flexneri*). One can find the statement that bacteria of the *Sh. dysenteriae* group have strongly developed type antigens and only weakly developed group antigens, and the opposite statement about strains from *Sh. flexneri*. If these had been *Escherichia*, one would have said that strong O cross-reactions were found within the *Sh. flexneri* group (and all the *flexner* serotypes might indeed have been found in one O group with subgroups (ab, ac, abc, etc.)), whereas the *Sh. dysenteriae* strains would have been regarded as individual O groups among which few cross-reactions could be detected.

2. K antigens

Several authors have noted that non-heated *Shigella* cultures might not agglutinate in the homologous O serum and it has been reported that this inagglutinability could be overcome by heating. Madsen (1949), who examined such O inagglutinable strains, applied the methods developed by Kauffmann and his collaborators for K antigen determination (see Kauffmann, 1966) and found that several subgroup A and C serotypes and some *Sh. flexneri* 6 cultures contained so-called B antigens. Ewing in fact described K antigens of the B type in cultures belonging to most serotypes of both groups A and C, and in some group B and D strains. He also gave numbers to these presumed K antigens (Edwards and Ewing, 1972). However, the O antigens and the postulated K antigens were always found in the same combinations. The establishment of these antigens was based on O inagglutinability. It was not possible to produce pure K sera by absorption by heat-treated culture, as no other cultures existed with the same O antigen and a different K antigen. Several of these supposedly B antigen-containing *Shigella* strains gave O antigen cross-reactions with *Escherichia*, but whenever attempts were made to absorb a *Shigella* OB serum by the cross-reacting *Escherichia* strain, all agglutinins were removed (Edwards and Ewing, 1972).

We have not personally worked with such *Shigella*, but from our experience with similar *Escherichia* strains earlier described as containing K antigens of the B type, e.g. *E. coli* O111, O26 or O124—the last mentioned has a strong cross-reaction with *Sh. dysenteriae* 3—we would not suppose that many such strains have special polysaccharide K antigens, biochemically and genetically independent of the LPS O antigens. A definite explanation of the O inagglutinability of such strains remains to be given, but O inagglutinability alone hardly warrants the numbering of a special K antigen which (1) need not be determined because it does not add any epidemiologically significant information to the O antigen determination, and (2) does not exist as a serological entity.

3. *Variation phenomena*

Within the *Sh. flexneri* O group antigens variational phenomena have been extensively described in the older literature. Chemical structural analysis of the lipopolysaccharides in recent years has, to a large extent, elucidated the chemical background for these differences. The reader is referred to Edwards and Ewing (1972) and to the original papers of Simmons (1971) and Seltmann (1972).

4. *Other surface structures important for serotyping*

In the *Sh. flexneri* group the same fimbrial antigen is found in all types, except *Sh. flexneri* 6. With the use of properly produced O sera (100°C, $2\frac{1}{2}$ h) antifimbrial antibody will generally not interfere with serotyping. In the case of other types of sera, of course, the fimbrial antigen may interfere with the typing results.

5. *Antigenic scheme*

For details of the development of the antigenic scheme of *Shigella* groups, see Edwards and Ewing (1972) and Slopek (1968a).

6. *Cross-reactions*

Many cross-reactions between O antigens of *Shigella* and *Enterobacteriaceae* genera have been reported. Most important are cross-reactions to the closely related *Escherichia* group (Edwards and Ewing, 1972; Kauffmann, 1966; Slopek, 1968; Kampelmacher, 1959; Ørskov *et al.*, 1977).

F. Klebsiella

The main biochemical characters are shown in Table II. "Bergey's Manual (1974)" describes three species, *K. pneumoniae*, *K. ozaenae* and *K. rhinoschleromatis*. In Table II the indole and gelatin reactions are recorded as "d", which means that some *K. pneumoniae* strains are positive in these tests. Jain *et al.* (1974) have examined four indole-gelatin positive *Klebsiella* strains and their results clearly indicate that these strains (the *Oxytoca* group) represent a distinct DNA homology group which could be classified separately from *Klebsiella* as a new genus.

Klebsiella strains possess O and K (capsular) antigens both of which can be typed. Serological type determination, however, is based on examinations of K antigens which can be determined by a variety of methods; this is in contrast to the other genera of *Enterobacteriaceae*, where O antigen determination is the backbone of the serotyping procedure.

1. *K antigens*

The method most frequently used for K antigen determination has been the capsule quellung reaction. By addition of specific K (capsular) antiserum, a precipitin reaction occurs at the bacterial surface. This makes the capsule highly refractile and thus easily visible microscopically. Normally, *Klebsiella* strains grow as large moist colonies which are more or less mucoid. The larger colonies usually appear more translucent and have larger capsules. India ink preparations are recommended to determine capsular size. This is carried out as follows: A small loopful of India ink and a large loopful of a dilute bacterial suspension (about 10^7 cells/ml) are mixed on a large slide. A large coverslip is placed over the mixture and the preparation examined in the light microscope. For comparison it is recommended that two further, but small preparations be made on the same slide, one of them with and the other without addition of antiserum. In the India ink preparation, the capsule will be revealed by an India ink-free zone around the bacterial body. By comparison of the preparations it will be seen that addition of antiserum to very large capsules seems to cause their shrinkage, while small capsules may look larger after addition of antiserum.

Some *Klebsiella* strains are very stringy and produce abundant intercellular capsular specific substance demonstrable as bands after precipitation with antiserum. In other strains, the mucoid substance is not really stringy; in these, detached capsular substance will be precipitated in clumps by antiserum as a finely granulated material. The younger the culture, the less substance will be solubilised. Most often cells—quelled or not—will be seen entangled within the bands of precipitated intercellular substance.

Strains of unknown capsule types can be serotyped as follows: The culture is grown on a medium considered useful for development of capsules. If the culture appears poorly capsulated after the ordinary incubation period at 37°C, it may be advantageous to leave it at room temperature for an additional 24 h. A dense suspension ($10^{10}-10^{11}$ cells/ml) is prepared in formolised saline. Small droplets are mixed with large droplets of antiserum on a glass slide. Positive agglutination reactions appear quickly. If serum pools are used, single sera of the positive pools are tested. Positive findings are confirmed by the capsular quellung test. For that test. the heavy suspension is diluted (about 10^{10} cells/ml) and a loopful mixed with a loopful of the serum giving a positive slide agglutination test. Several cultures agglutinate on the slide in most antisera, probably because of a fimbria–antifimbria reaction. In these cases, the slide agglutination technique should be abandoned and the culture examined by the specific quellung reaction only. If the culture gives a quellung reaction in several antisera with known relationships, absorbed antisera—if available—should be tested; otherwise, the culture is examined with serial dilutions of the antisera in order to see in which serum it reacts to homologous titre. If the culture reacts with two antisera, not known to be related, it is assigned to two types, e.g. 3, 37. It should be emphasised that the labelling of a strain as a certain capsule type does not mean that the capsular antigen is identical with that of the test strain, unless the identity is confirmed by cross-absorption.

A few comments should be made regarding production of antisera against capsular antigens (OK antisera). Cultures that produce capsules of moderate size are better immunogens than those producing large capsules. Huge capsules or abundant extracellular substance may cause immunological tolerance, previously called immunological paralysis (Ørskov, 1956). In that case, a better antibody response can be obtained if most of the solubilised substance is removed by centrifugation or, even better, if the culture is heated at 100°C for e.g. 30 min before centrifugation, in order to increase the amount of capsular substance solubilised from the cells. If the antibody content is low (titre $< 1 : 16$) after five to six injections of the rabbit, it is hardly worth while to proceed further. The titre of a *Klebsillea* K antiserum is rarely higher than $1 : 128$, usually about $1 : 32$.

The following capsular types have been described in the *Klebsiella* group: K1–K72 (see Kauffmann, 1966), K73–K76 (Maresz-Babczyszyn, 1962), K77–K80 (Durlakowa *et al.*, 1963), K81 (Nimmich and Münter, 1975) and K82 (Ørskov and Fife-Asbury, 1978). K73 and K75–K78 could not be confirmed by us as separate types for the following reasons; K73 is motile and labelled as *E. aerogenes*, while K75 to K78 are found to be closely related to, or identical with, already known types.

K. ozaenae belongs to K4, K5 and K6, *K. rhinoschleromatis* to K3, while all types will be found in *K. pneumoniae*.

The number of strong relationships between *Klebsiella* K antigens is so high that use of cross-absorbed antisera is desirable for more precise antigen determination. Since the number of cross-reactions varies from rabbit to rabbit, from time to time and from laboratory to laboratory, it is impossible to give a detailed table of such cross-reactions. In our laboratory some of the most frequently found K cross-reactions are as follows: 1–6, 2–69, 2–13–30, 3–68, 5–46, 6–46, 7–10, 8–74, 11–21, 11–33, 14–64, 15–78, 18–44, 20–51, 22–37, 24–43, 26–74, 27–46, 28–46, 29–42, 30–69, 33–35, 41–79, 41–81, 65–80 and 70–80.

2. O antigens

As mentioned above, *Klebsiella* strains possess O antigens which can also be typed. However, two facts make the determination of these antigens less valuable: (1) the number of O groups is small compared with the number of capsular types and (2) the determination is difficult because of capsular heat stability. An encapsulated culture heated at 100°C will not agglutinate in an O antiserum, in contrast to an autoclaved culture.

The best way to determine the O antigen is to isolate non-encapsulated mutants and test boiled cultures in O antisera produced with non-encapsulated mutants. If no such antiserum exists, an ordinary OK antiserum against the capsulated form may be used, as it contains O in addition to K antibodies. However, in that case it should be pointed out that it is of the utmost importance that the culture is non-encapsulated because if capsular substance is left, the agglutination may be a mixed O and K agglutination.

Twelve O groups have been described for *Klebsiella*: O1–O5 (see Kauffmann, 1966), O6–O9, O11, O12 (Durlakowa *et al.*, 1960) and O10 (Maresz-Babczyszyn, 1962), but it is somewhat doubtful whether all 12 should be recognised. O4 is identical with O11 and there are findings suggesting that O1 and O6 should be combined and that O8 and O9 should be regarded as O2 (see Nimmich and Korten, 1970, and for comparison of LPS structures, see Jann and Westphal, 1975).

G. Enterobacter

The main characteristic features are seen in Table II from "Bergey's Manual", 8th edition (1974), where two species are recognised: *En. cloacae* and *En. aerogenes*. The inclusion of *Hafnia alvei* in the *Enterobacter* group has been suggested by Sakazaki (1961) and Ewing (1963). Edwards and Ewing (1972) include not only *H. alvei* but also *E. liquefaciens* into the *Enterobacter* group.

1. O antigens

Only *E. cloacae* has been analysed serologically. 53 O groups have been described by Sakazaki and Namioka (1960). According to Sedlak and Matejovska (1958) and Sedlak (1968), 19 O groups were established by Sedlak and his co-workers. No comparison of the O grouping systems seems to exist.

2. H antigens

E. cloacae strains are monophasic. Sakazaki and Namioka (1960) have characterised 57 H antigens.

3. K antigens

Sakazaki and Namioka (1960) reported on the occurrence of K antigens and α antigen. Edwards and Ewing (1972) state that the majority of capsulated *En. cloacae* and *En. aerogenes* yield quellung reactions in *Klebsiella* antisera (see below).

4. Antigenic scheme

Several O and H antigenic systems have been described, but apparently very little serotyping of *Enterobacter* has been carried out. No generally accepted antigenic scheme exists.

5. Cross-reactions

Some O antigen relationships to both *E. coli*, *Salmonella* and *Shigella* have been described (see Sedlak, 1968).

K antigen relationships to *Klebsiella* capsular types 8, 11, 21, 26 and 69 have been reported by Edwards and Ewing (1972) (see above). These authors emphasise that *Enterobacter* cultures rarely react in a single *Klebsiella* antiserum. We would assume that the above-mentioned cross-reactions are caused by the M antigen, as these *Klebsiella* types are exactly those types which cross-react with the M antigen (see page 65).

H. Hafnia

Details of the main biochemical characters are shown in Table II. Only one species, *H. alvei*, is described. The taxonomic position of this genus is disputed and Sakazaki (1961) has suggested that it should be moved to the genus *Enterobacter*—a point of view supported by Edwards and Ewing (1972).

1. O antigens

Antiserum production and antigen determination follow the general rules. Sakazaki (1961) has described 29 O groups among which the O antigens described earlier by Eveland and Faber (1953) (the 32011 group) are included. Matsumuto (1963, 1964) examined further O antigens increasing the total number to 68 O groups (cited from Edwards and Ewing, 1972).

2. K antigens

Sakazaki (1961) mentions, without giving details, that many cultures were inagglutinable in O serum and he therefore assumes that K antigens probably slime antigens, exist.

3. H antigens

Antiserum production and antigenic determination are performed according to general procedures. *Hafnia* cultures are motile, but motility is often best developed in cultures grown at 20°C. Only monophasic cultures have been described. Sakazaki (1961) and Matsumuto (1963, 1964) described a total of 34 H antigens.

4. Antigenic scheme

The reader is referred to the paper by Sakazaki (1961). K antigens were not included in the scheme.

5. Cross-reactions

Antigenic cross-reactions between *Hafnia* and *Shigella*, *Citrobacter* and *Salmonella* have been recorded (see v. Oye, 1968). Sakazaki (1961) reported O antigenic cross-reactions between *Hafnia* and *En. cloacae* and *Escherichia coli*.

I. Serratia

The main biochemical characters can be found in Table II. Only one species, *Sr. marcescens*, is described. The delimitation of this genus to other groups of *Enterobacteriaceae*, especially *En. liquefaciens*, and to the bacterial group called *Bacterium rubidaeum* is discussed by Ewing *et al* (1972).

1. O antigens

Antiserum production and antigen determinations are carried out according to general procedures. However, Edwards and Ewing (1972) suggested the use of tube agglutination for the preliminary O determination. Davis and Woodward (1957), Ewing *et al.* (1959) and Sedlak *et al.* (1965) (see Sedlak, 1968c) have described a total of 10 O antigens and Edwards and Ewing (1972) extended the number to 15 O antigens showing few cross-reactions.

2. K antigens

No detailed examination of *Serratia* K antigens has been carried out, but Ewing *et al.* (1959) mention the detection of some encapsulated strains and furthermore the possible existence of thermolabile K antigens causing inagglutinability of non-heated cultures in O serum. The reader is referred to the general discussion of K antigens (see pages 20–25).

3. H antigens

Antiserum production and antigen determination follow the general procedures. However, Edwards and Ewing (1972) point out that many strains have to be passaged through semisolid agar to improve flagellation. The use of 0·5% phenolised saline is recommended instead of the usual formalin to preserve H antigen suspensions.

4. *Antigenic scheme*

Ewing *et al.* (1959) published the first provisional antigenic scheme and Sedlak *et al.* (1965) extended this to 11 O groups. It comprises 13 H antigens.

5. *Cross-reactions*

O antigen cross-reactions between *Serratia* and *E. coli* were described by Kauffmann (1949) and an antigenic relationship to *Enterobacter* and *Hafnia* O groups was described by Ewing *et al.* (1959).

J. Proteus

The main biochemical characters of this genus are recorded in Table II. In "Bergey's Manual (1974)" it is subdivided into five species, two of which, *P. mirabilis* and *P. vulgaris*, could be regarded as two biotypes of one species. The three other species are *P. rettgeri*, *P. morganii* and *P. inconstans*. The status of these species has been under consideration for a long time and the interested reader is referred to the discussion by Edwards and Ewing (1972) and by Sedlak and Rauss (1968) where further references can be found. *P. inconstans* is the same as genus *Providencia* (Edwards and Ewing, 1972).

Extensive serological analyses have been performed on strains from the single species, but little comparative work on antigens from different species within the genus has been carried out. Exceptions are *P. mirabilis* and *P. vulgaris* which were examined as one unit by Kauffmann and Perch (1947) and Perch (1948).

1. *Proteus vulgaris* and *Proteus mirabilis*

(a) *O antigens.* The general principles for serum production and antigen determination are applicable. Perch (1948) detected some differences in colony morphology described as "dull" and "bright" colony forms, where the bright form consisted of actively motile forms, while the dull form consisted of long filaments and rods, sluggishly motile. After one passage on 0·1% phenol agar, the bright form did not give any H agglutination, while the dull form gave H agglutination even after seven passages. After 2½ h boiling the dull form gave antisera containing both O and H agglutinins, while the bright form gave a pure O serum. The bright form should therefore be used for O serum production. If such forms cannot be obtained, washing of the boiled culture will be helpful. It is stressed that O determinations of *Proteus* should be carried out only with cultures heated at 100°C for 1 h, in order to prevent the O inagglutinability effect of the flagellar antigen. For further details, see Kauffmann (1966). Kauffmann and Perch (1947) described 49 O antigens and later Perch (1948) gave a more detailed analysis of 26 of these O groups (see Kauffmann, 1966).

(*b*) *K antigens* have not been demonstrated. Recently Larsson *et al.* (1973) have applied immunodiffusion techniques to the analysis of some *P. vulgaris* and *P. mirabilis* strains. No K antigens could be detected.

(*c*) *H antigens*. Antiserum production and antigen determination are performed according to general procedures. Slide or tube agglutination of live or formolised cultures can be used. Tube agglutinations should be read after 4 h at 50°C instead of the usual 2 h incubation. 19 different H antigens have been described. All strains are monophasic.

(*d*) *Antigenic scheme*. A simplified antigenic scheme developed according to the principles used for the *Escherichia* scheme has been described by Kauffmann and Perch (1947) (see Kauffmann, 1966). This scheme contains 49 O groups and 19 H antigens. A more elaborate scheme comprising only the first 26 O groups, but in which both O groups and H antigens have been subdivided into O and H antigenic factors, has been worked out by Perch (1948) (see Kauffmann, 1966).

(*e*) *Cross-reactions*. Kauffmann and Perch (1948), Perch (1948) and Frantzen (1950) found several O antigen cross-reactions to *Escherichia* and *Salmonella*. Further cross-reactions involving *P. vulgaris* and *P. mirabilis* strains are mentioned by Sedlak and Rauss (1968).

2. *Proteus morganii*

(*a*) *O antigens*. O serum production and antigen determination are performed according to general procedures. Rauss and Vörös (1967) have described 32 O groups (numbered O1 to O34) showing few cross-reactions.

(*b*) *K antigens*. Rauss (1957) and Ralovich and Vörös (1967) have described two thermolabile K antigens, K1 and K2. The *P. morganii* strain O29: K1 was identical with *E. coli* O112ac: K66(B) and *P. morganii* O34: K2 with *E. coli* O111: K58(B). In the light of our own examinations of *E. coli* K antigens of the B type (see page 24) we would suggest that these strains have no special acidic polysaccharide K antigens, but that the two K antigens described most likely are part of the respective lipopolysaccharides. Obviously a closer examination of these K antigens should be attempted.

(c) *H antigens*. Antiserum production and antigen determination are according to general procedures and may be carried out as tube or slide agglutination. 25 H antigens have been established. For details, see Sedlak and Rauss (1968).

3. *Proteus rettgeri*

(a) *O antigens*. Antiserum production and antigen determination according to general procedures. 34 O antigens were described by Namioka and Sakazaki (1958). Penner and Hinton (1973) described 18 O antigens found in Canadian strains. They did not compare these antigens with those of Namioka and Sakazaki, and they changed the traditional technique on two points using (1) antisera produced with non-heated motile cells and (2) indirect haemagglutination instead of bacterial agglutination technique. Penner and Hinton (1976) have later described the practical application of their serological system.

(b) *K antigens*. Namioka and Sakazaki (1959) described a partly heat-labile antigen in *P. rettgeri* and other *Proteus* strains. They considered it to be a K antigen different from the L and B antigens of *E. coli* and from the common α and β antigens (see page 30); they termed it the C antigen. Only a limited number of strains were examined. We consider it likely that the C antigen is a fimbrial antigen. Penner and Hinton (1973) described a K antigen which they termed K1 and which, according to their examinations, was heat stable at 100°C but was destroyed by autoclaving. It was a common antigen of *P. rettgeri* cultures and found in several serologically different strains. It was not compared with the C antigen of Namioka and Sakazaki (1958).

(c) *H antigens*. Antiserum is produced and antigen determined according to general procedures. Namioka and Sakazaki (1958) numbered 26 flagellar antigens. All cultures were monophasic. Penner and Hinton (1973) listed 12 H antigens and did not compare them with those of Namioka and Sakazaki (1958).

(d) *Antigenic schemes*. Namioka and Sakazaki (1958) published a provisional antigenic scheme containing 34 O antigens, 26 H antigens and a total of 45 O: H types. Penner and Hinton (1973) published an antigenic scheme comprising 18 O antigens, 1 K antigen and 18 H antigens. The scheme was not compared with that of Namioka and Sakazaki (1958).

(e) *Cross-reactions*. Sedlak and Rauss (1968) have found O antigen cross-reactions to several strains from the *P. vulgaris-mirabilis* group.

4. *Proteus inconstans* = genus *Providencia*

Most authors agree to establish this group as a special genus (see Edwards and Ewing, 1972).

(*a*) *O antigens.* O serum production and antigen determination may be carried out according to the general procedure. Ewing *et al.* (1954) described 56 O antigen groups and the number was later extended to 62 (Edwards and Ewing, 1972). A limited number of cross-reactions between the groups were detected.

(*b*) *K antigens.* Many cultures are inagglutinable or poorly agglutinable in O serum. Ewing *et al.* (1954) described two K antigens. The first one, labelled K1, is the same as the α antigen of Stamp and Stone (1944). This antigen was found in strains belonging to O groups 14, 15, and 46. The K2 antigen was found in the test strain of O group 25; this strain showed capsular quellung in homologous OK antiserum and absorption results indicated that it was a K antigen of the L type. No details were given.

(*c*) *H antigens.* Antiserum production and antigen determination follow general procedures. Ewing *et al.* (1954) described 28 H antigens and later brought the number to 30 (Edwards and Ewing, 1972).

(*d*) *Antigenic scheme.* Ewing *et al.* (1954) have published an antigenic scheme comprising 56 O groups and 28 H antigens.

(*e*) *Cross-reactions.* A number of reciprocal O antigenic relationships between *Providencia* and *E. coli* were described by Ewing (see Edwards and Ewing, 1972). Further O antigen relationships between *Providencia* and *Salmonella* subgroup III (*Arizona*) and *Shigella* are recorded in Edwards and Ewing (1972) who also mention O and H relations to *P. morganii* and *P. rettgeri* strains.

K. *Erwinia*

Erwinia has a special position within the *Enterobacteriaceae*. This genus consists of plant pathogens which until the last decade were excluded from *Enterobacteriaceae* and therefore have been independently grouped.

Only recently has it been shown that *Erwinia* belongs to the *Enterobacteriaceae* (for a review, see Starr and Chatterjee, 1972). Agglutination has been used for routine detection of single species, but no attempt has been made to carry out a systematic antigenic analysis of bacteria of this group and less has been done to examine the possible relationships between antigens of *Erwinia* and those found in other *Enterobacteriaceae*.

L. Yersinia

See chapter by Sten Winblad in volume 12.

X. SEROLOGICAL CROSS-REACTIONS WITH ANTIGENS OUTSIDE THE *ENTEROBACTERIACEAE* FAMILY

A. Cross-reacting antigens found in micro-organisms

The most famous example is probably the cross-reactivity between certain *Proteus* O antigens and antigens in *Rickettsia prowazekii* which was exploited for the Weil–Felix reaction for detection of antibodies in epidemic typhus.

In the following a few similar examples are given, emphasis being on polysaccharide capsular antigens. Avery *et al.* (1925) described cross-reactions between *Klebsiella* type 2 and pneumococcus type 2, and Kauffmann (1966) has shown that *Klebsiella* types 8 and 9 also cross-react with pneumococcus type 2. Cross-reactions of *E. coli* K30 polysaccharide in antipneumococcus sera types II and V and of K42 polysaccharide in antipneumococcus sera types I and XXV were reported by Heidelberger *et al.* (1968). Grados and Ewing (1970) detected cross-reactions between *E. coli* K1 and *Neisseria meningitidis* group B. A more detailed analysis of this relationship has been carried out by Kasper *et al.* (1973). Glode *et al.* (1977) report the cross-reactivity of polysaccharide capsule substance from *E. coli* K92 and *Neisseria meningitidis* type C. Bradshaw *et al.* (1973) described the antigenic relationship between the polysaccharide capsular substance found in *Haemophilus influenzae* type b and in *E. coli* K100. Cross-reactions between *Streptococcus pneumoniae* type 3 polysaccharide and the *E. coli* K7 antigen were reported by Robbins *et al.* (1975) who at the same time discussed the possible immunological importance of such cross-reactions. Winkle *et al.* (1972) demonstrated O antigen relationships between *Vibrio cholerae* and *Salmonella*, *Citrobacter* and *E. coli* strains.

B. Cross-reacting antigens found outside the microbial world

Cross-reactions between antigens of the *Enterobacteriaceae* and animal tissues (particularly blood group antigens) have been intensively studied. Thus the strong cross-reaction between *E. coli* O86 and human blood group B has been noted. For an extensive review written by one of the pioneers in this field, see Springer (1971).

REFERENCES

Aleksic, V., Rohde, R., and Aleksic, S. (1973). *Ann. Microbiol. Inst. Pasteur*, **124B**, 587–590.
Anderson, E. S., and Rogers, A. H. (1963). *Nature, Lond.*, **198**, 714–715.

Anderson, T. F. (1949). In "The Nature of the Bacterial Surface" (Eds. A. A. Miles and N. W. Pirie), p. 76. Oxford Univ. Press, New York.

Andrewes, F. W. (1922). J. Path. Bact., 25, 505–521.

Avery, O. T., Heidelberger, M., and Goebel, W. F. (1925). J. exp. Med., 42, 709–725.

Axelsen, N. H., Krøll, J., and Weeke, B. (Eds.) (1973). "Quantitative Immunoelectrophoresis". Universitetsforlaget, Oslo.

Axelsen, N. H. (1975). "Quantitative Immunoelectrophoresis". Universitetsforlaget, Oslo.

Baker, E. E., Whiteside, R. E., Basch, R., and Derow, M. (1959). J. Immunol., 83, 680–696.

Barry, G. T., and Goebel, W. F. (1957). Nature, Lond., 179, 206.

Bergey's Manual of Determinative Bacteriology, 8th edition (1974), pp. 290–340. (Eds. R. E. Buchanan, and N. E. Gibbons). The Williams and Wilkins Company, Baltimore.

Bettelheim, K. A., Bushrod, F. M., Chandler, M. E., and Trotman, R. E. (1975). Zbl. Bakt. I. Abt. Orig. A, 230, 443–451.

Beutner, E. H. (1961). Bact. Rev., 25, 49–76.

Boyden, S. V. (1951). J. exp. Med., 93, 107–120.

Bradshaw, M., Schneerson, R., Parke, J. C., Jr, and Robbins, J. B. (1971). Lancet, i, 1095–1096.

Brenner, D. J., Fanning, G. R., Skerman, F. J., and Falkow, S. (1972). J. Bact., 109, 953–965.

Brinton, C. C. (1965). Trans. N.Y. Acad. Sci. Ser. II., 27, 1003–1054.

Buczowski, Z. (1968). In "Enterobakteriaceae Infektionen" (Eds. J. Sedlak, and H. Rische), 2nd edn., pp. 251–273. VEB Georg Thieme, Leipzig.

Chandler, M. E., and Bettelheim, K. A. (1974). Zbl. Bakt. I. Abt. Orig., 229, 74–79.

Clark, W. R., McLaughlin, J., and Webster, M. E. (1958). J. biol. Chem., 230, 81–89.

Davis, B. R., and Woodward, J. M. (1957). Canad. J. Microbiol., 3, 591–597.

Duguid, J. P. (1959). J. gen. Microbiol., 21, 271–286.

Duguid, J. P., Anderson, E. S., and Campbell, I. (1966). J. Path. Bact., 92, 107–138.

Duguid, J. P., and Campbell, I. (1969). J. med. Microbiol., 2, 535–553.

Duguid, J. P., Smith, I. W., Dempster, G., and Edmunds, P. N. (1955). J. Path. Bact., 70, 335–348.

Durlakowa, I., Marecz-Babczyszyn, J., Przondo-Hessek, A., and Lusar, Z. (1963). Arch. Immunol. Ther. exp., 11, 549–562.

Durlakowa, I., Przondo-Hessek, A., and Maresz-Babczyszyn, J. (1960). Arch. Immunol. Ther. exp., 8, 645–654.

Edwards, P. R., and Ewing, W. H. (1972). "Identification of Enterobacteriaceae", 3rd edn. Burgess Publishing Company, Minneapolis.

Evans, D. G., Silver, R. P., Evans, D. J., Jr, Chase, D. G., and Gorbach, S. L. (1975). Inf. and Imm., 12, 656–667.

Eveland, W. C., and Faber, J. E. (1953). J. infect. Dis., 93, 226–236.

Ewing, W. H. (1963). Int. Bull. bact. Nomencl., 13, 95–110.

Ewing, W. H., Davis, B. R., and Reavis, R. W. (1959). "Studies on the Serratia Group". CDC Monograph, Atlanta.

Ewing, W. H., and Davis, B. R. (1971). "Biochemical Characterization of Citrobacter freundii and Citrobacter diversus". CDC Monograph, Atlanta.

Ewing, W. H., Davis, B. R., and Fife, M. A. (1972). "Biochemical Characterization of *Serratia liquefaciens* and *Serratia rubidaea*". CDC Monograph, Atlanta.

Ewing, W. H., Tanner, K. E., and Dennard, D. A. (1954). *J. infect. Dis.*, **94**, 134–140.

Felix, A. (1938). *J. Hyg. (Lond.)*, **38**, 750–772.

Felix, A., and Pitt, R. M. (1934). *Lancet*, **ii**, 186–191.

Fey, H., and Wetzstein, H. P. (1975). *Med. microbiol. Immunol.*, **161**, 73–78.

Frantzen, A. (1950). *Acta path. microbiol. scand.*, **27**, 647–649.

Frederiksen, W. (1970). *Publications of the Faculty of Science of the University J. E. Purkinye, Brno*, **47**, 89–94.

Freer, J. H., and Salton, M. R. J. (1971). *In* "Microbial Toxins" (Eds. G. Weinbaum, S. Kadis, and S. J. Ajl), Vol. IV: Bacterial Endotoxins, pp. 67–126. Academic Press, New York and London.

Gard, S. (1938). *Z. Hyg. Infekt.-Kr.*, **120**, 615–619.

Garegg, P. J., Lindberg, B., Onn, T., and Sutherland, I. W. (1971). *Acta chem. scand.*, **25**, 2103–2108.

Gemski, P., Jr, Koeltzow, D. E., and Formal, S. B. (1975). *Inf. Imm.*, **11**, 685–691.

Gillies, R. R., and Duguid, J. P. (1958). *J. Hyg. (Lond.)*, **56**, 303–318.

Glode, M. P., Robbins, J. B., Liu, T. Y., Gotschlich, E. C., Ørskov, I., and Ørskov, F. (1977). *J. infect. Dis.*, **135**, 94–102.

Goebel, W. F. (1963). *Proc. nat. Acad. Sci. (Wash.)*, **49**, 464–471.

Grabar, P., and Williams, C. A. (1953). *Biochim. biophys. Acta*, **10**, 193–194.

Grados, O., and Ewing, W. H. (1970). *J. infect. Dis.*, **122**, 100–103.

Gross, R. J., and Rowe, B. (1975). *J. Hyg. (Lond.)*, **75**, 121–127.

Guinée, P. A. M., Agterberg, C. M., and Jansen, W. H. (1972). *Appl. Microbiol.*, **24**, 127–131.

Guinée, P. A. M., Jansen, W. H., and Achterberg, C. M. (1976). In preparation.

Guinée, P. A. M., Veldkamp, J., and Jansen, W. H. (1977). In preparation.

Heidelberger, M., Jann, K., Jann, B., Ørskov, F., Ørskov, I., and Westphal, O. (1968). *J. Bact.*, **95**, 2415–2417.

Heidelberger, M., and Jann, K. (1975). Unpublished, see Jann, K., and Westphal, O. (1975).

Hellerquist, C. G., Lindberg, B., Svensson, S., Holme, T., and Lindberg, A. A. (1969). *Acta chem. scand.*, **23**, 1588–1596.

Hellerquist, C. G., Lindberg, B., Lönngren, J., and Lindberg, A. A. (1971). *Carbohyd. Res.*, **16**, 289–296.

Henriksen, S. D. (1949). *Acta path. microbiol. scand.*, **26**, 903–916.

Henriksen, S. D. (1950). *Acta path. microbiol. scand.*, **27**, 107–109.

Henriksen, S. D. (1954). *Acta path. microbiol. scand.*, **34**, 271–275.

Hungerer, D., Jann, K., Jann, B., Ørskov, F., and Ørskov, I. (1967). *Europ. J. Biochem.*, **2**, 115–126.

Iino, T. (1969). *Bact. Rev.*, **33**, 454–475.

Iino, T. M., and Lederberg, J. (1964). *In* "The World Problem of Salmonellosis" (Ed. E. van Oye), pp. 111–142. Dr W. Junk Publishers, The Hague.

Isaacson, R. E., Nagy, B., and Moon, H. W. (1977). *J. infect. Dis.*, **135**, 531–539.

Iseki, S., and Sakai, T. (1953a). *Proc. Japan. Acad.*, **29**, 121–126.

Iseki, S., and Sakai, T. (1953b). *Proc. Japan. Acad.*, **29**, 127–131.

Jain, K., Radsak, K., and Mannheim, W. (1974). *Int. J. system. Bact.*, **24**, 402–407.

Jann, K., Jann, B., Ørskov, F., Ørskov, I., and Westphal, O. (1965). *Biochem. Z.*, **342**, 1–22.

Jann, K., and Westphal, O. (1975). In "The Antigens" (Ed. M. Sela), Vol. III, pp. 290–340. Academic Press, New York and London.

Jude, A., and Nicolle, P. (1952). C. R. Acad. Sci., 234, 1718–1720.

Julianelle, L. A. (1926). J. exp. Med., 44, 113–128.

Kabat, E. A., and Mayer, M. M. (1961). "Experimental Immunochemistry", p. 124. Charles C. Thomas Publ., Springfield.

Kampelmacher, E. H. (1959). Antonie van Leeuwenhoek, 25, 289–324.

Kasper, D. L., Winkelhake, J. L., Zollinger, W. D., Brandt, B. L., and Artenstein, M. S. (1973). J. Immunol., 110, 262–268.

Kauffmann, F. (1935). Z. Hyg. Infekt.-Kr., 116, 617–651.

Kauffmann, F. (1936). Z. Hyg. Infekt.-Kr., 117, 778–791.

Kauffmann, F. (1940). Acta path. microbiol. scand., 17, 135–144.

Kauffmann, F. (1943). Acta path. microbiol. scand., 20, 21–44.

Kauffmann, F. (1944). Acta path. microbiol. scand., 21, 20–45.

Kauffmann, F., and Vahlne, G. (1945). Acta path. microbiol. scand., 22, 119–137.

Kauffmann, F. (1949). Acta path. microbiol. scand., 26, 879–881.

Kauffmann, F. (1956). Acta path. microbiol. scand., 39, 299–304.

Kauffmann, F. (1957). Acta path. microbiol. scand., 40, 343–344.

Kauffmann, F. (1966). "The Bacteriology of Enterobacteriaceae". Munksgaard, Copenhagen.

Kauffmann, F., and Perch, B. (1947). Acta path. microbiol. scand., 24, 135–149.

Kauffmann, F., and Perch, B. (1948). Acta path. microbiol. scand., 25, 608–620.

Ketyi, I., Vertényi, A., and Financsek, I. (1971). Acta microbiol. acad. sci. hung., 18, 47–54.

Knipschildt, H. E. (1945a). Acta path. microbiol. scand., 22, 44–64.

Knipschildt, H. E. (1945b). "Undersøgelser over Coligruppens Serologi". Nyt Nordisk Forlag, Arnold Busck, Copenhagen.

Knipschildt, H. E. (1946). Acta path. microbiol. scand., 23, 179–186.

Kunin, C. M., Beard, M. V., and Halmagyi, N. E. (1962). Proc. Soc. exp. Biol. (N.Y.), 160–166.

Lachowicz, K. (1968). In "Enterobacteriaceae Infektionen" (Eds. J. Sedlak, and H. Rische), pp. 445–512. VEB Georg Thieme, Leipzig.

Larsson, P., Hanson, L. Å, and Kaijser, B. (1973). Acta path. microbiol. scand. B, 81, 641–649.

Laurell, C.-B. (1966). Analyt. Biochem., 15, 45–52.

Le Minor, L. (1968a). Ann. Inst. Pasteur, 109, 505–515.

Le Minor, L. (1968b). Int. J. system. Bact., 18, 197–201.

Le Minor, L., Fournier, P. J., and Eliachar, E. (1962). La Semaine des Hôpitaux, 38, 3073–3086.

Le Minor, L., Rohde, R., and Rowe, B. (1974). Ann. Microbiol. Inst. Pasteur, 125A, 427–432.

Le Minor, L., Rohde, R., and Rowe, B. (1975). Ann. Microbiol. Inst. Pasteur, 126B, 63–68.

Lüderitz, O., Jann, K., and Wheat, R. (1968). In "Compr. Biochem." 26A, 105–228 (Eds. M. Florkin, and E. H. Stotz), Elsevier Publ. Co., Amsterdam.

Lüderitz, O., Staub, A. M., and Westphal, O. (1966). Bact. Rev., 30, 192–255.

Lüderitz, O., Westphal, O., Staub, A. M., and Nikaido, H. (1971). In "Microbial Toxins" (Eds. G. Weinbaum, S. Kadis, and S. J. Ajl), vol. IV, pp. 145–224. Academic Press, New York and London.

McWhorter, A. C., Ewing, W. H., and Sakazaki, R. (1967). Bact. Proc., p. 89.

Madsen, Sten. (1949). "On the Classification of the Shigella Types". Munksgaard, Copenhagen.
Mäkelä, P. H., and Stocker, B. A. D. (1969). *Ann. Rev. Genet.*, 3, 291–322.
Maresz-Babczyszyn, J. (1962). *Arch. Immunol. Ther. exp.*, 10, 589–617.
Markovitz, A. (1964). *Proc. natn. Acad. Sci.* (*Wash.*), 51, 239–246.
Matsumoto, H. (1963). *Jap. J. Microbiol.*, 7, 105–114.
Matsumoto, H. (1964). *Jap. J. Microbiol.*, 8, 139–141.
Mushin, R. (1949). *J. Hyg.* (*Lond.*), 47, 227–234.
Mushin, R. (1955). *J. Hyg.* (*Lond.*), 53, 297–304.
Namioka, S., and Sakazaki, R. (1958). *Ann. Inst. Pasteur*, 94, 485–499.
Namioka, S., and Sakazaki, R. (1959). *J. Bact.*, 78, 301–306.
Neter, E., Bertram, L. F., Zak, D. A., Murdock, M. R., and Arbesman, C. E. (1952). *J. exp. Med.*, 96, 1–15.
Neter, E., Westphal, O., Lüderitz, O., and Gorzynski, E. A. (1956). *Ann. Rev. N.Y. Acad. Sci.*, 66, 141–156.
Neufeld, F. (1902). *Z. Hyg. Infekt.-Kr.*, 40, 54–72.
Nimmich, W., and Korten, G. (1970). *Path. Microbiol.*, 36, 179–190.
Nimmich, W., and Münter, W. (1975). *Z. allg. Mikrobiol.*, 15, 127–129.
Novgorodskaya, E. M. (1968). *Pasteur Institute of Epidemiology and Microbiology Leningrad*, 33, 7–16.
Old, D. C., and Payne, S. B. (1971). *J. med. Microbiol.*, 4, 215–225.
Ørskov, F. (1954). *Acta path. microbiol. scand.*, 35, 179–186.
Ørskov, F. (1976). *Acta path. microbiol. Scand.*, B84, 319–320.
Ørskov, F., and Ørskov, I. (1962). *Acta path. microbiol. scand.*, 55, 99–109.
Ørskov, F., Ørskov, I., Jann, B., and Jann, K. (1971). *Acta path. microbiol. scand.*, 79, 142–152.
Ørskov, F., and Ørskov, I. (1972). *Acta path. microbiol. scand. B*, 80, 905–910.
Ørskov, F., and Ørskov, I. (1975). *Acta path. microbiol. scand. B*, 83, 595–600.
Ørskov, F., Sharma, V., and Ørskov, I. (1978a). In preparation.
Ørskov, F., Ørskov, I., Schneerson, R., Egan, W., Sutton, A., and Robbins, J. B. (1978b). In preparation.
Ørskov, I. (1956). *Acta path. microbiol. scand.*, 38, 375–384.
Ørskov, I., Ørskov, F., Sojka, W. J., and Leach, J. M. (1961). *Acta path. microbiol. scand.*, 53, 404–422.
Ørskov, I., and Ørskov, F. (1970). *Acta path. microbiol. scand. B*, 78, 593–604.
Ørskov, I., Ørskov, F., Smith, W. S., and Sojka, J. M. (1975a). *Acta path. microbiol. scand B*, 83, 31–36.
Ørskov, I., Ørskov, F., Bettelheim, K. A., and Chandler, M. E. (1975b). *Acta path. microbiol. scand. B*, 83, 121–124.
Ørskov, I., Sharma, V., and Ørskov, F. (1976). *Acta path. microbiol. Scand.*, B84, 125–131.
Ørskov, I., and Fife-Asbury (1978). *Int. J. Syst. Bacteriol.* (In press).
Ørskov, I., Ørskov, F., Jann, B., and Jann, K. (1977). *Bact. Rev.*, 41, 667–710.
Ouchterlony, O. (1958). *Progr. Allergy*, 5, 1–78.
Oye, E. van (1968). In "Enterobacteriaceae Infektionen" (Eds. J. Sedlak, and H. Rische), 2nd edn., pp. 603–606. VEB Georg Thieme, Leipzig.
Penner, J. L., and Hinton, N. A. (1973). *Canad. J. Microbiol.*, 19, 271–279.
Penner, J. L., and Hinton, N. A. (1976). *J. clin. Microbiol.*, 3, 385–405.
Perch, B. (1948). *Acta path. microbiol. scand.*, 25, 703–714.
Perch, B. (1950). *Acta path. microbiol. scand.*, 27, 565–571.

Popoff, M., and Richard, C. (1975). *Ann. Microbiol. Inst. Pasteur*, **126B**, 17-23.
Ralovick, B., and Vörös, S. (1967). *Acta microbiol. hung.*, **14**, 189-194.
Rauss, K., and Vörös, S. (1967). *Acta microbiol. hung.*, **14**, 195-198.
Refai, M., and Rohde, R. (1975). *Zbl. Bakt. I. Abt. Orig. A*, **233**, 171-179.
Richard, C. (1975). *Bull. Inst. Pasteur*, **73**, 357-381.
Robbins, J. B., Schneerson, R., Liu, T-Y., Schiffer, M. S., Schiffman, G., Myerowitz, R. L., McCracken, G. H., Ørskov, I., and Ørskov, F. (1975). *In* "The Immune System and Infectious Diseases" (Eds. E. Neter, and F. Milgrom), pp. 218-241. S. Karger, Basel.
Robbins, P. W., and Uchida, T. (1962). *Biochem.*, **1**, 323-335.
Rohde, R., Aleksic, S., Müller, G., Plavsic, S., and Aleksic, V. (1975). *Zbl. Bakt. I. Abt. Orig. A*, **230**, 38-50.
Roschka, R. (1950). *Klin. Med.*, **5**, 88-93.
Rowe, B., Gross, R. J., and Woodroof, D. P. (1977). *Int. J. Syst. Bacteriol.*, **27**, 15-18.
Sakazaki, R. (1961). *Jap. J. med. Sci. Biol.*, **14**, 223-241.
Sakazaki, R. (1967). *Jap. J. med. Sci. Biol.*, **20**, 205-212.
Sakazaki, R., and Namioka, S. (1960). *Jap. J. med. Sci. Biol.*, **13**, 1-12.
Sakazaki, R., Tamura, K., and Saito, M. (1967). *Jap. J. med. Sci. Biol.*, **20**, 389-399.
Sarvas, M., and Mäkelä, P. H. (1965). *Acta path. microbiol. scand.*, **65**, 654-656.
Scheidegger, J. J. (1955). *Int. Arch. Allergy*, **7**, 103-110.
Schlecht, S., and Westphal, O. (1966). *Zbl. Bakt. I. Abt. Orig.*, **200**, 241-259.
Schlecht, S., and Westphal, O. (1967). *Zbl. Bakt. I. Abt. Orig.*, **204**, 335-355.
Schlecht, S., and Westphal, O. (1968a). *Zbl. Bakt. I. Abt. Orig.*, **205**, 487-501.
Schlecht, S., and Westphal, O. (1968b). *Zbl. Bakt. I. Abt. Orig.*, **207**, 317-333.
Sedlak, J. (1968a). *In* "Enterobacteriaceae Infektionen" (Eds. J. Sedlak, and H. Rische), pp. 521-530. VEB Georg Thieme, Leipzig.
Sedlak, J. (1968b). *In* "Enterobacteriaceae Infektionen" (Eds. J. Sedlak, and H. Rische), pp. 595-601. VEB Georg Thieme, Leipzig.
Sedlak, J. (1968c). *In* "Enterobacteriaceae Infektionen" (Eds. J. Sedlak, and H. Rische), pp. 607-612. VEB Georg Thieme, Leipzig.
Sedlak, J., Dlabac, V., and Motlikova, M. (1965). *J. Hyg. Epidem. (Praha)*, **9**, 45-53.
Sedlak, J., and Matejovska, D. (1958). *J. Hyg. Epidem. Microbiol. Immun.*, **2**, 285-299.
Sedlak, J., and Rauss, K. (1968). *In* "Enterobacteriaceae Infektionen" (Eds. J. Sedlak, and H. Rische), pp. 559-594. VEB Georg Thieme, Leipzig.
Sedlak, J., and Rische, H. (Eds.) (1968). "Enterobacteriaceae Infektionen", 2nd edn. VEB Georg Thieme, Leipzig.
Sedlak, J., and Slajsova, M. (1966). *Zbl. Bakt. I. Abt. Orig.*, **200**, 369-374.
Seltmann, G. (1971). *Zbl. Bakt. I. Abt. Orig.*, **A219**, 324-335.
Seltmann, G. (1972). *Z. allg. Mikrobiol.*, **12**, 497-520.
Seltmann, G., and Beer, W. (1974). *J. Hyg. Epidem. (Praha)*, **18**, 397-404.
Seltmann, G., and Reissbrodt, R. (1975). *Zbl. Bakt. I. Abt. Orig.*, **A231**, 122-125.
Semjen, G., Ørskov, I., and Ørskov, F. (1977). *Acta path. microbiol. Scand.*, **B85**, 103-107.
Shands, J. W., Jr (1971). *In* "Microbial Toxins" (Eds. G. Weinbaum, S. Kadis, and S. J. Ajl), vol. IV: "Bacterial Endotoxins", pp. 127-144. Academic Press, New York and London.
Simmons, D. A. R. (1971). *Bact. Rev.*, **35**, 117-148.

Slopek, S. (1968a). *In* "Enterobacteriaceae Infektionen" (Eds. J. Sedlak, and H. Rische), pp. 375–444. VEB Georg Thieme, Leipzig.
Slopek, S. (1968b). *In* "Enterobacteriaceae Infektionen" (Eds. J. Sedlak, and H. Rische), pp. 531–557. VEB Georg Thieme, Leipzig.
Smith, D. (1927). *J. exp. Med.*, **46**, 155–166.
Smith, H. W., and Linggood, M. A. (1971). *J. med. Microbiol.*, **5**, 243–250.
Smith, T., and Bryant, G. (1927). *J. exp. Med.*, **46**, 133–140.
Smith, T. (1927). *J. exp. Med.*, **46**, 141–154.
Springer, G. F. (1970). *Ann. N.Y. Acad. Sci.*, **169**, 134–152.
Springer, G. F. (1971). *Progr. Allergy*, **15**, 9–77.
Stamp, L., and Stone, D. M. (1944). *J. Hyg. (Lond.)*, **43**, 266–272.
Starr, M. P., and Chatterjee, A. K. (1972). *Ann. Rev. Microbiol.*, **26**, 389–426.
Stocker, B. A. D., and Mäkelä, P. H. (1971). *In* "Microbial Toxins (Eds. G. Weinbaum, S. Kadis, and S. J. Ajl), vol. IV, pp. 369–438. Academic Press, New York and London.
Thomason, B. M., Cherry, W. B., and Edwards, P. R. (1959). *J. Bact.*, **77**, 478–486.
Tinelli, R., and Staub, A. M. (1960a). *Bull. Soc. Chim. biol. (Paris)*, **42**, 583–599.
Tinelli, R., and Staub, A. M. (1960b). *Bull. Soc. Chim. biol. (Paris)*, **42**, 601–610.
Toenniessen, E. (1915). *Zbl. Bakt. I. Abt. Orig.*, **75**, 329–336.
Toenniessen, E. (1920). *Zbl. Bakt. I. Abt. Orig.*, **85**, 225–237.
Topley, W. W. C., and Ayrton, J. (1924). *J. Hyg. (Lond.)*, **23**, 198–222.
Vahlne, G. (1945). *Acta path. microbiol. scand., Suppl.*, **62**, 1–127.
Weibull, C. (1950). *Acta chem. scand.*, **4**, 268–276.
West, M. G., and Edwards, P. R. (1954). *Publ. Hlth Monogr. No. 22.*
Winblad, S. (1978). *In* "Methods in Microbiology" (Eds. T. Bergan and J. Norris), Vol. 12, Academic Press, London, New York and San Francisco.
Winkle, S., Refai, M., and Rohde, R. (1972). *Ann. Inst. Pasteur*, **123**, 775–781.

CHAPTER II

Bacteriocin Typing of *Enterobacteriaceae*

R. R. GILLIES

Professor of Clinical Bacteriology, Department of Microbiology,
The Queen's University of Belfast

I. INTRODUCTION

The need for laboratory techniques allowing epidemiological characterisation of micro-organisms isolated from patients, carriers or the environment is primarily dictated by the popularity or severity of the infection caused by the species concerned.

Thus serological typing of members of the genera *Shigella* and *Salmonella* is well established; similarly bacteriophage typing of certain commonly occurring serotypes of salmonellae is invaluable in epidemiological studies; again biochemical types within some serotypes of salmonellae can be recognised and have merit for epidemiological purposes. In this latter regard a recently published biotyping scheme for *Salmonella typhimurium* (Duguid *et al.*, 1975) allows reliable differentiation of such strains by the normally equipped diagnostic bacteriological laboratory without the need for submission of isolates to Reference Laboratories for bacteriophage typing.

Although Gratia (1925) first reported bacteriocin activity in the family *Enterobacteriaceae* more than half a century ago, utilisation of this biological activity for epidemiological purposes was long delayed and even now we are only beginning to appreciate the merits of bacteriocin typing of strains within the several genera of this large family.

Although several authors, and in particular Fredericq, had investigated extensively many of the properties of colicins (the name "colicin" for bacteriocins produced by *Escherichia coli* was introduced by Gratia and Fredericq, 1946) the first publication demonstrating the epidemiological relevance of bacteriocin typing was by Shannon (1957) who reported his method of typing *E. coli* O55: B5 strains by means of colicin production.

He identified six colicin types of that particular serotype of *E. coli* and in investigating 15 separate incidents of infection demonstrated that there was a close relationship between biochemical, antigenic and colicin types but that occasionally it was possible to subdivide a biochemical and antigenic type on the basis of colicin production.

It is in the author's opinion regretable that Shannon's technique in typing *E. coli* has not attracted wide attention although other methods (e.g. *McGeachie*, 1965) have been reported.

Similarly an early attempt at bacteriocin typing of *Proteus* species (Cradock-Watson, 1965) received little attention but a significant resurrection of interest in the bacteriocin typing of such species was recently made by Al-Jumaili (1975a, b). The most significant bacteriocin typing technique of species within the *Enterobacteriaceae* concerns *Shigella sonnei*; this serologically homogeneous species is now the most common cause of bacillary dysentery in many parts of the world, both temperate and tropical, and this high prevalence undoubtedly attracted worldwide interest in the method of typing introduced by Abbott and Shannon (1958) and subsequently modified by Gillies (1964).

II. COLICIN TYPING OF *SHIGELLA SONNEI*

A. General principles

The method basically relies on the demonstration of colicin production by the test strain against a battery of stock indicator strains; the detection of such colicin production is made by streaking the test strain diametrically across the surface of a suitable solid medium and, after incubation and removal of the growth of the test strain, the indicator strains are inoculated on to the surface of the medium at right angles to the original inoculum area. Further incubation will reveal which of the indicator strains have been inhibited by the colicins produced by the test strain. This simple statement of the basic principles of testing for colicin production must be amplified if reliable results are to be obtained and attention must be given to such factors as the medium used, the temperature and duration of incubation and the preparation of the inocula of the indicator strains. These factors vary depending on the genus under investigation and the optimal conditions which will now be noted pertain only to *S. sonnei.*

B. The medium

Abbott and Shannon (1958) favoured the use of layered blood agar plates made with nutrient agar and incorporating 10% horse blood. However Gillies (1964) reported that over a four-year period he had satisfactorily used unlayered tryptone soya agar incorporating 5% horse blood; McGeachie and McCormick (1963) investigated six different media (including two different batches of tryptone soya agar) and demonstrated the superiority of digest nutrient agar prepared from Hartley's broth with 1·2% agar added. These latter authors also noted that, of the media tested, only tryptone soya agar showed variation between batches in colicin production. Such batch variation is safeguarded against, in my laboratories, by obtaining samples of several batches from the manufacturer and testing these for suitability before placing a bulk order for delivery.

C. Incubation temperature

Although a temperature of 37°C is favoured for colicin production by *E. coli* (Shannon, 1957; Linton, 1960; McGeachie and McCormick, 1963) the optimum temperature for the production of colicins by *S. sonnei* is 35–36°C (Abbott and Shannon, 1958; Gillies, 1964). However Aoki *et al.* (1967) follow Hart's (1965) instruction in incubating between 35–37°C; this increase in the upper temperature range reduces the value of the method since certain strains which produce colicins at 35°C do not do so at 37°C; thus for several years it has been our practice to incubate the producer strains at 35°C (± 0.5°C).

Secondary incubation of plates after inoculation with the set of indicator strains is at 37°C since colicins are stable at this temperature during the period required for growth of the indicator strains.

D. Duration of incubation

Gillies (1964) recommended that the potential producer strain should be incubated for 24 h in comparison with the three day incubation period suggested by Abbott and Shannon (1958). Hart (1965) and Aoki *et al.* (1969) use a two day incubation for the producing strain but our selection of 24 h receives backing from McGeachie and McCormick (1963) and continues to be adequate. Any modification which speeds up the issue of typing results must make the method more attractive to the clinician and epidemiologist.

Faster reporting is also facilitated by reducing the period of exposure to

4

chloroform from the 5 h recommended by Abbott and Shannon (1958) to the 15 min used by Gillies (1964) who reported also that the macroscopic growth of the producer strain could safely be removed with a glass slide without previous exposure to $CHCl_3$ provided the growth was jettisoned on to cellulose wadding soaked with $CHCl_3$. Thus the 15 min exposure to that reagent is solely to effectively sterilise the microscopic remnants of growth on the surface of the medium.

E. Maintenance and preparation of indicator strains

The indicator strains (15 in number) routinely used in British and most other laboratories, comprise five wild strains of *S. sonnei*, and eight more strains derived from these wild strains by the selection of resistant variants harvested from the zones of inhibition of the parent strain. The fourteenth indicator is *Shigella dysenteriae* 2 (NCTC 8218) and finally *E. coli* Row donated by Fredericq and derived by him from *E. coli* K12.

There appears to be uniformity among workers in maintaining the indicator strains on Dorset's egg slopes in the dark and at room temperature (ca. 22°C); subcultures are made every 3–4 months. Weekly subculture is made to MacConkey's agar plates and inocula from these are made into peptone water tubes and incubated for 4 h at 37°C before the indicator strains are applied to prepared test plates. A useful multiple loop inoculator described by Brown (1973) greatly facilitates application of the indicator strains and has been used in my laboratories for almost a decade; the simplicity of construction and its cheapness confirms that this multiple loop inoculator can be strongly recommended for use in any bacteriocin typing technique which involves the cross-streaking method of applying indicator strains. Patterns of inhibition can be read and the colicin type determined after 8 or more hours of incubation at 37°C following inoculation of the indicator strains.

F. The method of colicin typing of *Shigella sonnei*

Because of the need to expose plates to chloroform it is usually necessary to use all-glass or glass and metal Petri dishes but an alternative technique allowing the use of plastic Petri dishes will be referred to.

(1) The strain of *S. sonnei* to be tested for colicin production (the producer strain) is inoculated in a diametric streak on the surface of a tryptone soya blood agar plate (5% horse blood) containing 15 ml of medium. Incubation is carried out for 24 h at 35°C (± 0.5°C).

(2) Thereafter the macroscopic growth of the producer strain is removed with a microscope slide; 3–5 ml of $CHCl_3$ is placed in the lid of the Petri dish and the inverted medium-containing portion is replaced for 15 min;

the plate is then opened and the residual CHCl$_3$ decanted into a beaker and retrieved for future use.

(3) The medium is exposed to the air for a few minutes and the 15 indicator strains applied at right angles to the original line of growth; eight strains are placed on the left and the other seven strains on the right half of the plate.

(4) The plate is reincubated for 8–18 h at 37°C which allows growth of the indicator strains and recognition of patterns of inhibition, and thus allocation of the producer strain to one or other colicin type. When for any reason glass Petri dishes cannot be used it is possible to use plastic dishes and then exposure to CHCl$_3$ vapour is made by inverting the medium containing portion of the plate (9 cm dia.) over the lid of a Honey pot (7 cm dia.) containing the CHCl$_3$; this is an irksome procedure.

G. Reliability of the method

Before any laboratory technique can be accepted as a valid method of distinguishing types within a given bacterial species its reliability must be established; indices of reliability at the *in vitro* level are that (1) the type of a given isolate must remain constant on repeated testing and (2) that the type should be maintained after prolonged storage of the isolate.

In vivo reliability should be assessed by firstly demonstrating that replicate isolates from the same patient (or carrier) should all be of the same type and secondly there should be uniformity of type in all cases in a given epidemic situation.

In regard to the *in vitro* indices our experience over more than a decade has demonstrated their fulfilment although one or two strains of the several hundreds kept in storage have shown weakening or partial loss of colicin production after 5+ years, a feature which is analogous to partial loss of antigenicity of strains, e.g. of salmonellae, in serological tests. We have had many years experience of judging the *in vivo* indices of reliability and our findings remain as in the first report (Gillies, 1964); 95·2% of excretors show uniformity of colicin type when consecutive isolations are made during the course of their infection and/or their attainment of carrier status. In such studies it is important to type several colonies from each isolation plate since only then can one demonstrate the minority of excretors shedding more than one colicin type of *S. sonnei* simultaneously.

Because of the ubiquitous nature of Sonne dysentery one would be surprised if a minority of individuals did not harbour more than one colicin type and in longitudinal studies (Gillies, 1965) we have surveyed families and shown that when subjected to a second attack of Sonne dysentery individuals carrying a particular colicin type from a previous episode acquire infection with a different colicin type and may continue to excrete both types for some time.

Similar results are obtained when studying the uniformity of colicin type in epidemics of Sonne dysentery; here approximately 3·8% of epidemics may show lack of uniformity and frequently only one individual in an outbreak in a residential home or other semi-closed community will not conform to the colicin type being excreted by the majority.

Such minority differences need not detract from the reliability of the method particularly when we accept that people may simultaneously harbour a salmonella and shigella strain or two different serotypes of salmonellae without invalidating the serological typing of salmonellae.

In addition to the reliability of this or other methods for the epidemiological characterisation of bacterial species, the discriminatory value of such techniques may not be satisfactory especially if the number of recognised types is small or even if when a large number of types can be detected a large proportion of strains falls into one type. This discriminating ability may vary from time to time as exemplified by the work of Cook and Daines (1964) who reported that in four consecutive years the percentage of untypable strains of *S. sonnei* was respectively 32, 44, 38 and 13%; this variability in discriminating ability may disenchant workers unfortunate enough to use colicin typing only for a short period of time when a particular type predominates.

III. EPIDEMIOLOGICAL VALUE OF COLICIN TYPING

In any species which has been subjected to bacteriocin typing by a reliable method the usefulness of the method in relating source cases or carriers to fresh cases of infection is undoubted and this is perhaps most obvious in pyocin typing of *Pseudomonas aeruginosa* in the nosocomial infection situation.

The excitement generated by colicin typing of *S. sonnei* seems to have been confined to laboratory workers and not reflected by any deep or prolonged interest of Medical Officers of Health (Community Medicine Specialists) perhaps because of the essentially mild (although highly prevalent) nature of Sonne dysentery.

However long term studies (Gillies, 1968; Aoki *et al.*, 1969) suggest that the periodicity in the incidence of Sonne dysentery in a community is related to the emergence of a particular colicin type of *S. sonnei* which predominates for 5–7 years and is then replaced by a different colicin type; thus type 7 strains predominated in Edinburgh from 1957 to 1961 and in 1962 type 11 strains made an attempt to dominate the scene. However from 1963 to 1967 untypable strains accounted for the majority of isolates and were in turn usurped by colicin type 4 strains. These periodicity cycles were mirrored by the waxing and waning of infection in Edinburgh suggesting the development of herd immunity to the emerging and then dominant colicin type of *S. sonnei*. Such cyclical associations merit investigation and might lead to some understanding of immunity to Sonne dysentery.

IV. BACTERIOCIN TYPING OF OTHER ENTEROBACTERIA

Reference has already been made to McGeachie's (1965) report on typing of *E. coli* strains from urinary tract infections where he employed sensitivity of his isolates to stock bacteriocin preparations; earlier, Linton (1960) had used the more orthodox method of assessing colicin production of such isolates in an epidemiological investigation of post-operative urinary tract infections. Both authors were satisfied as to the reliability of their typing procedures.

Again in *Proteus* species Craddock-Watson (1965) reported only three significant types in strains of *P. mirabilis* and *P. vulgaris* although this was later extended to nine types (Pitt, 1975 quoted by Al-Jumaili, 1975b). Al-Jumaili (1975a, b) uses 12 standard bacteriocin preparations to type *Proteus* strains and recognises 48 different sensitivity patterns among the 1000 isolates thus typed; no strain was untypable and a high level of discrimination is obtained since Al-Jumaili's most commonly encountered type (2) accounted for only 9·8% of 204 isolates. Further investigation of this most recently reported bacteriocin typing technique for enterobacteria should be encouraged especially in the light of possible improvements by using optimal media and cultural conditions (George, 1975).

REFERENCES

Abbott, J. D. and Shannon, R. (1958). *J. clin. Path.*, **11**, 71–77.
Al-Jumaili, I. J. (1975a). *J. clin. Path.*, **28**, 784–787.
Al-Jumaili, I. J. (1975b). *J. clin. Path.*, **28**, 788–792.
Aoki, Y., Naito, T., Fujise, N., Miura, K., Iwanaga, Y., Ikeda, A., Jinnouchi, K., Morino, T. and Miyahara, M. (1969). *Trop. Med.*, **11**, 57–75.
Aoki, Y., Naito, T., Matsuo, S., Fujise, N., Ikeda, A., Miura, K. and Yakushiji, Y. (1967). *Jap. J. Microbiol.*, **11**, 73–85.
Brown, D. O. (1973). *Med. Lab. Tech.*, **30**, 351–353.
Cook, G. T. and Daines, C. F. (1964). *Mon. Bull. Min. Hlth.*, **23**, 81–85.
Craddock-Watson, J. E. (1965). *Zbl. Bakt. I. Orig.*, **196**, 385–388.
Duguid, J. P., Anderson, E. S., Alfredsson, G. A., Barker, Ruth and Old, D. C. (1975). *J. med. Microbiol.*, **8**, 149–166.
George, R. H. (1975). *J. clin. Path.*, **28**, 25–28.
Gillies, R. R. (1964). *J. Hyg., Camb.*, **62**, 1–9.

Gillies, R. R. (1965). *Mill. Mem. Quart.*, **43**, 224–232.

Gillies, R. R. (1968). *Arch. Immunol. et Therap. Exper.*, **16**, 410–420.

Gratia, A. (1925). *C.R. Soc. Biol., Paris*, **93**, 1040–1041.

Gratia, A. and Fredericq, P. (1946). *C.R. Soc. Biol., Paris*, **140**, 1032–1033.

Hart, Judith M. (1965). *Zbl. f. Bakt. I. Orig.*, **196**, 364–369.

Linton, K. B. (1960). *J. clin. Path.*, **13**, 168–172.

McGeachie, J. and McCormick, W. (1963). *J. clin. Path.*, **16**, 278–280.

McGeachie, J. (1965). *Zbl. f. Bakt. I. Orig.*, **196**, 377–384.

Shannon, R. (1957). *J. med. Lab. Tech.*, **14**, 199–214.

CHAPTER III

Phage Typing of *Escherichia Coli*

HEDDA MILCH

National Institute of Hygiene, Budapest, Hungary

I. INTRODUCTION

A. Pathogenicity

The pathogenic role of *Escherichia coli* has been a subject of increasing interest. According to the type of the associated disease it is usual to speak of enteric and extra-enteric infections. Based on the symptoms, epidemiological characteristics and the type of pathogenic agent *human enteric diseases* are classified into two main groups.

(i) The cholera-like syndrome includes infantile diarrhoea and traveller's diarrhoea. The former affects children up to 5 years of age and can give rise to outbreaks. The disease is usually associated with toxic symptoms, the outcome is often fatal. Traveller's diarrhoea affects adults causing sporadic or epidemic diarrhoeal disease. Toxigenic *E. coli* strains of certain serogroups or serotypes are responsible for the syndrome.

(ii) The dysentery-like syndrome may affect children or adults. It occurs in sporadic or epidemic form. Often the disease is water-borne or food-borne. Non-toxigenic, invasive *E. coli* serotypes are the causative agents.

Human extra-enteric diseases due to *E. coli* include pyelonephritis, cholecystitis, cholangitis, appendicitis, endocarditis, meningitis, otitis, sinusitis. In addition, *E. coli* strains have been incriminated for local abscesses in a variety of organs, diseases of the respiratory tract and sepsis. Definite serotypes have been demonstrated as etiological agents in these conditions.

Enteric and extra-enteric diseases of animals (pigs, calves, lambs) manifest in diarrhoea or in generalised infection are also caused by some definite serotypes of *E. coli*.

Based on symptoms and the type of the pathogenic agent pig diseases are grouped as follows:

(i) Neonatal *E. coli* diarrhoea of piglets.
(ii) Weaning *coli* bacillary diarrhoea occurring at the time of weaning.
(iii) Oedema disease also frequently occurring soon after weaning.

In calves, (i) an intestinal form and (ii) a generalised form with bacteraemia, in lambs (i) enteric disease, (ii) localised infections, and (iii) generalised bacteraemia are distinguished.

Microbiological and epidemiological observations have indicated that *E. coli* can also cause human diseases and the affected animals can serve as reservoirs of human enteritis.

In addition to the known *E. coli* serogroups and types involved in human and animal enteric disease, the existence of new pathogenic or potentially pathogenic serological groups and types has to be considered. Some strains cannot be grouped into the rigid limits of the above classification.

The ubiquity of *E. coli* strains lends to the differentiation between pathogenic and non-pathogenic strains greater importance than is the case with obligate pathogens. An increasing number of serological groups and types have been isolated in enteritis outbreaks and the same groups and types could be isolated from the intestinal tract of healthy individuals. The often contradictory data have given impetus to intensive research and many attempts have been made at defining more precisely strains of the same serological group by biochemical-, sero-, colicin- and phage-typing.

The development of relevant laboratory methods has facilitated epidemiologic studies of infantile enteritis. Sporadic cases and outbreaks of enteritis in adults have been investigated more extensively only recently. Epidemiological studies on extra-enteric *E. coli* infections have been carried out mainly in urinary tract diseases. Edwards and Ewing (1972), Sakazaki *et al.* (1962), Costin (1966), Lachowicz (1968), Novgorodskaya *et al.* (1970), Ørskov *et al.* (1971) and Sojka (1972) have reviewed the *E. coli* serotypes known to be pathogenic for man and animals. The actual importance of *E. coli* infections can only be established if the results of laboratory diagnosis and typing are related to epidemiological data.

B. Classification and cultural characters: minimum requirement for diagnosis

Bacteria of the *Escherichia* group are usually motile, and seldom capsulated. The organisms ferment mannitol, usually with gas production. Most strains acidify lactose promptly and form gas from it at 37°C and 44°C. They produce indole, give a positive methyl-red reaction, a negative Voges–Proskauer reaction and do not grow in Koser's citrate medium. No growth occurs in Møller's KCN medium. The organisms do not liquify gelatin, they possess lysine and glutamic acid decarboxylases.

E. coli strains grow readily on simple nutrient media. The strains are aerogenic or semi-aerogenic, they grow at temperatures between 18 and 42°C, some grow well at 44°C. Colonies on nutrient agar are 2–3 mm in diameter, circular, low convex, smooth or rough with a mucous shiny surface or dry. The M-S-R morphological variation can also be observed in subcultures of the same strain. On blood agar some strains form β-haemolysin. The colonies are violet in colour and have a metallic lustre on Eosin-Methylene blue agar, red on Endo agar and on MacConkey's lactose agar.

Edwards and Ewing (1972) recommend the examination of the antigenic structure as the first step of subdividing *E. coli* strains. This is based on the reactions obtained in an agglutination test with O and OK sera. If the results are positive, the agglutination tests are carried out on boiled bacterial suspensions. Simultaneously the biochemical tests according to Edwards and Ewing (1972), and Kauffmann (1966) can be performed. (See this volume, Chapter 1.)

A number of methods have been developed for assessing pathogenicity of *E. coli* strains. These are based on the type of the antigen, the toxins produced (haemolysin, necrotoxin, enterotoxin), ability of the strains to multiply, penetrate and to resist phagocytosis, or on the combination of these characters. Kauffmann (1948) observed that strains isolated from septic infections of man were virulent for the mouse, belonged to the haemolytic type, agglutinated erythrocytes, caused necrosis when injected into the rabbit's skin and had an antigen of the L type. Medearis and Kenny (1968) associated the differences between mouse virulence of strains with their ability to resist phagocytosis. Medearis *et al.* (1968) found that mutants with incomplete O side-chains had greatly reduced virulence. Wolberg and de Witt (1969) also observed a relation between mouse virulence and the presence of L antigen. Glynn and Howard (1970) showed that there was a direct relation between the presence of the K polysaccharide and resistance to complement.

Most but not all of the strains responsible for the oedema disease of piglets possess the protein antigen K88 the production of which is governed by the presence of a transmissible plasmid (Smith and Linggood, 1971).

Smith (1963) found that most strains of animal origin but less than 20% of strains of human origin were haemolytic on sheep blood agar. He recognised two different haemolysins: α-haemolysin and β-haemolysin. Smith and Halls (1967) showed that the ability to form α-lysin was transferable from one strain to another.

Taylor *et al.* (1958) were the first to use the dilation of ligated rabbit ileal loop test for the demonstration of enterotoxin from enteropathogenic *E. coli* (EEC). Enteropathogenic strains of human origin have no dilatation effect on calf, sheep and swine ileum. Bettelheim and Taylor (1970) demonstrated a substance different from O and K antigen. It precipitated with the O-antiserum. Ørskov *et al.* (1971), Bettelheim and Taylor (1971) using electrophoresis demonstrated antigens responsible for enteropathogenicity. Methods recently developed for the demonstration of heat labile enterotoxin are: the production of oedema in the mouse lung (Avdeeva *et al.*, 1973) and the adrenal cell tissue culture test (Donta *et al.*, 1974). Neish *et al.* (1975) developed an *in vitro* assay system to measure bacterial adhesion to the mucosa of human foetal small intestine. The non-toxigenic strains able to penetrate epithelial cells cause kerato-conjunctivitis in guinea pigs (Serény, 1957) or have a cytopathogenic effect on HeLa cell monolayers (La Brec *et al.*, 1964).

Among the methods of rapid diagnosis fluorescence microscopy has found wide acceptance. Examining faecal samples from patients with enteritis and from symptom-free carriers poly- and monovalent sera can be used.

C. Cell wall composition—phage receptor

Extracts from *E. coli* envelopes have been shown to contain the receptor activity (Burnet, 1934; Levine and Frisch, 1934; Weidel, 1953). Chemical analysis of the cell wall has revealed that it is made up of two layers (Weidel, 1953). The outer is composed of lipoprotein containing amino-acids, the inner is mainly lipopolysaccharide in nature, it comprises a network of units built from phospholipid, glucose, glucosamine and L-gala-D-mannoheptose, linked together by units containing muramic acid, alanine, glutamic acid and diaminopimelic acid (Rogers, 1965). Cell walls from which the outer lipoprotein layer has been extracted by phenol, no longer adsorb phages T2, T5, T6 although this treatment actually enhances the adsorption of phages T3, T4 and T7. The receptors for T2, T5 and T6 must therefore reside in the outer layer and those for T3, T4 and T7 in the inner (Weidel *et al.*, 1954, 1958; Weidel and Primosigh, 1958). The role of defective lipopolysaccharides in antibiotic resistance and phage adsorption was proved by Tamaki *et al.* (1971) in experiments with *E. coli* mutants. In some cases the somatic antigen itself is the phage receptor. Highly purified bacterial antigens consisting of lipid, protein and polysaccharides inactivate phages. Adsorption is much influenced by environmental conditions such as temperature, ionic composition, the presence of L-tryptophan, and cationic requirements.

Jackson *et al.* (1967) suggested that the phages T2, T3, T4, T6 and T7 had a common or closely linked attachment locus on the bacterial surface. Bayer (1968) found that T phages seemed to adsorb to the cell envelope areas opposite to cell wall-membrane adhesions. Male-specific phages adsorb to certain type of pili (Brinton et al., 1964).

Randall-Hazelbauer and Schwartz (1973) isolated the bacteriophage lambda receptor from E. coli. Ryter et al. (1975) showed how integration of the receptor for bacteriophage lambda takes place in the outer membrane of the cell wall.

The correlation between phage sensitivity and antigen was studied by Kauffmann and Vahlne in 1945. The non-capsulated forms of strains belonging to serogroup O9 are phage sensitive, while the capsulated ones are phage resistant and strains of the same serotype exhibited different sensitivity to phages. Toft (1947) also investigated the relationship between capsular antigen and phage sensitivity in L+, L−, A+ and A− strains of the O9 serogroup. Based on sensitivity, phages attacking non-capsulated forms (commonest), only capsulated forms (rarest) and both forms could be distinguished. Stirm (1968, 1971) described 13 specific K phages. Stirm et al. (1972) and Fehmel et al. (1957) demonstrated that the spike organelles of phages played a role in adsorption and penetration through the polysaccharide gel functioning as receptor. The surface polysaccharides of the E. coli Bi 161/42 (O9:K29V:H−) strain have been extensively studied; its cell wall lipopolysaccharide was found to consist of a core oligosaccharide of R1 type (Schmidt et al., 1969; Schmidt, 1972). Schmidt (1972) demonstrated that the lipopolysaccharide can be characterised by phage pattern. Nhan et al. (1971) and recently Choy et al. (1975) investigated the hexasaccharide repeating units and determined the structure of these units.

A similarity in the enzymatic activity of E. coli K phages and Vi phage II was shown by Stirm (1968). Chemical analysis of cell wall lipopolysaccharides (Weidel et al., 1954; Taylor and Taylor, 1963; Bernard et al., 1965a, 1965b) support this similarity. Borisov et al. (1970) found a correlation between the phage and colicin receptors and the OB antigens of group O26 and O111 strains.

II. PHAGE TYPING

A. Principles and methods

Serological methods have been successfully used to reveal pathogenicity of E. coli strains. However, biochemical and serological techniques have proved less suited for clarifying the route and source of nosocomial infections. Biochemical methods are of limited value for type sub-division, while serological methods are time consuming and can be reliably performed in but a few laboratories. These difficulties have led to the elaboration of phage typing methods.

E. coli belonging to different serological groups or untypable by serological methods are divided into phage types or phage patterns using a
suitable phage-set. In general, phages isolated from different sources and
selected empirically are used. Occasionally phages produced by the
adaptation of an isolated phage are utilised. In another phage typing
method, the effect of prophages carried by lysogenic strains on selected
indicator strains is examined.

First, phage typing methods for *E. coli* causing human enteric infections
were elaborated. Typing of strains derived from human extra-enteric
infections and of strains pathogenic for animals have also been developed.

B. Phage typing systems

In the following more important phage typing methods, attempts to
elaborate such methods will be reviewed, since no international standard
technique exists.

1. *Phage typing of* E. coli *from human enteric infections*

(a) Nicoll *et al.* (1952a) differentiated the strains O111:K58(B4) into
7, the strains O55: K59(B5) into 9, the strains O26: K60(B6) into 5 phage
types. A set of 30 phages was used in the studies. Some were specific for
O111 others for O55, others lysed strains of both these groups or of all
three.

Nicolle *et al.* (1952b, 1954, 1958, 1960) classified more than 12,000
strains from different geographical areas and reported good results from an
epidemiological viewpoint. However, they stated that the phage types of the
strains under study were liable to degeneration. Later they partly extended
and simplified the method. Using 25 typing phages strains O111: K58(B4)
were grouped into 11, strains O55: K59(B5) into 7, and strains O26: K60
(B6) into 4 phage types (Rische, 1968).

Nicolle *et al.* (1964) directed attention to certain associations between
phage sensitivity and other characters.

(i) Relation was found between phage sensitivity, antigen H, and
biochemical characters (β-phenylpropionic acid reaction, (PPR), described
by D'Allessandro and Comez, 1952) in the strains O111:K58(B4) and
O55: K59(B5).

(ii) In studies on O111: K58(B4) Nicolle *et al.* (1963) found that strains
of the phage type Sèvres did not behave uniformly with respect to lysogeny.

(iii) A relationship was suggested between changes in phage type of the
strains O111, O55 and O26, and the antibiotic therapy applied (Nicolle
et al., 1960, Viallier *et al.*, 1963).

(b) Eörsi *et al.* (1953, 1954) using four typing phages differentiated the nosocomial strains O111: K58(B4) and O55:K59(B5). With the exception of one phage, the typing phages were specific for the corresponding serological groups. Using a set of 4 phages, strains of serological group O111: K58(B4) were classified into 4, the strains O55: K59(B5) into 3 types. The phage typing method was further developed by Milch and Deák (1961) and used for the typing of strains belonging to serological group O26:K60(B6). In subsequent work, the set of typing phages was supplemented with newly isolated or adapted phages. Strains of group O111 could be differentiated into 7, strains of O55 into 5, the strains of O26 into 4 phage types. Simultaneously H antigen, antibiotic sensitivity, biochemical characters (Kauffmann and Dupont, 1960), and PPR (D'Alessandro and Comes, 1952) were determined. The epidemiological use of the method was tested on strains derived from various hospital infections, on strains obtained from healthy infants in Hungary and on strains received for typing from France, the GDR and Romania. Based on the results of phage typing, serological typing and biochemical characterisation the following conclusions could be drawn:

(i) In agreement with the findings of Nicolle and co-workers, the strains O111: K58(B4) showed a close association between phage type, serotype and the result of the PPR.

(ii) Antibiotic sensitivity tests carried out in 1952 showed that phage types 111/1 of the non-flagellate strains were resistant to streptomycin, the rest were sensitive.

(iii) No association was found between lysogeny and phage type, although temperate phages could be demonstrated in 61·8, 11·1 and 0%, from the strains of the O111, O55 and O26 serogroups, respectively. The fact that typing phages are unrelated to prophages carried by lysogenic strains may account for the finding.

(c) Adzharov (1966) elaborated a method for typing the strains O111:K58(B4). Using a set of 13 phages he could differentiate 12 phage types and 2 variants and reported that the method could be successfully applied in epidemiological studies. The author used the phage set for the differentiation of the serogroups O111 and O55 and recommended phage typing also for diagnostic purposes indicating that phage sensitivity tests can support serological and morphological results.

(d) In addition to strains of O111, O55 and O26 serogroups, EEC O127: K63(B8) has been encountered most frequently in Europe and North America. Serological types and biochemical characters of these strains showed heterogeneity (Le Minor et al., 1962). Most of the strains were non-flagellated and strains having the same antigen showed uniform fermentation characteristics. Ackermann et al. (1962) using a set of 9 phages differentiated 9 phage types and 2 variants. In contrast to the strains belonging to serogroups O26, O55, O111 and O119 examined by Nicolle *et al.* (1960) and Kasatiya (1962) only a small number of the O127 strains proved lysogenic. A close correlation was found between lytic pattern, flagellar antigen and biochemical properties. The authors maintained that phage typing was more rapid and simple than were serological and biochemical differentiation.

(e) Strains of the *E. coli* serogroup O114 often give rise to enteric outbreaks. Kayser (1964) studied the biotypes of the strain and ¦elaborated a method of phage typing using 7 non-specific phages isolated from sewage and faeces. He differentiated 10 phage types and observed clear association between biotype and phage type. The epidemiological value of the method has been checked on a limited number of strains.

(f) Chistovich and Matyko (1967) developed a method for the phage typing of EEC O9 and O26 strains. The typing phages were isolated from stool samples of infants with enteritis and from sewage. Using a set of 5 phages, the O26 group could be differentiated into 10 phage types. By the help of a further 6 typing phages the O9 serogroup could be classified into 24 phage types. The value of the method in epidemics was checked on 112 strains.

(g) Several authors (Edwards and Ewing, 1972; Ewing, 1963; Costin, 1966) reported on *E. coli* strains of the O1 to O25 serogroups associated with infantile enteritis. The most common O serogroups were 4, 6, 8, 18, 20 and 25. With the help of 57 typing phages isolated from sewage, Bercovici *et al.* (1969, 1975) differentiated strains of the O1 to O25 serogroups obtained from stool samples of infants with enteritis, further strains of animal origin. A set of 4–12 typing phages were used for each of the serogroups, 80% of the strains were typable by phages. The serogroups O4, O5 and O15 proved to be the most heterogenic having 7–20 phages types.

2. *Phage typing of E. coli from human extra-enteric infections*

Attempts at differentiation have been mainly directed at phage typing of strains isolated from patients with urinary tract infections. Serological typing carried out by Ujváry (1957), Rantz (1962), Kunin (1966) revealed that some serogroups different from those found in infantile enteritis were responsible for these infections. The following O serological groups were found: 1, 2, 4, 6, 7, 11, 15, 21, 22, 25, 48, 57, 62, 65, 75, 112 and 118. Brown and Parisi (1966) used 8 typing phages isolated from sewage for the differentiation of *E. coli* strains belonging to 7 serogroups or non-identifiable by serological methods. Of the 90 strains obtained from urinary infections 69·4% proved to be typable. Adding another 5 typing phages to the former set of 8, Parisi *et al.* (1969) differentiated 717 *E. coli* strains derived from urinary tract infections and other sources; 50·3 and 34·4%, respectively, of the strains were found typable, 41 phage types could be distinguished. Specificity of the typing phages was tested on other *Enterobacteriaceae*, beside *E. coli* only *Shigella* strains were lysed. These results indicate that phage typing is an easy and rapid method suitable for epidemiological studies in *E. coli*-associated urinary tract infections.

3. *Phage typing of* E. coli *pathogenic for animals*

Smith and Crabb elaborated a method in 1956 for the phage typing of bovine faecal *E. coli* strains. With the help of the phage set, human, ovine, porcine and avian strains could be differentiated. A total of 71 phage types were found.

Sterne *et al.* (1970) examined 571 *E. coli* strains isolated from healthy and diarrhoeal pigs. Typing was carried out by the method of Brown and Parisi (1966) and Parisi *et al.* (1969). Sixty-one per cent of the strains were typable. Diarrhoeal pigs had phage type patterns similar to those found in normal pigs. Thus, pathogenicity of the strains could not be assessed by phage typing. Several *E. coli* strains of porcine origin were examined, a higher percentage of strains isolated from swine were typable than was the case with strains obtained from piglets.

4. *Phage typing of enteric and extra-enteric* E. coli *strains*

Milch and Gyenes (1972) investigated phage sensitivity of different *E. coli* serogroups. The strains under study were isolated from infants and adults with enteritis, from premature babies with interstitial pneumonia and from other pathological conditions. Comparative studies of O antigen and phage sensitivity revealed that some phage patterns or phage groups were characteristic for a given serological group. Other serogroups again could be subdivided on the basis of phage sensitivity. In the period 1969–1975, a total of 2657 serologically identified and 789 non-identified *E. coli* strains were examined for phage pattern, colicinogeny and lysogeny. The studies indicated that serotyping, a more labour demanding procedure could be dispensed with if the strains isolated from the same source had an identical phage pattern.

Strains non-identifiable by phage typing should be re-examined at elevated temperature (45°C), when most strains assume phage sensitivity. This technique was adopted for phage typing of *E. coli* by Marsik and Parisi (1971). Several theories and hypotheses were offered for the explanation of phage sensitivity following heat treatment. These include inactivation of restrictive enzymes (Schell and Glover, 1966) changes in O antigen (Rhodes and Fung, 1970) and/or changes in the lysogenic state (Cavallo, 1951; Lieb, 1953; Zichichi and Kellenberger, 1963).

III. TECHNICAL METHODS OF DIFFERENT TYPING SYSTEMS

A. Strains associated with human enteric infections

1. *Phage typing system according to Nicolle* et al. *for* E. coli O111: K58(B4), O55: K59(B5) *and* O26: K60(B6) *strains*

(*a*) *Typing phages and propagating strains.* Of the 200 phages isolated, first 30 (Nicolle *et al.*, 1960), later only 25 (Rische, 1968) were used. Three phages of unknown origin (designated 3, 4 and 6) came from E. Wollman's collection (Services des Bactériophages, Institut Pasteur, Paris). A further 3 phages (1, 2 and 5) were isolated from sewage in Clichy, one phage (24) was isolated from a lysogenic strain. The rest were isolated from stool samples of infants suffering from gastroenteritis. With the exception of phage 24, the phages were propagated on the strains used for their isolation. The typing phages can be obtained from Service des Bactériophages, Institut Pasteur, Paris.

Table I illustrates the typing phages and their propagating strains.

(*b*) *Culture media.* The strains to be tested are maintained at 4°C in small tubes containing agar medium or Dorset's medium. The culture medium used for phage propagation is:

Vaillant peptone 5 B	20 g
NaCl	6 g
Tap water	1000 ml

For typing the same medium is used containing 1·7% agar.

(*c*) *Propagation of phages* is carried out in the above mentioned liquid medium. The phages are used undiluted with the exception of phages 2, 11, 14 and 21. The first is applied in a dilution of 1: 25, the rest in 1: 5. The filtered phage preparations contain 10^6–10^8 plaque-forming units (PFU)/ml. The preparations are stored at 4°C in closed bottles. Before use the undiluted or diluted phage is transferred into another bottle mounted with a drip nozzle.

TABLE I

Typing phages and propagating strains
according to Nicolle *et al.* (1960, 1968)

Phages	Propagating strains	Serogroups
1	E.c.1330	O111
2	E.c. 47	O111
3	E.c. 333 c.10	O111
4	E.c. 333 c.10	O111
5	E.c. 47 c. 4	O111
6	E.c. 47 c. 4	O111
7	E.c. 76	O55
8	E.c.1941 P1.m.	O55
9	E.c.B	non-identified
10	E.c. 27	O128
11	E.c. 110	non-identified
12	E.c. 117 c.blue	non-identified
13	E.c. 116	non-identified
14	E.c. 24	non-identified
15	E.c. 116	non-identified
16	E.c. 112	O111
17	E.c. 112	O111
18	E.c. 111	O26
19	E.c. 123	O55
20	E.c. 129	O26
21	E.c. 28	O111
22	E.c. 28	O111
23	E.c. 28	O111
24	E.c. 309	*S. paratyphi B* Phage type 1
25	E.c.1233	O111

(*d*) *Phage titration.* The phages can be used in the concentrations indicated if PFU on the propagating strain is 10^6–10^8/ml.

The propagating strain is inoculated into broth and incubated at 37°C for 4–6 h. From the growing culture 4 drops are transferred to agar and spread with the help of a glass rod. The place where the phage is to be dropped should be marked in advance. Ten-fold dilutions are made from the propagated phage and 0·02 ml is placed on the plate with the aid of a glass dropper. After drying, the plates are incubated at 37°C overnight and lysis is then recorded.

(*e*) *Method of typing.* A 5-h culture of the strain to be tested is used. The same technique is applied as in titration. The results are read after incubation at 37°C for 8 h.

(*f*) *Reading and evaluation of results.* Lysis is characterised according to its intensity. The lytic patterns corresponding to the degree of lysis are shown in Tables II, III and IV. The tables illustrate also the serotypes and the results of PPR.

Nicolle *et al.* (1960) investigated strains from various countries of Europe, Africa, Asia and America. Certain phage types, especially those obtained from regions where no antibiotics had been used, proved to be relatively stable. The introduction of phage typing made it possible to follow up hospital infections, to trace the sources of infections, to differentiate hospital infections from newly introduced ones and from non-nosocomial outbreaks (Buttiaux *et al.*, 1956a, 1956b). Phage typing was more useful epidemiologically than were biotype or serotype determinations

Nicolle *et al.* (1957, 1960) found that changes of phage sensitivity occurred quite frequently *in vitro* and *in vivo*. It involved either sensitivity to new phages, or resistance to some phages. Without going into details on the underlying causes, the recommendations of Nicolle *et al.* (1960) for reducing variability are listed as follows:

(i) Typing should be started immediately after receiving the strain.

(ii) Several colonies should be checked separately.

(iii) Phage typing after intraperitoneal mouse passage, in case of changes in phage sensitivity.

(iv) Simultaneous examination of several biological characters by serotyping, PPR and LDC (lysine decarboxylase reaction) should be performed.

(v) Examination of colicinogeny, identification of the colicin produced by the strains, i.e. colicin typing should be carried out.

(vi) Examination of lysogeny should be performed.

2. *Phage typing system according to Eörsi* et al. (1953, 1954), *Milch and Deák* (1961) *for* E. coli O111: K58(B4), O55: K59(B5) and O26: K60 (B6)

(*a*) *Typing phages and propagating strains.* The typing phages were isolated from stool specimens of patients (mainly infants) with enteritis. Specimens were suspended in 6–8 ml yeast broth and the suspension centrifuged at 2000 rev/min for 10 min. The supernatant was incubated at 60°C in a water bath for 30 min. then kept at 4°C until tested for phage. Various *Enterobacteriaceae* were used as indicator strains, e.g. *Salmonella typhi* phage type A, *Shigella flexneri* serotype 2a, *Shigella sonnei* phase II, *E. coli* O111: K58(B4), O55: K59(B5), O26: K60(B6) and *E. coli* cultivated from that faecal sample from which attempts at phage isolation were being made. Before phage typing the indicator strains were grown in broth for 2–3 h.

TABLE II

Phage typing scheme of O111: K58(B4) according to Nicolle et al. (1960, 1968)

Phage types	H antigen	PPR	1	2	3/4	5/6	7/19	8	9	10	11/12	13	14	15	16	17/18/20	21	22/23	24	25
Montparnasse	2	+	+	+	+	+	−	−	−	−	−	−	−	−	+	+	+	+	−	−
Sèvres	2	+	+	+	−	+	−	−	−	±	−	−	−	−	+	+	+	+	−	−
Sèvres var. lyon	2	+	+	+	−	+	−	−	−	±	−	−	±	−	+	+	−	+	−	−
Tourcoing	2	+	+	−	−	+	−	−	−	±	−	−	−	−	+	+	+	+	−	−
Vienne	2	+	+	−	−	+	−	−	−	−	−	−	−	−	−	+	−	+	−	−
Bretonneau	12	−	+	−	−	−	−	−	−	−	+	−	−	−	−	+	−	−	+	+
Dorf	12	−	+	−	−	−	−	−	−	−	+	−	+	−	−	+	−	−	+	−
Lille	12	−	+	−	−	−	−	−	−	−	−	−	−	−	−	+	−	−	+	−
Israel	4	−	+	−	−	−	−	−	−	+	−	−	−	−	−	+	−	−	−	−
Palerme	21	−	+	−	−	+	−	−	−	−	+	−	−	−	−	+	+	−	+	+
Indonésie	12	−	+	+	−	+	−	−	−	−	+	−	−	−	+	+	+	+	−	−
Japon (Nagoya)	40	−	+	+	−	+	−	−	−	+	+	−	−	−	+	+	+	+	+	+

+ = confluent lysis.

± = semiconfluent lysis or plaques.

PPR = β-phenyl-propionic acid reaction.

TABLE III

Phage typing scheme of O55: K59 (B5) according to Nicolle et al. (1960, 1968)

Phage types	H antigen	PPR	1	2	3 / 4	5 / 6	7 / 19	8	9	10	11 / 12	13	14	15	16	17 / 18 / 20	21	22 / 23	24	25
																			Typing phages	
St. Christopher	2	+	−	−	−	+	+	−	−	−	−	−	±	−	+	−	+	+	−	−
Weiler	7	−	−	−	−	−	+	−	−	−	v	−	+	−	+	−	+	+	−	+
Flandre	2, 6, 7, 11, 27, 32	−	−	−	−	−	+	v	−	+	−	−	−	−	−	−	−	−	+	−
Graz	6	−	−	−	−	−	+	+	−	v	−	−	+	−	−	−	v	−	+	+
Londres	2, 6, 7	−	−	−	−	−	+	+	−	−	−	−	+	−	−	−	−	−	v	−
Lomme	2, 6, ?	−	−	−	−	−	+	+	−	−	−	−	−	−	−	−	−	−	v	−
Béthune	6, 21	−	−	−	−	+	+	+	−	+	v	−	+	−	+	−	+	+	−	−
Jérusalem	32	−	−	−	−	+	+	−	−	+	−	−	+	−	+	−	+	+	−	−
Finlande	4	−	−	−	−	+	+	−	−	+	−	−	+	−	+	−	+	+	−	−

+ = confluent lysis.
± = semiconfluent lysis or plaques.
PPR = β-phenyl-propionic acid reaction.
v = variable.

TABLE IV

Phage typing scheme of O26: K60 (B6) according to Nicolle et al. (1960, 1968)

Phage types	H antigen	PPR	Typing phages																								
			1	2	3	4	5	6	7	19	8	9	10	11	12	13	14	15	16	17	18	20	21	22	23	24	25
Birmingham	11	—	—	—	—	—	+	—	+	—	—	—	+	+	+	—	—	—	+	+	+	—	±	+	—	—	—
Zürich	—	—	—	—	—	—	+	—	—	—	—	—	—	+	+	—	—	—	—	+	+	—	—	—	—	—	—
Warwick	—	—	—	—	—	—	—	—	—	—	—	—	+	+	+	—	—	—	+	+	+	+	+	—	—	—	—
Liège	—	—	—	—	—	—	+	—	—	—	—	—	—	—	—	—	—	—	+	+	+	+	+	+	—	—	—

+ = confluent lysis.

± = semiconfluent lysis or plaques.

PPR = β-phenyl-propionic acid reaction.

From each culture drops 15 mm in diameter were placed on to agar plates. After drying the plates, the supernatant was pipetted on to the drops. Results were read following incubation at 37°C for 8 h. From the several hundred phages that were isolated 5 were chosen for typing. Phages "a1" and "aF42" were obtained by adaptation from phage "a". The typing phages and the propagating strains are shown in Table V.

TABLE V

Typing phages and propagating strains according to
Eörsi *et al.* (1953, 1954) and Milch & Deák (1961)

| Phages | Propagating strains | |
	Phage type	H antigen
a	111/1	—
a1	111/5	—
aF42	26/1	11
b	111/2	2
c	26/1	11
k	55/4	6
l	55/1	6

First undiluted phages were used, since 1968 they have been used at routine test dilution (RTD), that is at the highest dilution of which 0·02 ml gives a semi-confluent lysis with the propagating strain. Typing phages and propagating strains can be obtained from H. Milch, National Institute of Hygiene, Budapest, Hungary.

(*b*) *Culture media.* For the propagation of phages Hartley's broth is used. It is prepared as described by Craigie and Felix (1947), with the modification that instead of beef digest, horse flesh digest is employed.

Hartley's agar (Hartley's broth + 1·9% agar) is used for typing.

(*c*) *Propagation of phages.* Hartley's agar plates are swabbed with a 2 to 3-h broth culture of the strain used for propagation, so as to obtain an even spread on the surface of the medium. Excess suspension is pipetted off and the plates are allowed to dry with the lids open for approximately 30 min at room temperature. Subsequently 12 places are marked on the agar medium with the help of a sterilised test-tube. Using a loop 4 mm in diameter one drop of the phage to be propagated is applied on each of the previously marked places. After drying, the plates are incubated for 16 h at 37°C. The lytic areas together with the surrounding bacterial growth and the agar medium are cut out circularly with the help of a loop, and placed into tubes containing 4·5 ml broth. The tubes are shaken for 4 h at 37°C. Thereafter the content is removed and centrifuged for 15 min at 3000 rev/min. The supernatant is diluted and titrated against the propagating strain. If a 10^{-6} to 10^{-8} dilution still results in complete lysis, the phage is filtered through a G5 glass filter and titrated again.

(d) *Method of typing.* Plating of the strain and application of typing phages is carried out as described in paragraph (c). The results are read following incubation at 37°C for 6 h.

(e) *Reading and evaluation of results.* Using phages at RTD the scheme of interpretation used by Eörsi *et al.* (1953, 1954), and Milch and Deák (1961) is shown in Tables VI, VII and VIII. Lysis is characterised according to the intensity of phage effect (+ indicates complete, or semi-complete lysis, ± separate plaques).

The use of the method is based on experience obtained in the phage typing of 3250 strains; 1860 strains belonged to serogroup O111, 1238 strains to group O55, 152 strains to group O26. The strains were isolated from hospital epidemics, sporadic cases of enteritis and from healthy infants in Hungary, or were received for typing from the GDR and Romania.

It is often found that following repeated subculture or on storage the phage patterns of some strains have changed. This variability *in vitro* makes it important that typing should be carried out immediately after isolation and as far as possible all the strains isolated during a given epidemic should be examined at the same time.

TABLE VI

Phage typing scheme of O111: K58 (B4) according to Eörsi *et al.* (1953, 1954) and Milch & Deák (1961)

Phage type	H antigen	PPR	Typing phages				
			a	a1	aF42	b	b2
111/1	—	+	+	±	+	+	+
111/5	—	+	—	+	—	—	±
111/7	—	+	±	—	—	+	+
111/2	2	+	—	—	—	+	+
111/4	2	—	—	—	—	+	+
111/3	12	—	—	—	—	±	±
111/6	12	—	—	—	—	—	—

+ = confluent lysis.
± = plaques.
PPR = β-phenyl-propionic acid reaction.

TABLE VII

Phage typing scheme of O55: K59 (B5) according to
Eörsi *et al.* (1953, 1954) and Milch & Deák (1961)

Phage type	H antigen	PPR	Typing phages		
			k	1	aF42
55/1	6	—	+	+	—
55/2	6	—	+	—	—
55/3	6	—	—	+	—
55/4	2 or —	+	+	±	+
55/5	6	—	—	—	—

+ = confluent lysis.
± = plaques.
PPR = β-phenyl-propionic acid reaction.

TABLE VIII

Phage typing scheme of O26: K60 (B6) according to
Eörsi *et al.* (1953, 1954) and Milch & Deák (1961)

Phage type	H antigen	PPR	Typing phages	
			aF42	c
26/1	11	—	+	+
26/2	11 or ?	—	—	+
26/3	11	—	+	—
26/4	1 or ?	—	—	—

+ = confluent lysis.
± = plaques.
PPR = β-phenyl-propionic acid reaction.

Phage type stability *in vivo* was examined by repeated typing of a strain isolated on several occasions from stool samples of one and the same infant. Of the many specimens from 98 hospitalised infants, the same phage type could be isolated in 87 cases. In 5 patients type 111/2 changed to 111/1, in 6 type 55/1 to 55/4. However, in these 11 cases the possibility of a nosocomial cross infection could not be excluded.

In vitro and *in vivo* studies indicated that phage types of *E. coli* O111, O55 and O26 were less stable than was the case with *S. typhi*. It is believed that the main reason of this finding was the frequency of phages acting upon enteric *E. coli* strains and the fact that these phages may lysogenise the strains under study. In spite of the changes in phage sensitivity, phage typing proved to be a useful tool in studying outbreaks, complementing the results of serotyping and PPR.

3. Comparative trials of the typing phages used by Nicolle et al., Eörsi et al., and further by Milch and Deák for the typing of O111: K58(B4), O55: K59(B5) and O26: K60(B6)

(a) Phages and test strains studied. Nicolle's phages necessary for differentiation are: 2, 4, 6, 7, 8, 11, 14, 16, 18, 21, 24.

Phages isolated by Eörsi et al., Milch and Deák: a, al, aF42, b, c, k, l.

The test strains used in the comparative trials are shown in Tables IX to XI.

Lytic and serological characters were compared. Anti-phage sera were produced against each phage and the serological relation between the phages was determined by cross neutralisation (Milch, 1967).

(b) Results. In Hungary the most common phage type of the serogroup O111 is 111/2. Using Nicolle's phage set it was found identical with the Tourcoing type. Phage type 111/1 isolated from severe hospital epidemics corresponded to Sèvres ubiquitaire. Types 111/3 and 111/7 different only in the intensity of lytic action proved to be identical with phage type Bretonneau and Tourcoing, respectively. The two types had different H antigen and have different PPR (see Table VI). Nicolle's phage set did not lyse phage type 111/5. Phage types 111/1 and 111/3 could be subdivided with Nicolle's set, the Tourcoing type with that of Eörsi, Milch and Deák (see Table IX). Phage types 55/1 and 55/2 corresponded to Nicolle's Bethune type. The less common type 55/4 could be identified as Lomme degr. Nicolle's phage types of world-wide distribution (Béthune, Lomme) corresponded to phage type 55/1, the types St. Christopher and Weiler to 55/4.

The comparative serological studies of phage types are summarised in Table XII.

Based on the neutralisation of anti-phage sera Nicolle's phages and those of Eörsi, Milch and Deák could be divided into serological groups and subgroups. Letters indicate the groups, figures the subgroups. There was serological identity between phage "a" (Eörsi–Milch–Deák) and Nicolle's phages 6 and 24, all of them belonging to subgroup Al. Nicolle's phages 12, 18 and 7 also proved identical serologically (B1 subgroup).

No close correlation could be observed between lytic action and serological characteristics. Thus, for example, lytic activity of the serologically identical phages 12 and 18 varied greatly.

The results of the comparative trials as shown in Tables IX, X and

TABLE IX

Comparison of the O111: K58(B4) phage types of Nicolle et al. (1960) and Eörsi et al. (1953, 1954) and Milch and Deák (1961)

Phage types Eörsi et al. (1953, 1954) Milch and Deák (1961)	H antigen	Eörsi et al. (1953, 1954) Milch and Deák				Nicolle et al. (1960)									Phage types Nicolle et al.
		a	al	aF42	b	2	4	6	11	14	16	18	21	24	
111/1	–	+	±	+	+	+	–	+	–	–	+	+	+	–	Sèvres ubiquitaire
111/5	–	–	+	–	+	+	–	–	–	–	+	+	–	–	Non typable
111/7	–	–	–	–	+	–	–	+	–	–	+	+	+	–	Tourcoing
111/2	2	–	–	±	+	–	–	+	–	–	±	±	+	–	Tourcoing
111/3	12	–	–	–	±	–	–	–	–	–	–	±	–	+	Bretonneau
111/6	12	–	–	–	–	–	–	–	–	–	–	–	–	±	Non typable

Phage types Nicolle et al. (1960)	H antigen	Eörsi et al. (1953, 1954) Milch and Deák				Nicolle et al. (1960)									Phage types Eörsi et al. (1953, 1954) and Milch and Deák (1961)
		a	al	aF42	b	2	4	6	11	14	16	18	21	24	
Montparnasse	2	+	+	+	+	+	+	+	–	±	+	+	+	–	111/1
Sèvres ubiqu.	2	+	±	+	+	+	–	+	–	–	+	+	+	–	111/1
Sèvres lyon.	2	+	±	+	+	+	–	+	–	–	+	+	–	–	111/1
Tourcoing	2	–	–	–	+	–	–	+	+	–	+	+	+	–	111/2
Lille	12	–	–	–	+	–	–	–	+	+	–	+	–	+	111/3
Dorf	12	–	–	–	+	–	–	–	+	–	–	+	±	±	111/3
Bretonneau	12	–	–	–	+	–	–	–	–	–	–	+	–	+	111/3

+ = confluent lysis.
± = plaques.

TABLE X

Comparison of the O55: K58(B4) phage types of Nicolle *et al.* (1960) and Eörsi *et al.* (1953, 1954) and Milch and Deák (1961)

Phage types Eörsi *et al.* (1953, 1954) Milch and Deák (1961)	H antigen	Eörsi *et al.* (1953, 1954) Milch and Deák (1961)		Nicolle *et al.* (1960)								Phage types Nicolle *et al.*
		k	1	6	7	8	11	14	16	18	24	
55/1	6	+	+	−	+	−	−	−	−	−	+	Béthune
55/2	6	+	±	−	+	−	−	−	−	−	−	Béthune
55/3	6	−	+	−	−	−	−	−	−	−	−	Non typable
55/4	6	+	−	−	+	+	−	−	−	−	−	Lomme degr.
55/5	6	−	−	−	−	±	−	−	−	−	−	Degr.

Typing phages according to

Phage types Nicolle *et al.* (1960)	H antigen	k	1	6	7	8	11	14	16	18	24	Phage types Eörsi *et al.* (1953, 1954) and Milch and Deák (1961)
St. Christopher	2	+	±	+	−	−	−	+	+	−	−	55/4
Lomme	2 or 6	+	+	−	+	+	−	−	−	±	±	55/1
Béthune	6 or 21	+	+	+	+	−	±	−	−	±	±	55/1
Weiler	7	+	−	+	+	−	+	+	+	−	−	55/4

+ = confluent lysis.

± = plaques.

TABLE XI

Comparison of the O26: K60(B6) phage types of Nicolle *et al.* (1960) and Eörsi *et al.* (1953, 1954) and Milch and Deák (1961)

Phage types Eörsi *et al.* (1953, 1954) Milch and Deák (1961)	H antigen	Typing phages according to							Phage types Nicolle *et al.* (1960)
		Eörsi *et al.* (1953, 1954) and Milch and Deák (1961)		Nicolle *et al.* (1960)					
		aF42	C	6	11	16	18	21	
26/1	11	+	+	+	+	+	+	+	Birmingham
26/2	11 or ?	−	+	−	+	−	−	−	Non typable
26/3	11	+	−	−	+	−	−	−	Non typable
26/4	11 or ?	−	−	−	+	−	+	−	Warwick

Phage types Nicolle *et al.* (1960)	H antigen	aF42	C	6	11	16	18	21	Phage types Eörsi *et al.* (1953, 1954) and Milch and Deák (1961)
Birmingham	11	+	+	+	+	+	+	+	26/1
Warwick	—	−	+	−	+	−	+	−	26/2

+ = confluent lysis.
± = plaques.

TABLE XII

Neutralisation of the *E. coli* O111, O55 and O26 typing phages of Nicolle *et al.* (1960) and Eörsi *et al.* (1953, 1954) and Milch and Deák (1961)

Typing phages	Serogroup of typing phages	6	a	al	24	aF42	2	16	21	14	k	18	7	8	b	1	c	4	11
6	A1	●	●	●	●	●	●	●	●	○									○
a		●	●	●	●	●	●	●	●	○								○	
al	A2	●	●	●	●	●	●	●	●	○								○	
24	A3	●	●	●	●	●	●	●	●	●									
aF42		●	●	●	●	●	●	●	●	○									
2	A4	○	○	○	●	●	●	○	●										
16	A5	●	●	●	●	●	○	●	●										
21		●	●	●	●	●	●	●	●										
14	A6	○	●	●	●	●	●	●	●										
k	B1								○		●	●	●						
18											●	●	●						
7											●	●	●						
8	B2	○					○		○	○		○	●	●					○
b	C1		○				○		○	○	○	○	●	●	●	○		○	
1	C2			○							○	○	○		○	●			
c	D1																●		
4	E1																	●	
11	F1																		●

● = neutralisation in antiphage-sera 1 : 1000 or 1 : 10,000.
○ = neutralisation in antiphage-sera 1 : 100 or 1 : 10.
without sign = negative reactions.

XI indicate that Nicolle's phages 4, 6, 11, 16, 18 and 24 should be added to the phage set of Eörsi, Milch, Deák.

4. *Phage-typing according to Adzharov* (1966) *for* E. coli O111: K58(B4)

(*a*) *Typing phages.* Thirteen phages were recovered from rivers, sewage and lysogenic strains or were developed by adaptation. The phages were used at RTD ($10^{-1} - 10^{-9}$). No information is given on the strains used for propagation.

(*b*) *Methods of propagation and typing, culture media, reading and evaluation of results.* Methods and culture media were the same as those employed by Nicolle *et al.* The results of phage typing were read after 6–8 h when incubated at 37°C, after 16–18 h when incubated at 30°C. The phage patterns are illustrated in Table XIII. Twelve phage types and 2 variants were differentiated. Efficacy of the method for epidemiological purposes was tested on 700 strains (Adzharov 1966, 1968, 1969).

(*c*) *Comparative trials with Nicolle's and Adzharov's phages for* E. coli *O111 typing.* Adzharov performed trials with 39 strains. Phage types 3 and 4 corresponded to Nicolle's Sèvres degr., the types 195, 7, 11 and 35 to Sèvres ubiquitaire, type 224 to Bretonneau, type 21 to phage type Israel.

5. *Phage typing system according to Ackermann* et al. (1962) *for* O127: K63(B8)

(*a*) *Typing phages and propagating strains.* Of the 9 typing phages 7 were isolated from sewage in Paris, one was Felix–Callow's O1, one Eörsi–Milch's. The strains used for propagation were not mentioned.

(*b*) *Propagation of phages and culture media.* The culture media were the same as used by Nicolle *et al.* (1960). Five drops of 18 h bacterial suspension are added to 20 ml broth, incubated for 3 h at 37°C, filtered through a Chamberland L3 filter and the phage filtrates titrated. The typing phages are used at RTD.

(*c*) *Method of typing and evaluation of results.* Nicolle's scheme is shown in Table XIV. Strains containing no H antigen can be divided into 3 phage types. In the flagellate strains a characteristic phage pattern corresponds to each H antigen. Ackermann and co-workers maintained that phage typing could substitute for serological and biochemical differentiation. A total of 1115 strains have been studied, 90% were typable. The most common phage type, marked A, had 2 variants. Phage types of non-flagellate strains were not variable. Data are scarce on flagellate strains containing H antigen.

TABLE XIII

Phage typing scheme of O111: K58(B4) according to Adzharov (1966)

Phage type	Typing phages													H antigen	Salicin	Saccharose
	1	2	3	4	5	6	7	8	9	10	11	12	13			
A-1	CL	CL	CL	—	—	—	—	—	—	—	—	CL	—	2	+	+
2	—	CL	CL	SCL	CL	—	—	—	—	—	—	CL	—	2	+	+
3	—	SCL	CL	CL	CL	CL	—	—	—	—	—	CL	CL	2	+	+
4	—	—	CL	CL	—	CL	—	—	—	—	—	CL	—	2	+	+
5	—	—	—	CL	CL	CL	—	—	—	—	—	CL	—	2	+	+
5a	—	—	—	SCL	CL	—	CL	—	—	—	—	CL	—	2	+	+
6	—	—	—	SCL	CL	CL	—	—	—	—	—	CL	—	2	+	+
7	—	—	—	—	CL	CL	CL	—	—	—	—	CL	—	2	—	+
8	—	—	—	—	—	CL	CL	CL	—	—	—	CL	CL	2	+	+
9	—	—	—	—	—	—	—	—	CL	—	—	CL	CL	2	+	+
B-1	—	—	—	CL	CL	CL	—	—	—	CL	—	—	—	12	+	+
2	—	—	—	—	—	—	—	—	—	CL	CL	—	—	12	+	—
2a	—	—	—	—	—	—	—	—	—	—	—	—	—	12	+	—
C-1	—	—	—	—	—	—	—	CL	CL	—	—	—	—	4	—	+

CL = confluent lysis
SCL = semiconfluent lysis

TABLE XIV

Phage typing scheme of O127: K63 (B8) according to Ackerman *et al.* (1962)

Phage types	\multicolumn{9}{Typing phages}								
	1	2	3	4	5	6	7	8	9
A	CL	CL	CL	—	×	×	×	×	—
B	CL	CL	CL	—	CL	×	×	×	—
C	CL	CL	CL	—	×	++	×	CL	—
H4	×	—	—	CL	SCL	—	—	—	—
H6	CL	CL	CL	—	×	—	CL	×	—
H11	CL	CL	CL	CL	v	+	—	v	CL
H21	CL	CL	CL	CL	v	CL	—	v	—
H26	SCL	CL	—	—	CL	—	—	CL	—
H40	CL	CL	CL	CL	—	v	CL	v	—
A var. A1	—	CL	CL	—	×	×	×	×	—
A var. A2	CL	SCL	SCL	—	×	×	×	×	—
H21-a	CL	v	CL	—	CL	CL	—	v	—
H21-40	CL	CL	CL	CL	×	×	—	×	—

CL = confluent lysis.
SCL = semiconfluent lysis.
v = variable reactions: CL or negative.
+, ++ = rising number of plaques.
× = inconstant, weak reaction.
— = no lysis.

6. Phage typing system according to Kayser (1964) for O group 114

The 7 typing phages were recovered from passages of single plaques of strains isolated from sewage and from stool samples of infants with enteritis. The phages were used at RTD. Nicolle's phage typing system was adopted. Ten phage types could be distinguished, type I being the most common. The phage patterns are shown in Table XV.

B. Phage typing of *E. coli* from human extra-enteric infections

1. Phage typing system according to Brown and Parisi (1966), Parisi et al. (1969) for E. coli strains causing urinary infections

(*a*) *Typing phages.* Quantities (10 ml) of raw sewage were inoculated into 90 ml of Brain Heart Infusion (BHI, Difco) broth, incubated at 37°C for 24 h, centrifuged, and filtered through a 0·45-nm membrane filter (Millipore). First 8, then 13 phages (A through M) were used for typing. To distinguish between phage activity and colicin activity, the method of Baily and Glynn (1961) was applied.

5

TABLE XV
Phage typing scheme of serogroup O114 according to Kayser (1964)

Phage type	Typing phages						
	1	2	3	4	5	6	7
I	CL	—	—	—	—	—	—
II	CL	CL	—	—	—	—	—
III	SCL-CL	—	CL	—	—	—	—
IV	—	—	—	CL	CL	—	—
V	CL	—	—	—	—	CL	—
VI	CL	—	—	—	CL	—	CL
VII	CL	—	—	—	CL	—	—
VIII	—	—	—	—	+++	—	CL
IX	CL	—	—	CL	CL	—	—
X	—	—	—	CL	CL	—	CL

CL = confluent lysis.
SCL = semiconfluent lysis.
+ + + = plaques.

(b) *Culture media*. Brain Heart Infusion (BHI, Difco) broth was used for phage propagation, BHI agar plates for typing.

(c) *Propagation of phages*. The filtrates were placed on agar plates previously swabbed with indicator strains grown in broth for 18 h at 37°C. Filtrates showing phage activity were diluted serially in 0·1 M ammonium acetate (Bradley, 1963) and 0·2 ml of the diluted filtrate mixed with 0·1 ml of the indicator strain in 2·5 ml of broth containing 0·75% agar. This mixture was then layered on the surface of a BHI agar plate. After incubation at 37°C for 4 to 6 h, discrete plaques were picked and washed into 1 ml of ammonium acetate. This suspension of phages was then used for propagating by the agar layer method (Swanstrom and Adams, 1951). The phages were used at RTD. No details are given of the indicator strains used.

(d) *Method of typing*. Cultures to be typed were grown in BHI broth overnight at 37°C and then swabbed on a BHI agar plate. Each phage was placed with the phage applicator originally described by Zierdt *et al.* (1960), and phages were applied to the surfaces of the inoculated plates. Plates were incubated at 37°C for 4 to 6 h and then examined for lysis. The plates were read again after 12 h at room temperature.

(*e*) *Interpretation and evaluation of results.* Phage types lysed by the typing phages A through M were indicated. Several phage types could be distinguished within the serological groups (Table XVI). Forty-one different phage types were identified; 81·2% of the strains proved typable. The ease and speed of phage typing justified its use as a new tool in epidemiological studies of urinary tract infections.

Urinary *E. coli* originating from various geographical regions were sero-typed and phage-typed by Marsik and Parisi (1971). The method of Brown and Parisi (1966) and Parisi *et al.* (1969) was employed for phage typing. Of the 454 strains 66·1% were typable with the help of phage-sets, 65·2% with 48 sera. Phage typing was successful in 80·2% of the most common serological groups: 4, 6, 75, 1, 50, 7 and 25. Different phage

TABLE XVI

Phage typing of bacteriuric *E. coli* according to Brown and Parisi (1966) (comparison of serological groups and phage types)

No. of cultures	Serological group	Phage types	No. of strains with phage type
15	N.T.*	AH, BCD, BCDEFGH, BDE, BE, DE	1
		D	2
		BCDE	3
		ABCDE	4
5	Auto-aggl.	ABCDE, BCDE, F	1
		BDE	2
1	O1 and O6	ABCD	1
6	O4	BCD	1
		ABCDE	3
		BCDE	2
4	O7	DE	4
1	O11	BE	1
3	O25	ABCDE, BDE, DE	1
5	O75	CD	1
		BCDE, DE	2

* Non typable.

types could be distinguished within one and the same serological group. No relationship was found between O antigen and phage type.

In cases of untypable strains, the plates were incubated at elevated temperature (45°C for 18 to 24 h) or ethidium bromide was applied as curing agent in a final concentration of 100×10^{-9}; the culture was incubated at 37°C for 18 to 24 h (Marsik and Parisi, 1971).

C. Phage typing of E. coli pathogenic for animals

1. *Phage typing of* E. coli *strains of bovine origin according to Smith and Crabb* (1956)

(*a*) *Typing phages*. Specimens of sewage and faeces were incubated in Difco broth at various temperatures between 20 and 37°C for 6 to 24 h, sometimes with the addition of cultures of *E. coli* hitherto regarded as untypable. They were then centrifuged at 3000 rev/min for 45 min, the supernatants either heated to 58°C for 30 min or treated with chloroform and then spotted on to plates of nutrient agar previously spread with broth cultures of *E. coli* strains. The plates were incubated at 28°C for 18 h. A group of 16 different phages, A, B, C, D, E, G, H, L, M, O, T, X, Z_1, Z_2, Z_3, and Z_4 were chosen for typing strains of bovine origin. Another 8 phages Z_5–Z_{12} were developed for the typing of strains isolated from other species of animals.

(*b*) *Culture medium*. The nutrient agar used was prepared by solidifying Difco Nutrient Broth No. B3 with 1·5% (w/v) New Zealand agar.

(*c*) *Propagation of phages*. Lytic areas were picked from the plates and mixed with a broth culture of the susceptible strains of *E. coli*, spread over a plate of nutrient agar and incubated at 28°C for 18 h. A discrete plaque was picked and replated with the propagating strain. (The strains used for phage propagation were not indicated.) This process was repeated twice. Finally, a discrete plaque with some of the surrounding bacterial growth was picked into broth and incubated at 28°C until lysis occurred. More susceptible bacterial culture was then added and the process repeated until a high-titre phage preparation was obtained. Any bacteria present were then killed by heating to 58°C for 30 min. Heat-sensitive phages were sterilised by shaking with a few drops of chloroform.

(*d*) *Phage titration*. Ten-fold dilutions of the phage preparation were made in phosphate buffer (KH_2PO_4, 3·4 g; Na_2HPO_4, 6 g; distilled water, 1000 ml). A broth culture of the susceptible strain was spread evenly over the surface of an agar plate, allowed to dry, and then spotted with one drop (1/150 ml) of each dilution of the phage preparation. Plates were incubated at 28°C for 18 h, and then read. The highest dilution which produced a large number of plaques that were semi-confluent or nearly so was chosen as the critical test dilution. The highest dilution producing confluent lysis was not used since a few of the preparations would inevitably contain colicin causing confusing results.

(e) *Method of typing.* A modification of the method devised by Wilson and Atkinson (1945) for the phage typing of *Staphylococcus aureus* was employed. Nutrient agar plates were dried at 37°C for 2 h with their lids partly open. Four drops (0·08 ml) of an 18-h broth culture of the *E. coli* strain to be typed were spread over the surface of one of the nutrient agar plates by means of a glass spreader. When these had dried, the phage preparations at their critical test dilution were spotted on to each plate by means of a dropping pipette. When the drops had been absorbed the plates were incubated at 28°C for 18 h and read. The lytic patterns for each phage type are shown in Table XVII.

D. Phage typing of enteric and extra-enteric *E. coli* strains

1. *Phage typing system according to Milch and Gyenes* (1972)

(a) *Typing phages and propagating strains.* The typing phages were isolated from stool samples of infants and adults with enteritis. *E. coli* strains of known antigenic structure and pathogenic according to data in the literature were used as indicator strains. Of the many hundred phages 30 were selected on a basis of the lytic patterns obtained with the indicator strains. Twenty-two were isolated by the authors, six by Eörsi *et al.* (1953, 1954) and Milch and Deák (1961) from enteritis, and two were adapted to *E. coli* O78 strains causing enteritis and meningitis. The indicator strains used for isolation were also used for phage propagation. The method of isolation and propagation is described on p. 13. The typing phages and propagating strains are shown in Table XVIII. In addition to *E. coli* phages, the *Salmonella* phage pools: Vi phage I + IV and "PO" (the latter containing phages O1, O2 and O3) were used.

Typing phages and propagating strains can be obtained from H. Milch, National Institute of Hygiene, Budapest, Hungary.

(i) *Serological examination of typing phages.* Anti-phage sera were produced against 28 typing phages. The neutralising effect of the anti-phage serum was determined for every phage, and in the presence of neutralisation, the K value of the serum was estimated according to Burnet *et al.* (1937). Classification of the 28 typing phages according to their serological properties is shown in Table XIX. The typing phages could be divided into seven serogroups including 14 subgroups. The serological groups were set up according to the neutralising action of the anti-phage sera produced

TABLE XVII

Phage typing reactions of the more common types of *E. coli* in calves and cows according to Smith and Crabb (1956)

Phage type*	A	B	C	D	E	G	H	L	M	O	T	X	Z₁	Z₂
1	−	−	−	−	−	+	−	−	−	−	+	−	−	−
2	−	−	−	−	−	−	−	−	−	−	−	−	+	−
3	−	−	−	+	−	+	+	−	+	−	+	−	−	−
5	−	−	−	−	−	+	+	−	+	−	+	−	+	−
6	−	+	−	−	−	+	+	+	+	−	+	−	−	−
8	−	−	−	+	+	+	+	−	+	−	+	−	−	−
11	−	+	+	+	−	+	+	+	+	+	+	−	−	−
13	−	+	+	−	−	−	−	+	−	+	−	−	−	−
18	−	−	+	+	−	+	+	+	+	+	+	−	−	−
22	−	−	+	+	−	−	+	−	+	−	+	−	−	−
26	−	−	−	+	+	+	+	+	+	+	+	−	−	−
31	−	+	+	+	−	+	+	−	+	−	+	−	−	−
36	−	−	−	+	−	+	+	+	+	−	+	−	−	−
37	−	−	−	−	−	−	−	−	−	−	−	+	−	−
38	+	+	+	+	−	+	+	−	+	−	+	−	−	−
39	−	+	+	−	−	+	+	+	+	+	+	−	−	−
40	−	+	+	+	−	+	+	−	−	−	+	−	−	−
41	−	−	−	+	−	+	+	+	−	+	−	−	−	−
43	−	−	−	+	−	+	−	−	+	−	+	−	−	−
45	−	−	−	−	−	+	+	−	−	−	+	−	−	−
48	−	+	+	+	−	+	+	+	+	−	+	−	−	−
58	−	+	+	−	−	+	+	−	−	−	+	−	−	−
60	−	−	−	−	−	+	−	+	+	−	−	−	−	−
63	−	+	+	−	−	−	+	−	−	−	+	−	−	−
64	−	−	−	−	−	+	+	−	+	−	+	−	−	−
65	−	−	−	+	−	−	+	−	+	+	+	−	−	−
71	−	−	−	−	−	+	+	+	−	+	+	−	−	−
72	−	−	−	+	−	+	+	−	−	−	+	−	−	−
77	−	+	+	−	−	+	+	−	+	−	+	−	−	−
80	−	+	+	−	−	−	−	−	−	−	−	−	−	−
83	−	−	−	+	−	−	−	+	−	−	−	−	−	−
88	−	−	−	−	−	+	+	+	+	+	+	−	−	−
90	−	−	−	−	−	−	+	−	−	−	+	−	−	−
91	−	−	−	−	−	−	+	−	−	−	−	−	−	−
95	−	−	−	−	−	+	−	−	−	−	−	−	−	−
96	−	−	−	−	−	+	−	+	−	+	−	−	−	−
101	−	−	−	+	−	+	+	+	+	+	+	−	−	−
106	−	−	+	−	−	+	+	−	+	−	+	−	−	−
108	−	+	−	+	−	−	+	+	+	−	+	−	−	−
125	−	−	−	−	−	+	−	+	+	+	+	−	−	−
128	−	−	−	−	−	+	+	−	−	−	−	−	−	−

TABLE XVII (continued)

Phage type*	A	B	C	D	E	G	H	L	M	O	T	X	Z₁	Z₂
130	—	—	—	—	—	—	—	+	—	+	—	—	—	—
142	—	—	—	—	—	—	+	—	—	+	—	—	—	—
143	—	—	—	—	—	—	—	—	—	—	+	—	—	—
162	—	+	—	—	—	+	+	—	+	—	+	—	—	—
167	—	+	+	—	—	—	+	+	+	+	+	—	—	—
176	—	+	—	—	—	—	—	—	—	—	—	—	—	—
179	—	—	—	—	—	+	—	—	—	—	—	—	—	—
188	—	—	—	—	—	+	—	—	—	—	—	—	—	+
192	—	+	—	+	—	+	+	+	—	+	+	—	—	—
195	—	—	—	+	+	+	—	+	+	+	+	—	—	—
198	—	+	—	+	—	+	+	—	—	—	+	—	—	—
199	—	—	—	+	—	—	+	—	—	—	+	—	—	—
201	—	+	+	—	—	—	+	—	+	—	+	—	—	—
205	—	+	—	—	—	+	+	—	—	—	+	—	—	—
206	—	—	—	—	—	+	—	—	+	—	—	—	—	—
215	—	+	—	—	—	—	+	—	—	—	+	—	—	—
Untypable	—	—	—	—	—	—	—	—	—	—	—	—	—	—

+ = lysis.
* The types listed in this table are only those that formed no less than 1% of the total number of cultures examined from at least one of the five sources. Phages Z₃, Z₄ are not included in the table as none of the cultures upon which they acted fell into this category.

against the typing phages. Table XIX shows the highest dilution of the sera completely neutralising lysis.

(ii) *Lytic patterns of the typing phages.* For the determination of the lytic pattern the typing phages were used at RTD. The results are shown in Table XX.

(iii) *Plaque morphology of typing phages.* Plaque-types differing from each other in shape and size were identified. Some characteristic appearances are shown in Plate I(a)–(c). Typing phages 6 and 15 formed clear plaques of pin-point size with sharp edges, typing phages 8 and 26 turbid plaques of pin-point size with sharp edges.

(b) *Culture media.* Infusion broth made from fresh meat and broth and agar made from various dehydrated products, such as Difco "Bacto" Nutrient broth and Oxoid Nutrient broth No. 2 were used. The same media were used for propagation and typing.

(c) *Propagation of phages.* The methods of propagation were the same as described in the typing system according to Eörsi *et al.* (1953, 1954) and

TABLE XVIII

Type phages, propagating and lysis-spectrum strains
according to Milch and Gyenes (1972)

Phages	Propagating strains	Antigenic structure
F1	C1	85: . : −
F2	C29	19, 133: . : 16
F3	C31	Rough: . : ?
F4	O8	8: . : .
F4a	R816	78: . : .
F4b	R984	78: . : .
F5	O5	9: . A: 9
F6	C23	25: . : ?
F7	C32	79: . : 2
F8	F42	26: 60B: .
F9	C26	20: . : −
F10	111/1	111: 58B: −
F11	111/5	111: 58B: −
F12	C35	125: . : 25
F13	C8	18ac: . : −
F14	C15	77: . : .
F15	C20	35: . : 9
F16	C19	18ac: . : −
F17	O58	18ac: 77B: −
F18	O58	18ac: 77B: −
F19	55/1	55: 59B: 6
F20	55/4	55: 59B: 6
F21	111/2	111: 58B: 2
F22	C3	4: . : 5
F23	C7	15: . : −
F24	O18	18ab: . : 14
F25	C9	18ab: . : 14
F26	C17	1: 1: 7
F27	F42	26: 60B: .
F28	C24	Rough: . : 2

Milch and Deák (1961). The phages were used at RTD estimated by titration in advance. The highest dilution of phages causing semi-confluent lysis with the propagating strains was regarded as RTD.

(*d*) *Method of typing: interpretation of results.* The *E. coli* strains to be typed were inoculated from 24-h agar plates into broth and after incubation for $2\frac{1}{2}$ h the cultures were transferred to agar plates, the excess being discarded. The plates were allowed to dry, then the typing phages were dropped on to the plates either by a pipette used specially for this purpose or by means of an applicator (Biddalph and Co., Lingard St.,

TABLE XIX

Neutralisation titre of antiphage sera of phages used in the typing scheme of Milch and Gyenes (1972)

Sero-group	Typing phage	A1			A2				A3				A4		B1		B2				B3	B4	C1	D1		E1	F1	F2	G1
		1	2	3	4	5	6	7	8	9	10	11	12	13	14	15	16	17	18	19	20	21	22	23	24	25	26	27	28
A1	1	10^{-2}	10^{-1}	10^{-2}	10^{-3}	10^{-3}	10^{-2}	10^{-2}	10^{-2}	10^{-1}	10^{-1}	10^{-1}																	
	2	10^{-1}	10^{-3}	10^{-4}	10^{-2}	10^{-2}	10^{-1}	10^{-2}																					
	3	10^{-1}	10^{-3}	10^{-5}	10^{-4}	10^{-1}	$10^{-7}/10^{-2}$	10^{-2}	10^{-1}																				
A2	4	10^{-1}		10^{-2}	10^{-4}				10^{-1}																				
	5	10^{-1}	10^{-2}	10^{-2}	10^{-3}	10^{-3}	10^{-3}	10^{-1}	10^{-3}																				
	6	10^{-1}	10^{-1}	10^{-4}	10^{-4}	10^{-2}	10^{-2}	10^{-2}	10^{-1}																				
	7	10^{-1}	10^{-3}	10^{-4}	10^{-3}	10^{-2}	10^{-2}	10^{-3}																					
A3	8	10^{-1}	10^{-1}	10^{-4}	10^{-2}	10^{-3}			10^{-3}	10^{-2}	10^{-2}	10^{-3}																	
	9	10^{-1}	10	10^{-1}	10^{-2}	10^{-3}			10^{-3}	10^{-2}	10^{-2}	10^{-2}																	
	10	10^{-1}	10	10^{-2}	10^{-2}	10^{-3}			10^{-3}	10^{-2}	10^{-2}	10^{-3}																	
	11	10^{-1}		10	10^{-2}	10^{-1}			10^{-2}	10^{-1}	10^{-2}	10^{-3}																	
A4	12	10^{-1}											10^{-3}	10															
	13													10^{-2}															
B1	14														10^{-2}														
	15				10^{-1}										<10	10^{-3}													
B2	16																10^{-4}	10^{-2}	10^{-2}										
	17																10^{-3}	10^{-2}	10^{-2}										
	18																		10^{-3}	10^{-2}									
	19									10								10^{-1}	10^{-1}	10^{-2}									
B3	20																			10	10^{-4}								
B4	21																	<10				10^{-4}							
C1	22																						10^{-2}						
D1	23																							10^{-3}	10^{-1}				
	24																							10^{-2}	10^{-3}				
E1	25																									10^{-4}			
F1	26																										10^{-4}	10^{-1}	
F2	27																											10^{-2}	
G1																													10^{-4}

TABLE XX

Lysis spectrum of E. coli typing phages according to Milch and Gyenes (1972) (Typing phages at RTD)

Test strain	Antigenic structure	A1						A2			A3				A4			B1		B2		B3		B4	C1	D1		E1	F1	F2	G1
		1	2	3	4	4a	4b	5	6	7	8	9	10	11	12	13	14	15	16	17	18	19	20	21	22	23	24	25	26	27	28
C1	85::−	Cl																													
C29	19,133:·:16		Cl	sCl											Cl	++	+++	Cl								+	+++				
C31	Rough:·:?	++	++		Cl	Cl	++		+++						Cl	+	+++	Cl									+				
O8	8:·:·				Cl	Cl	++	Cl							Cl			Cl									+				
R816	O78::−				Cl	Cl	Cl																								
R984	O78:·:				Cl	Cl	sCl																								
C5	9:(A):9		±					Cl							ol		+														
C23	25:·:?		+++	Cl					+++	Cl									+++	+											
C32	79:·:2	+	Cl	Cl						+									+++	+										+++	
F42	26:60(B):·																	Cl													
C26	20:·:−	+++									Cl	sCl	Cl	++	sCl		+							+							
111/1	111:58(B):−										Cl	Cl	Cl	++	Cl						+++									±	Cl
111/5	111:58(B):−										Cl	Cl	Cl	Cl	Cl		+				+++										
C35	125:·:25		±						++										Cl	+					±	+					
C8	18ac:·:−				±		Cl				ol	ol	±		++	Cl															
C15	77:·:?											ol			Cl	Cl		Cl													
C20	35:·:9															Cl															
C19	18ac:·:−														ol	Cl		Cl	Cl												
O58	18ac:77(B):−				±	Cl									ol	Cl			Cl	+	+++										
55/1	55:59(B):6																					Cl	±	+							
R604	4:·:				±	Cl										•						Cl	Cl								
111/2	111:58(B):2															Cl							+++	+							
C3	4:·:5				±	Cl	Cl																				Cl				
C7	15:·:−				±	Cl	++																			Cl					
O18	18ab:·:14														Cl												Cl				
C9	18ab:·:14						++								Cl														Cl		±
C17	1:1:7																											Cl	Cl		
C24	Rough:·:2												ol					Cl													Cl

Cl = Confluent lysis   sCl = Semiconfluent lysis   ol = Opaque lysis   ± to +++ = Increasing numbers of discrete plaques

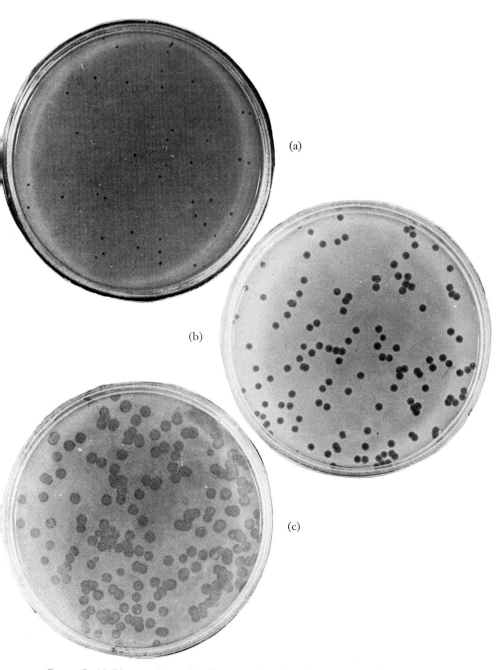

(a)

(b)

(c)

PLATE I. (a) Plaques 1 mm in diameter showing clear areas with entire edges (*E. coli* typing phages 2, 4, 4a, 4b, 5, 7, 9, 11 and 17). (b) Plaques 3 mm in diameter showing clear areas with entire edges (*E. coli* typing phages 3 and 18). (c) Plaques 4 mm in diameter showing turbid areas with entire edges (*E. coli* typing phages 12, 19 and 27).

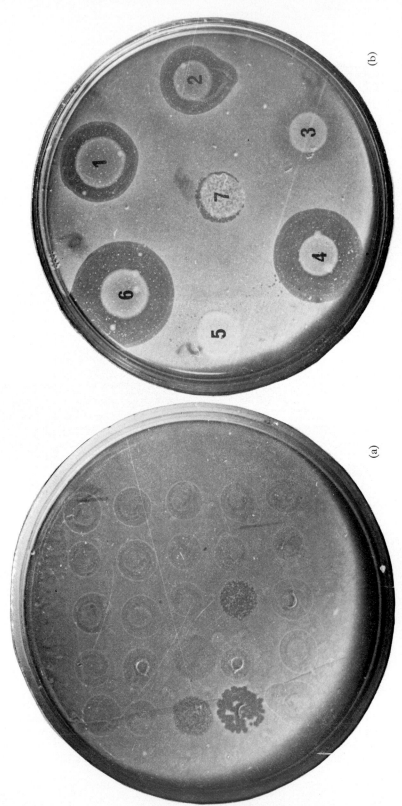

PLATE II. (a) Strain 18 ac; 77B: −; phage pattern: 12, 13, 16, 18; phage group: A4, B2. (b) Colicinogenic and lysogenic properties of 7 *E. coli* strains on *E. coli* K12 as indicator strain. Colicinogenic +: strains Nos. 1 and 2. Colicinogenic ±: strain No. 3. Colicinogenic ++: strains Nos. 4 and 6. Lysogenic: strains Nos. 5 and 7.

Manchester 15, England). After drying, the plates were incubated at 37°C for 6 h and then read. Lysis was recorded according to its intensity. The results are given in terms of lytic pattern. The phage pattern of the strains is characterised by listing the phages producing lysis (Arabic numerals). Phages producing lysis weaker than + + (less than 50–100 plaques) are indicated in brackets. Phage groups (capital letters) indicating the sero-groups of the phages producing lysis are also shown. A typical pattern is shown in Plate II(a).

(i) *Phage typing after incubation at elevated temperature.* Untypable strains were grown in broth at 45°C for 18 h, the inoculated plates were incubated at 45°C for 5 h, then lysis was read.

(ii) *Additional methods.* Beside phage sensitivity, colicinogenic and lyso-genic properties and PPR (D'Alessandro and Comes, 1952) were examined. (The methods for the examination of colicinogeny and lysogeny are described on pages 47–48.)

(*e*) *Phage pattern, phage groups, colicinogeny, lysogeny and PPR of the frequent* E. coli *serogroups.* 2657 *E. coli* strains belonging to different serogroups and 789 serologically unidentified strains were isolated in the period 1969 to 1975 by the National Institute of Hygiene, Budapest or by Regional Public Health Laboratories in Hungary.

I. Ørskov and F. Ørskov at the WHO International *Escherichia* Centre carried out serological typing of the propagating strains and of 231 *E. coli* strains belonging to different serogroups. The serogroups of 2426 *E. coli* strains were determined by E. Czirók, National Institute of Hygiene, Budapest (NIHB), F. Baron, Regional Public Health Station, Kaposvár and M. Lakatos, Regional Public Health Station, Debrecen. An additional 247 type strains maintained in the Culture Collection of the NIHB were analysed using the different typing methods. The results are presented according to the serogroups of the strains (Tables XXI, XXII).

The Tables comprise those strains which had been isolated with high frequency and were most probably pathogenic. Thus the results were believed sufficiently reliable to enable some general conclusions to be drawn. Since a great variety of phage patterns was observed, only the more frequent are shown here.

Examination of the Culture Collection strains revealed that the different K or H antigens were not related to the phage patterns, since phage patterns of strains with the same K or H antigens proved to be different (Table XXIII).

E. coli O4. Most of the 273 strains shown in Tables XXI and XXII were sensitive only to phage 4, while some of them were sensitive to phages 4, 22; 4, 20, 22 or 4, 12, 13, 22 and PO. The phage group characteristic for

TABLE XXI

Typability of the most frequent *E. coli* serogroups according to Milch and Gyenes (1972)

Serogroup	No. of strains examined	No. of phage patterns	Percentage of strains			
			Typable by phages	col+	ly+	PPR+
O2	51	22	74·5	84·3	3·9	—
O4	274	43	79·9	47·6	9·2	—
O6	101	20	89·1	41·6	11·9	1·0
O8	52	15	55·8	34·6	11·5	3·8
O18	302	52	75·5	26·8	35·1	—
O20	49	13	32·0	22·4	6·1	—
O25	76	22	61·8	28·9	6·6	2·6
O26	44	16	75·0	29·5	4·5	2·3
O55	73	30	63·0	46·6	8·2	1·4
O75	150	44	62·7	28·7	12·0	—
O78	183	6	99·7	68·3	10·4	—
O111	226	27	81·4	22·6	40·7	13·3
O114	28	4	32·2	60·7	17·9	—
O115	25	7	60·0	48·0	—	—
O124	621	57	94·5	30·8	44·3	10·0
O125	33	11	75·8	18·1	33·3	—
O143	89	12	82·0	21·3	18·0	26·3
O147	31	8	93·5	42·0	6·5	—
O149	33	4	75·8	72·7	6·0	—

serogroup O4 was A1, B3, C1. The phage patterns were manifold, totalling 43. Of the strains studied 79·9% were typable by phages, 47·6% were colicinogenic, 9·2% were lysogenic and none of them gave positive PPR.

E. coli O18. For the 101 strains listed in Tables XXI, XXII, 52 phage patterns were observed. The simultaneous occurrence of sensitivity to phages 4, 12, 13, 16, 17 and 18 was the most characteristic pattern. Typical phage groups were A1, A4, B2. 75·5% of the strains were typable by phages, 26·8% were colicinogenic, 35·1% lysogenic and all of them gave negative PPR.

E. coli O78. The 183 strains in this serogroup were almost uniform in their phage pattern. Differentiation was possible only by assessing colicinogenic properties. Of the strains 68·3% were colicinogenic, 10·4% lysogenic and none of them have positive PPR. In colicinogenic strains it is also advisable to determine the type of the colicins produced.

E. coli O111. The 226 strains shown in Tables XXII and XXIII had 27 different lytic patterns. The four most frequent phage patterns and phage groups were: 9; 21, 28; 28; 12, 21; A3; B4, G1; G1; A4, B4. The phage

TABLE XXII

Phage pattern, phage group, colicinogeny, lysogeny and PPR of the most frequent _E. coli_ serogroups according to Milch and Gyenes (1972)

Serogroup (total no. of strains)	The most frequent Phage patterns	Phage groups	Exam-ined	Coli-cino-genic	Lyso-genic	PPR+
O4	4	A1	41	22	6	—
(273)	4, 22	A1, C1	29	19	1	—
	4a, 4b, 20	A1, B3	19	—	—	—
	4, 20, 22	A1, B3, C1	12	11	—	—
	4, 12, 13, 22, PO	A1, A4, C1, S*	10	—	—	—
O6	4, 13, 15	A1, A4, B1	21	8	1	—
(101)	4, 13	A1, A4	20	10	—	4
	4, (12), 15	A1, A4, B1	8	4	2	—
O18	4, (12), 13, 16, 17, (18)	A1, A4, B2	24	—	16	—
(302)	12	A4	15	1	7	—
	13, 17	A4, B2	12	—	11	—
	13, 16, (17)	A4, B2	9	—	6	—
	16	B2	8	—	—	—
	12, 16, 17	A4, B2	7	—	7	—
O75	4	A1	23	7	2	—
(150)	4, 12	A1, A4	17	3	3	—
O78	4, 4a, 4b	A1	175	120	18	—
(183)	4a, 4b	A1	2	1	1	—
O111	9	A3	37	1	4	32
(226)	21, 28	B4, G1	29	1	19	29
	28	G1	20	—	5	5
	12, 21	A4, B4	17	16	—	—
	30	S*	14	14	—	—
	21	B4	11	—	11	10
	9, 21, 28	A3, B4, G1	7	—	7	7
	8, 9	A3	5	1	4	4
O124	2, 3, 4, 6, 7	A1, A2	147	48	92	80
(621)	2, 3, 4, 6, 7, 28	A1, A2, G1	83	32	36	51
	2, 3, 6	A1, A2	83	9	56	8
	7	A2	43	4	33	—
	2, 3, 6, 28	A1, A2, G1	37	7	14	14
	2, 3, 4, 6	A1, A2	24	18	21	19
	2, 3, 6, 7, 16, 17, 28	A1, A2, B2	20	18	—	20
	2, 3, 6, 7	A1, A2	15	6	7	9
	2, 3, 4, 6, 7, 16, 17, 18	A1, A2, A3, B2	13	9	—	13
O143	8, 9	A3	31	2	11	31
(89)	8	A3	27	1	10	27

* _Salmonella._
Phage pattern ( ) = weak and unstable lysis.

TABLE XXIII

Phage pattern, phage group, colicinogeny, lysogeny, PPR and haemolysis of *E. coli* serogroups 09, 018, 026, 055 and 0111 (Culture Collection strains) according to Milch and Gyenes (1972)

Antigenic structure	Phage pattern	Phage group	Colicinogen pattern	Lysogen pattern	P.P.R.	Haemolysis	Type strain
9: 9L: 12	Negative	Non typable	Phi, 17, K12	—	—	—	Bi 316/42
9a:26A: —	Negative	Non typable	—	—	—	—	Bi 449/42
9a,b: 28A: —	5	A2	—	—	—	—	K 14a
9: 29A: —	(2), 3, 4, (5), (7), 12	A1, A2, A4	Phi, 17, K12	—	—	+	Bi 161/42
9: 30A: 1	2, 3, 5, 6, 7	A1, A2	Phi, 17, K12	—	—	+	E 69
9: 31A: —	2	A1	—	—	—	—	Su 3973/41
9: 32A: 10	2, 3, 4, 5, 7, 12	A1, A2, A4	Phi, 17, K12	—	—	+	H 36
9: 33A: —	(3), 4, (5), (6), 7, (11), 12	A1, A2, (A3), A4	—	—	—	—	AP 289
9: 34A: —	Negative	Non typable	Phi, 17, K12	—	—	—	E 75
9: 34A: —	2, 3, 4, 5,12	A1, A2, A4	—	—	—	—	A 140 a
9: 36A: 19	4, (7)	A1, (A2)	—	—	—	—	A 198 a
9: 37A: —	2, 3, 9, 12, 15, 23, 24	A1, A3, A4, B1, D1	—	—	—	—	A 84 a
9: 38A: —	Negative	Non typable	—	—	—	—	A 262 a
9: 39A: 9	2, 3, 5, 6, 7, 12	A1, A2, A4	—	—	—	—	A 121 a
9: 55A: —	Negative	Non typable	—	—	—	—	N 24 c
9: 57B: 32	Negative	Non typable	Phi, 17, K12	—	—	—	H 509 d
9: ·: 19	2, 3, 4, 5, 6, 7, 12	A1, A2, A4	Phi, 17, K12	—	—	—	A 18 d
18ab: 76:14	7, 12, 24	A2, A4, D1	—	—	—	—	F 10018/41
18ac: 77:7	4, 12, 16, 17	A1, A4, B2	—	—	—	—	3219/54
26: 60B: —	1, 2, 6, 8, 9, 12, 20, 21, 27	A1, A2, A3, A4, B3, B4, F2	—	—	—	—	F 41
26: 60B:11	8, 15, 20, 27	A3, B1, B3, F2	Phi, 17, K12	—	—	—	H 311 b
26: 60B: 46	2, 12, 15	A1, A4, B1	—	—	—	—	5306/56
55: 59B: —	7	A2	Phi, K12	—	+	—	Su 3912/41
55: 59B: 6	20	B3	—	—	—	—	Aberdeen 1064
55: 59B: ·	Negative	Non typable	Phi, K12	—	—	—	972
111: 58B: —	1, 8, 9, 10, 21	A1, A3, B4	Phi, 17, K12	—	+	—	Stoke w

pattern of 81·4% of the strains was determined. 22·6% were colicinogenic, 40·7% lysogenic and 13·3% PPR positive.

E. coli O124. The majority of the 621 strains in this serogroup were fairly uniform in phage pattern, thus lysogeny complemented with the examination of colicinogeny provided a more adequate basis for differentiation. Of the strains 147 had the same phage pattern (2, 3, 4, 6, 7), 269 belonged to phage group A1, A2; 94·5% were typable by phages, 30·8% were colicinogenic, 44·3% lysogenic and 10·0% gave positive PPR.

In serogroups 78 and 124 it is mainly the colicinogenic and lysogenic properties that allow a further subdivision. In the most frequent serogroups (2, 4, 6, 8, 18, 20, 25, 26, 55, 75, 111, 114, 125, 143, 147, 149) phage pattern, colicinogeny, lysogeny and PPR were found useful for further subdivision.

Due to the small number of strains studied no general conclusions could be drawn concerning the other serogroups.

Of the 2657 strains belonging to different serogroups, 45 strains were lysed by *Salmonella* "O" or "Vi" phages. "O" phages lysed 48 strains (serogroups 4, 6, 18, 20, 28, 55, 111, 124), "Vi" phages lysed 6 strains (serogroups 19, 75, 79, 111).

Of the 789 serologically unidentified *E. coli* strains 45% were typable by phages, 31·2% were colicinogenic, 9·2% lysogenic and 3·6% were PPR positive.

IV. ADDITIONAL METHODS FOR SUBDIVIDING SERO- AND PHAGE-TYPES OF *E. COLI*

A. Biochemical typing methods

Jensen (1913) and Christiansen (1917) were the first to investigate biochemical characters of *E. coli*. Fermentation reactions of strains associated with infantile enteritis were first examined by Bahr (1912) and Bahr and Thomsen (1912). Adam (1923) described 6 biochemical types, of which A1 and A4 were the most common enteropathogenic organisms. Systematic studies of Kauffmann and Perch (1948) revealed that *E. coli* had characteristic biochemical properties. Based on the reactions with adonitol, dulcitol, inositol, rhamnose, salicin, sucrose and xylose, 10 biochemical types were distinguished.

1. *Differentiation of serogroups and serotypes by biochemical methods*

Kauffmann and Dupont (1950) examined the strains O111: K58(B4) and O55: K59(B5) on a basis of dulcitol, inositol and sucrose fermentation, and indole formation. Kauffmann and Ørskov (1956) accomplished the biochemical differentiation of the serogroups and serotypes O111: K58

(B4), O55: K59(B5), O26: K60(B6), O86: K61(B7), O25: K11(L), O25: K3(L), O44: K74(L), O111:K69(B14), O125: K70(B15), O126: K71 (B16), O127:K63(B8) and O128:K67(B12). Le Minor et al. (1964) distinguished 7 biochemical types (marked I through VII) of O111: K58 (B4), 9 types (A, B1, B2, C, D, E, F, G) of O55: K59(B5) and 4 types (Ørskov's types 1 to 3 plus one atypical) of O26: K60(B6). D'Alessandro and Comes (1952) complemented the biochemical typing system of Kauffmann and Dupont (1950) by the β-phenyl propionic acid reaction. Edwards et al. (1956), and Thibault and Le Minor (1957) found that decarboxylation of lysine (LDC reaction after Møller, 1955, and Carlquist, 1956) was of great taxonomic value. According to Ewing et al. (1960) 51% of E. coli demonstrated ornithine decarboxylase activity, (ODC).

Le Minor et al. (1962) described fermentation types corresponding to the serotypes of E. coli O127:K63(B8) as based on the reactions with adonitol, sorbose, sorbitol, dulcitol, sucrose, salicin and LDC. For the culture media used in fermentation tests the reader is referred to the review of Kauffmann and Ørskov (1956), for the method and value of LDC and arginine dihydrolase test (Moeller, 1955) to the monograph of Edwards and Ewing (1972). Nicolle et al. (1960) found the majority of the strains O111: K58(B4), O55: K59(B5), O26: K60(B6) LDC-positive. Most of the O127: K63(B8) strains are non-flagellate and LDC negative. Strains containing antigen H4 are LDC negative, those containing antigens H11, 21, 26 or 40 are LDC positive. LDC reaction varies in strains with the antigen H6 (Le Minor et al., 1962).

Richard (1966) studied different serotypes of O26: K60(B6), O55: K59 (B5), O86: K61(B7), O111: K58(B4), O119: K69(B14), O125: K70(B15), O126: K71(B16), O127: K63(B8) and O128: K67(B12). He established that the ODC reaction is of taxonomic and epidemiologic importance. An association was found between the presence of the enzyme and certain types of H antigen.

According to D'Allessandro and Comes (1952), E. coli O111: K58 (B4): H⁻ and O111: K58(B4): H12 are negative in the PPR. Strains of the serogroup O55 did not show such correlation between the result of PPR and the presence of H antigen. Milch and Gyenes (1972) could subdivide strains of the O serogroups 3, 9, 55, 59, 86, 111, 124 and 128 on the result in the PPR. In recent studies (Milch, unpublished data) the same was found for the O serogroups 8, 25 and 143.

2. Phage typing complemented by biochemical tests

Nicolle et al. (1960) reported that most phage types of O111:K58(B4) were uniform in the LDC reaction, except for some members (e.g. Lille).

Le Minor *et al.* (1954), Nicolle *et al.* (1960) Milch and Deák (1961) could demonstrate certain relations between serotype, phage type and PPR (Tables II, III, IV, VI, VII).

B. Colicin typing methods

Colicinogeny and sensitivity to colicins form the basis of two different kinds of differentiation, i.e. the assessment of the type of colicin produced and sensitivity to known colicins. Gardner (1950) found 11·6% of the strains obtained from human specimens to be colicin producers, Papavassiliou (1963) 32·1%. The same figures for porcine strains were 83·7% (Levine and Tanamito, 1960) and 43·7%, respectively (De Alwis and Thomlinson, 1973).

1. *Determination of colicin production and of colicin type for the division of serogroups and serotypes*

Frédéricq *et al.* (1956) were the first to suggest the epidemiologic use of colicin identification. They established that 28·11% of the strains O111: K58(B4), O55: K59(B5) and O26: K60(B6) were colicin producers. Shannon (1957) classified the strains O55: K59(B5) based on colicin production. Bavastrelli (1959) reported that on the average 17% of the O serogroups 25, 26, 44, 55, 86, 111, 112, 126 produced colicin, while none of the O serogroups 119 and 127 did so. In the study of Hamon (1959) 29% of the serogroups O111, O55 and O26 proved to be producers, the colicin types identified being I, E, B and G. Of the 40 strains in the O111 serogroup and of the 28 strains pathogenic for calves (serogroup O70: B80), 12 and 16 strains respectively were found colicinogenic by Papavassiliou (1961). Krajnovic (1961) demonstrated colicin production in 42·5% of the O55, in 27·45% of the O26, and in 50% of the O111 serogroups. In the study of Likhoded and Kudlai (1963) the same figures for the serogroups O111, O55 and O26 were 13·8, 50 and 47% respectively. The colicin types identified differed from those reported in the literature. Colicin I was produced by 68% of the O111, colicin G by 95% of the O55, colicin E, I, B and D by 46% of the O26 serogroup strains. Kasatiya (1963), Kasatiya and Hamon (1965) demonstrated colicin production in 40% of the strains of serogroup O119. The distribution of colicin types was different from that observed in other serogroups (E, V, E+N, V+B, I, V+N, V+I). It was mainly the frequency of colicin E and of the heat-resistant colicin V that differed from that of colicin types usual for EEC.

According to Frédéricq (1948) colicinogeny is a common character of other serogroups, too. Hamon (1958) demonstrated it in 28·8% of the O serogroups 1, 2, 4, 7, 8, 11, 12, 17, 18 and 20; the identified colicin types were I, V, E, D, B, C, G and H. Barry et al. (1962) showed that strains of the serotypes 25: . : 1, 4: . : 5, 6: . : 16 were often colicin producers; no association was found between colicin types (A, V, K, E, B, G, H, D) and H antigen. Colicin G + H was produced by all of the strains of the serogroup O25, by 90% of O4, and by 65% of O6. Linton (1960) studied colicin production of urinary E. coli. Using the method of Abbott and Shannon (1958) he could differentiate 30 colicin types. McGeachie and McCormick (1963), and McGeachie (1965) found that 36·3% of urinary strains were colicin producers, 49·8% sensitive to colicin, 14% producers and sensitive. They stated that colicin was a suitable marker for routine type differentiation. Vosti (1968) demonstrated colicins of different types and activities within the same serogroup.

A high percentage of porcine enteropathogenic E. coli proved to be producers in Vasenius (1967) series. De Alwis and Thomlinson (1973) could demonstrate colicinogeny only for the serogroup O141. According to Atanassova and Mladenova (1975) colicinogeny was frequent in strains of the O143 serogroup. They could demonstrate a transferable Col factor in strains of the O serogroups 28, 112a, 112b, 143 and 144. Milch (unpublished data) found that 36·5% of E. coli pathogenic for man and 56·3% pathogenic for animals produced colicin. Distribution of colicinogeny among strains belonging to the frequent serogroups is shown in Table XXI.

2. Phage typing complemented by determination of colicinogeny and colicin typing

Hamon (1959) investigated colicinogenic properties of phage types (determined after Nicolle) of the O serogroups 111, 55 and 26. No strains of the serogroup O55 phage types Lomme, Jérusalem and Londres were producers. Most of the other phage types could be characterised according to colicinogeny. It was frequent among the phage types Tourcoing, Bretonneau and Weiler. Phage types Sèvres ubiquitaire, Bretonneau and Weiler produced 5 to 7 types of colicin; St. Christopher, Sèvres var. lyonnaise and Béthune only 2 to 3 types. Hamon (1960) compiled in tables the distribution of sero-, phage-, bio- and colicin-types, colicin sensitivity and PPR for E. coli O111, O55 and O26. Chistovich and Matyko (1967) found differences in the colicinogenic properties of the different phage types of EEC O9 and O26. According to the phage typing method of Milch and Gyenes (1972) in serogroups 2, 4, 6, 8, 18, 20, 25, 26, 55, 75, 78, 111, 114, 124, 125, 143, 147 and 149 colicinogeny allowed a further subdivision.

3. *Determination of colicin sensitivity for the division of sero- and phage-types*

Colicin producing and non-producing strains can be classified according to their sensitivity to colicins (Hamon, 1960). Parr *et al.* (1960) studied sensitivity to 12 known colicins in different *E. coli* serotypes. No association could be demonstrated between sensitivity to colicins and serotype. Barbuti (1961) also failed to demonstrate any correlation. According to Papavassiliou (1962) colicin sensitivity is a less stable hereditary characteristic than is colicinogeny. Hamon and Brault (1958), and Hamon (1958) using the method of Frédéricq (1948) found differences among the various phage types of the O serogroups 111, 55 and 26 in their sensitivity to colicins. Chistovich and Matyko (1967) recommended the combined testing of EEC O9 and O26 strains for sensitivity to phages and colicins. Milch (unpublished data) found a difference in colicin sensitivity of *E. coli* O78 strains having the same phage pattern.

C. Tests for Lysogeny to supplement phage typing

1. *Principles and application of the method*

The method has its origin in the studies of Nicolle *et al.* (1960) who actually used it for subdividing strains of O111 phage type Sèvres, and therefore regarded it as a new phage typing technique (Nicolle *et al.*, 1963). The method is based on the following principle. Phages are isolated from strains of identical phage types. *E. coli* B strains are rendered lysogenic with the isolated phages. Lytic spectra of phages isolated from the lysogenic strains under study are examined using these *E. coli* B immune mutants. Nicolle *et al.* (1963) called the method "indirect phage typing" and recommended its trial on more extensive material.

Papavassiliou (1961) investigated lysogeny of the serogroup O111 pathogenic for man and of the serogroup O70: K80(B) pathogenic for calves. A total of 28·3% of the strains proved lysogenic. He maintained that the great number of indicator strains used accounted for the high frequency of lysogeny.

Krylova (1962) examined isolates from Norils (O111 serogroup) for lysogeny and could distinguish four types. The results were in good agreement with epidemiological data.

Kolta and Deák (1962), and Deák (1965, 1966) developed a method for the type differentiation of O124: K72(B17) strains. The isolated temperate phages were examined by a set of indicator strains.

Milch and Gyenes (1972) complemented phage typing of *E. coli* strains belonging to different serogroups and serotypes by studying lysogeny. Tables XXII, XXIII show that lysogenic properties allowed a further subdivision of almost every frequently occurring serological group.

2. Methods for examining lysogeny

(a) *Method of Nicolle* et al (1963).

Phage preparations isolated from strains of identical phage types are dropped on agar plates inoculated with growing cultures of *E. coli* B. The lysogenised colonies are transferred to agar slopes, then spread on agar plates. These lysogenic immune mutants are spread on agar medium. After drying, the phages isolated from the strains to be tested for lysogeny are dropped on to the plates. Lysis pattern of the strains originating from the same infectious source will be identical, whereas different lysis patterns will indicate different sources. Indirect phage typing has not yet been tested on sufficiently and the elaborate technique speaks against its routine use.

(b) *Method of Papavassiliou* (1961)

(i) *Indicator strains: E. coli* B, C and Y20. (*E. coli* strains obtained from P. Frédéricq).

(ii) *Test for lysogeny.* The indicator strain is grown in 5 ml broth for 4–5 h from an inoculum taken with a 5 mm loop from a 24-h broth culture. The indicator plate is prepared by pouring on the agar plate 0·1 ml of this culture diluted with 0·9 ml broth and then mixed with 4–5 ml soft agar melted and cooled to below 50°C. The plate is then dried by incubation at 37°C for 30–45 min with the lid open. Cultures of the strains to be tested grown for 3–4 h at 37°C in 5 ml broth from inocula of 0·1 ml taken from 24-h broth cultures are sterilised by chloroform (Frédéricq, 1950). After standing on the bench for at least 30 min, a 5 mm loopful is placed on the surface of an indicator plate, and the plate again dried for 30–45 min with the lid open. The plates are examined after incubation at 37°C for 20 h.

(c) *Method of Deák and Ádám* (1969) *for* E. coli O124: K72(B17)

(i) *Indicator strains: E. coli* Row, *E. coli* B, *E. coli* 36, *Sh. flexneri* No. 262 (serotype 1b), *Sh. sonnei* II, *E. coli* O124 (No. 138), *E. coli* O124 (No. 218).

(ii) *Colicin typing.* Colicinogenic strains are examined for inhibition spectrum against the indicator strains of Abbott and Shannon (1958). Inhibition patterns are characterised according to Hart (1965).

(iii) *Test for lysogeny* according to Papavassiliou (1961). (See paragraph b.)

(iv) *Interpretation of results.* On a basis of temperate phages, O124 strains isolated in Hungary were first classified into 7 (Kolta and Deák, 1962), and later into 11 types (Deák, 1965). Strains isolated from water samples collected from different geographical regions or from faecal specimens of patients proved to be stable *in vivo* and *in vitro*. Repeated testing showed that the lysogenic pattern of strains from carriers did not change for 14 months. Comparative trials for lysogeny, colicinogeny, H antigen and lactose fermentation were carried out later (Deák and Ádám, 1969). According to the patterns of lysogeny and colicinogeny 4 groups and 7 types could be characterised. For the differentiation between colicin and temperate phages the method of Hamon *et al.* (1965) was used (trypsin treatment, ionic requirement, sensitivity to heat and chloroform). The typing scheme is illustrated in Table XXIV. Type 1 producing colicin was the most common. The lysogenic Type 2 was non-flagellate and fermented lactose within a shorter time than did the other types.

TABLE XXIV

Typing scheme of serogroup O124 on the basis of lysogenity, colicinogenity, H antigens and lactose fermentation according to Deák & Ádám (1969)

| | | Lysis or inhibition spectrum | | | | | | | Lactose fermentation in days | |
| | | *E. coli* 36 | *Sh. sonnei* II | *E. coli* 138 | *E. coli* 218 | *Sh. flexneri* 262 | *E. coli* Row | H antigen | 2–3 | 4–17 |
Group	Type									
I	2	±	+	±	±	—	—	—	+	
	4	—	—	+	±	—	—	+		+
	6	—	—	—	—	+	—	v		+
II	1	×	—	—	—	—	×	+		+
	2b	×	×	×	×	×	×	+		+
III	1b	×	—	+	±	—	×	+		+
IV	NT	—	—	—	—	—	—	v		+

+ = lysis.
± = irregularly positive or negative.
× = inhibition.
v = with or without H antigen.

(d) *Method of Milch and Gyenes* (1972). The method is used to complement phage typing.

(i) *Indicator strains:* E. coli K12 Row, Sh. sonnei 17 (Abbott–Shannon), E. coli Phi (Gratia).

(ii) *Test for lysogeny and colicinogeny.* E. coli strains are plated on agar, inoculated into broth and incubated at 37°C for 24 h. One drop of each strain to be studied is dropped on to plates in triplicate. On every plate 7 strains are studied. After 16 h incubation at 37°C the plates are exposed to chloroform vapour for 30 min and then the chloroform is allowed to evaporate. Finally, 2-h broth cultures of 3 indicator strains are mixed with soft agar (0·5 ml with 4 ml soft agar) and then layered over the plates. Plaques and the inhibition zones of colicins are read after overnight incubation at 37°C. The phage isolated from the lysogenic strains is characterised by listing the strains surrounded by plaques. The colicin is characterised by listing the strains inhibited. (For the results see III. D.1d.) Typical reactions are shown in Plate II(b).

V. EPIDEMIOLOGY OF PATHOGENIC *E. COLI* STRAINS

A. Epidemiological use of the different typing procedures

Among infections associated with *E. coli*, infantile enteritis outbreaks, nosocomial or non-nosocomial, are of greatest importance. Water-borne or food-borne dysentery-like enteritis of children and adults are also causing increasing concern.

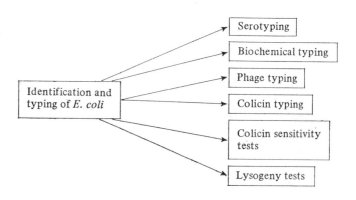

FIG. 1. Procedures used for identification and typing of *E. coli*

Braun (1956), Nicolle *et al.* (1957), Taylor (1960), Linzenmeier (1962), Rudnai and Adamis (1965) reported on the geographical distribution of the common serogroups (O111, O55, O26). In a study on water-borne epidemics in Hungary, Kubinyi (1965, 1971) directed attention to the pathogenic role of *E. coli*. Of the 74,283 enteritis cases 72·4% were of "unknown" etiology. 18·3 % were due to *E. coli* O124. These data stress the importance of the elaboration and application of diagnostic and epidemiological methods suited for assessing pathogenicity and spread of the microorganisms and for detecting new pathogenic strains.

Figure 1 illustrates the procedures used for the identification of *E. coli*. The methods can be used separately or in combination.

For use and comparison of the procedures the reader is referred to the relevant Chapters in this Volume.
Figure 2 illustrates methods approved in routine work.

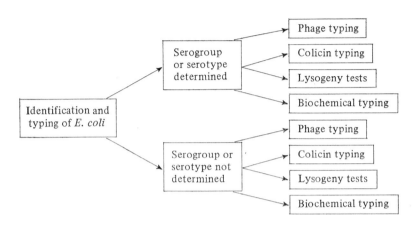

FIG. 2. Approved typing methods for use in routine work.

B. Spread of *E. coli* O111 phage types differentiated according to the method of Nicolle *et al.* (1960)

As shown in Table XXV Sèvres ubiquitaire, Tourcoing and Sèvres var. lyonnaise are the most common types in Europe. Due to the small number of strains examined from Africa, Asia and America, frequency of occurrence cannot be estimated. Nicolle and co-workers distinguished ubiquitous types and local types.

1. *Ubiquitous phage types*

Phage type *Montparnasse*. Uncommon, encountered in Europe, Asia and Africa. Believed not to be an independent type but a variant of Sèvres ubiquitaire.

Phage type *Sèvres ubiquitaire*. Within this group the most common type in several countries. Wide-spread distribution and frequency of degraded variants induced Nicolle and co-workers to subdivide the phage type by lysogeny.

Phage type *Tourcoing*. It ranks as the second commonest type in Europe.

TABLE XXV

Phage type distribution of O111: K58(B4) strains in Europe, Africa, Asia and America according to Nicolle et al. (1960)

Country	Total No. of strains	Mont-parnasse	Sèvres ubiqu.	Sèvres lyon.	Tour-coing	Vien-ne	Dorf	Lille	Breton-neau	Pa-lerme	Nago-ya	Israel	Non-typable
Europe													
Austria	262		138		40	38	3		26				17
			52·67		15·26	14·50	1·14		9·92				6·48
Belgium	452	6	289		132		1	1	10				13
		1·32	63·93		29·20		0·22	0·22	2·21				2·87
Bulgaria	68		34		23			1	3				7
			50		33·82			1·47	4·41				10·29
Denmark	7				4		2						1
Spain	12		11						1				
Finland	9		4					1	1				3
France	6068	154	2466	877	1901		16	8	260				386
		2·53	40·63	14·45	31·32		0·26	0·13	4·28				6·35
Great Britain	241		170		36		6	1	25				3
			70·53		14·93		2·48	0·41	10·37				1·24

DDR, GDR	837	2	370		323		29	8	75				30
		0·23	44·20		38·59		3·46	0·95	8·26				3·58
Greece	15				3				6				6
Hungary	35		14		16				2				3
Italy	30	1	6		1			1		21			
Roumania	83	3	74										6
Switzerland	2						1	1					
Yugoslavia	92	1	51		26				12				2
		1·07	55·43		28·26				13·04				2·17
Total	8213	167	3627	877	2505	38	58	22	421	21			477
		2·03	44·16	10·67	30·50	0·46	0·70	0·26	5·12	0·25			5·80
Africa													
Total	18		2	2	9				5				
Asia													
Total	130	3	77		5		8	1	17		4	6	9
America													
Total	260		171		58		1		15				15
			65·76		22·30		0·38		5·76				5·76

In some countries (Mexico, Morocco, Hungary) it is actually the most common. Differentiation from Sèvres ubiquitaire var. degr. is difficult, since the two differ only in their lytic reaction to typing phage 2. The fact that using the phage set of Eörsi–Milch–Deák Sèvres ubiquitaire var. degr. proved to be phage type 111/1, whereas Tourcoing corresponded to phage type 111/2 indicates two distinct phage types.

Phage type *Dorf.* Uncommon, encountered in several countries of Europe, in Israel, Canada and the U.S.A.

Phage type *Lille.* Found in several European countries and in Vietnam. More rare than phage type Dorf.

Phage type *Bretonneau.* Common in Europe and North America, found frequently also in Yugoslavia.

2. Local phage types

Phage type *Sèvres var. lyonnaise.* Isolated only in Lyon and its vicinity. Differs from Sèvres ubiquitaire only in sensitivity to typing phage 21.

Phage types *Vienne, Nagoya, Israel, Palerme, Indonésie.* Hitherto encountered only in Austria, Japan, Israel, Sicily and Indonesia, respectively.

Between 1952 and 1965, a total of 1249 strains belonging to the serogroup O111: K58(B4) were phage typed in Hungary. Of these 80·2% corresponded to phage type Tourcoing, 17·2% to Sèvres ubiquitaire, and 2·6% were untypable.

C. Spread of *E. coli* O55 phage types differentiated according to the methods of Nicolle *et al.* (1960)

Distribution of the phage types is shown in Table XXVI. As can be seen types Béthune, Lomme, Weiler and Londres are the most common in Europe.

1. Ubiquitous phage types

Phage type *Béthune.* Within the serogroup the most frequent type encountered in Europe, but has also spread to other countries.

Phage type *Lomme.* Most common in France.

Phage type *Weiler.* Fairly common in Europe, isolated mainly in Austria, Yugoslavia, England and France.

Phage type *Londres.* Common in France, rarely encountered in other European countries.

Phage type *Flandre.* Uncommon, often a degraded type.

2. Local phage types

Phage type *St. Christopher*. Differs from phage type Weiler only in the PPR and type of H antigen.

Page type *Graz*. Common in Austria, otherwise rare.

Phage type *Jérusalem*. Encountered only in Israel.

Phage type *Finlande*. Rare occurrence in Finland, the GDR and Hungary. Between 1958 and 1967 a total of 301 strains belonging to the serogroup O55: K59(B5) were phage typed in Hungary. Of these 88·9% corresponded to phage type Béthune, 3% to the phage type Lomme, and 8·1% were untypable.

D. Spread of *E. coli* O26 phage types differentiated according to the method of Nicolle *et al.* (1960)

In this group, the most common in Europe are the phage types Birmingham and Warwick.

1. Ubiquitous phage types

Phage type *Birmingham*. Common in Europe, mainly in Austria and Yugoslavia. Encountered also in Morocco, Tunisia, Senegal, Japan, Vietnam, the U.S.A., Mexico and Venezuela.

Phage type *Warwick*. Ranks as the second commonest type in Europe. Of the strains isolated in Denmark it is the most common. Encountered also in Africa, Asia, America.

2. Local phage types

Phage type *Zurich, Vietnam*. Hitherto encountered only in Switzerland and Vietnam.

In Hungary, 57 strains of the O26: K60(B6) serogroup have been phage typed. Of these 33 corresponded to phage type Birmingham, 8 to Warwick, and 16 were untypable.

E. Spread of *E. coli* O111 phage types differentiated by the method of Eörsi *et al.* (1953, 1954), Milch and Deák (1961)

The strains were isolated from hospital epidemics, or other sources in Hungary, France, the GDR and Roumania (Milch and Deák, 1961; Milch *et al.*, 1961; Milch, 1967).

TABLE XXVI

Phage type distribution of O55: K59(B5) strains in Europe, Africa, Asia and America according to Nicolle et al. (1960)

Country	Total No. of strains	St. Christopher	Weiler	Finlande	Graz	Flandre	Lomme	Béthune	Londres	Jéru-salem	Non typable
Europe											
Austria	77		42 / 54·54		10 / 12·98	2 / 2·59	13 / 16·88	6 / 7·79	2 / 2·59		2 / 2·59
Belgium	183		2 / 1·09				6 / 3·27	165 / 90·16	5 / 2·73		5 / 2·73
Bulgaria	10		7				1	1	1		
Denmark	1						1				
Spain	3		2				1				
Finland	11		4	3			2	1			1
France	1400		77 / 5·50		11 / 0·78	15 / 1·07	696 / 49·71	330 / 23·57	168 / 12		103 / 7·35
Great Britain	282	165 / 58·51	23 / 8·15		2 / 0·70	2 / 0·70	56 / 19·85	26 / 9·21	1 / 0·35		7 / 2·48

	Total									
DDR, GDR	707	1 / 0·14	21 / 2·96			1 / 0·14	32 / 4·52	615 / 86·93	10 / 1·41	27 / 3·81
Greece	1									
Hungary	35					1 / 2·85	21 / 60	13 / 37·14		
Italy	15						12			1
Sweden	2						2			
Roumania	1						1			
Switzerland	2						1			
Yugoslavia	49		12 / 24·48			2 / 4·08	26 / 53·06	8 / 16·32		1 / 2·04
Total	2779	166 / 5·97	193 / 6·94	3 / 0·10	23 / 0·80	24 / 0·86	871 / 31·34	1165 / 41·92	187 / 6·72	147 / 5·30
Africa Total	24	2	3				4	16		2
Asia Total	57	2	11			1	10	22		9
America Total	108	2 / 1·85	8 / 7·40			3 / 2·77	89 / 82·40	3 / 2·77	1 / 0·92	2 / 1·85

Table XXVII illustrates the geographical distribution of the phage types.

1. *Ubiquitous phage types*

Phage type 111/2. In this group the most common nosocomial and non-nosocomial type in Hungary. Of the 14 strains received from France 2, of the 30 from the GDR 23, and of the 5 from Roumania 4 belonged to this phage type.

Phage type 111/1. Ranks as the second most common isolated in Hungary or received for typing from the GDR. It was encountered also among the strains from France, and one of the Roumanian strains was also of this phage type.

2. *Local phage types*

Phage types 111/3 and 111/6. In order of frequency they were the third and the fourth between 1953–1956 in Hungary. Not encountered since 1957.

Phage type 111/7. The most common in Hungary in 1954, rarely isolated thereafter.

Phage type 111/5. Rare isolates in Hungary in 1952–1963, no occurrence thereafter.

F. Spread of *E. coli* O55 phage types differentiated according to Eörsi *et al.* (1954, 1953), Milch and Deák (1961)

1. *Ubiquitous phage types*

Phage type 55/2. In this group, the most common phage type in Hungary isolated from patients and symptom-free carriers. Of the 83 strains received from the GDR 71, and from the 85 from France, 12 belonged to this type.

Phage type 55/1. The second commonest type in Hungary. Possibly a variant of 55/2. Of the 83 strains from the GDR 5, and of the 85 from France 28 were of this phage type.

Phage type 55/4. Rare in Hungary and in the GDR. Of the 85 strains from France 33 belonged here.

2. *Local phage types*

Phage type 55/3. A rare type in Hungary. Of the strains received from France and isolated in Israel, Saigon, Vienna and Berlin respectively, 5 were of this type.

Phage type 55/5. Not lysed by any phage. Rare in Hungary. Six of the GDR and 4 of the French strains belonged to this type.

TABLE XXVII

Phage type distribution of O111: K58(B4) strains in different countries in 1952–1969 according to Eörsi et al. (1953, 1954) and Milch and Deák (1961)

Country	Total No. of strains	Phage types														Non-typable	
		111/1		111/2		111/3		111/4		111/5		111/6		111/7			
		No.	%	No.	%	No.	%	No.	%	No.	%	No.	%	No.	%	No.	%
Hungary	1796	304	17·1	1269	70·2	50	2·8	60	3·4	49	2·8	40	2·3	12	0·7	12	0·7
France	22	16		4		2		—		—		—		—		—	
GDR	33	8		23		2		—		—		—		—		—	
Roumania	6	1		4		1		—		—		—		—		—	
Belgium	2	—		—		2		—		—		—		—		—	
Bulgaria	1	1		—		—		—		—		—		—		—	
Total	1860	330	17·7	1300	70·0	57	3·1	60	3·2	49	2·6	40	2·2	12	0·6	12	0·6

G. Spread of *E. coli* O26 phage types differentiated according to Eörsi *et al.* (1953, 1954), Milch and Deak (1961)

In this group, phage types 26/1 and 26/2 were the most common in Hungary. Also encountered among the strains received from the GDR and France. Due to the small number of strains examined no epidemiological conclusions can be drawn.

H. Spread of *E. coli* O111 phage types differentiated according to the method of Adzharov (1966)

Ubiquitous phage types in Bulgaria: A5, A7, B1.
Local phage types: A1, A2, A3, A4, A6, A8, A9, B2, C1.

I. Spread of *E. coli* O127 phage types differentiated according to the method of Ackermann (1962)

Most strains of this serogroup are non-flagellate. Thus, serotyping allowing the differentiation of 6 types of H antigen (4, 6, 11, 21, 26 and 40) is unsuited for epidemiological purposes. Non-flagellate strains isolated in France belonged to phage types A, B, and C.

J. Spread of *E. coli* phage types differentiated according to the method of Brown and Parisi (1966)

Phage types A through H, and later I, J, L and M were the most common among the human isolates in Missouri (Parisi *et al.*, 1969). Sterne *et al.* (1970) examining strains from healthy and diarrhoeal pigs also in Missouri identified the same common phage types as did Parisi *et al.* (1969) from human specimens.

K. Spread of bovine *E. coli* phage types differentiated according to Smith and Crabb (1956)

Smith and Crabb (1956) performed phage typing of *E. coli* isolated from healthy calves and from calves suffering from white scours. Several types of *E. coli* were often found in the same faecal specimen in cases of scouring as well as in healthy animals reared in one, self-contained herd. Consecutive examinations revealed changes in the predominant phage type. It was suggested that change of phage type might be induced by phage in the intestinal tract.

L. Spread of phage patterns of pathogenic *E. coli* strains differentiated according to the method of Milch and Gyenes (1972)

The great majority of strains were isolated from hospitalised infants with enteritis or respiratory infections and from patients with urinary infections. Strains of the serogroups O4 and O6, isolated from the same hospital during the same period, each had identical phage patterns and colicinogenic properties.

Strains belonging to the serogroup O8 were isolated from patients with respiratory infections or from individuals with respiratory and enteric infections. The tracheal and faecal isolates collected in the same hospital were identical in phage pattern, colicinogeny and lysogeny.

Among strains of the serogroup O18 collected in the same hospital those isolated from pharyngeal or tracheal specimens of patients with respiratory infections and from faecal samples of patients with enteric disease had the same or nearly the same phage patterns (4, 12, 13, 16, 17 or 4, 12, 16, 17, 18). These strains were lysogenic and non-colicinogenic.

Madar *et al.* (1970), and later Czirók *et al.* (1973) performed phage typing of strains in the O18 serogroup and suggested their pathogenic role in enteric and respiratory diseases. The O18 strain recovered from cow's milk supported the assumption that enteritis in children can be transmitted by milk.

The pathologic role of strains of serogroup O59, can only be supposed for the time being. Strains O59: · : 19 were isolated from faecal samples of premature infants treated in the same hospital and from air samples of incubators. The clinical symptoms indicated enteritis and interstitial pneumonia. Phage patterns and colicinogenic properties were identical in every strain.

Strains belonging to the serogroup O111 and originating from the same hospital were identical in phage pattern, colicinogen, lysogeny and PPR. Strains at first untypable showed identical phage pattern after heat treatment when isolated from an identical source.

Czirók *et al.* (1975) reported an outbreak in an infants' ward. *E. coli* O4 and O114 strains were isolated. The responsibility of O114 for the epidemic was indicated by the finding that, unlike O4 strains, O114 isolates were uniform in phage pattern and colicinogeny, as well as by the fact that the latter produced exterotoxin while the former failed to do so.

Strains belonging to the serogroup O124 were derived from adult out-patients with enteritis or food poisoning. The patterns in this serogroup were fairly uniform, only colicinogeny provided a basis for differentiation. Strains isolated from the same family or from food poisoning cases were identical in phage pattern, colicinogeny, lysogeny and PPR. Phage pattern of the strains received from the GDR differed from the strains isolated in Hungary.

Studies of the lytic properties of strains belonging to different serogroups indicated that using the 30 typing phages of Milch and Gyenes (1972) identity of the strains could be deduced.

6

A total of 78·7% of serologically identified strains were typable by phages, and 36·5% were colicinogenic. Of the serologically untypable strains 45·0% were typable by phages, and 31·2% were colicinogenic.

Stability of phage pattern relates primarily to confluent or semi-confluent lysis. Weaker lytic reactions are less stable.

When a strain is re-typed, the choice of morphologically identical or non-identical colonies for examination can influence the result of the lytic reaction. Identical lytic reactions are experienced in about 90% of morphologically identical colonies. Interpretation of the results will not vary if carried out by different but well-trained persons.

In Hungary, the *E. coli* phage typing method of Milch and Gyenes (1972) has been applied by the Phage Typing Reference Laboratory and by one Regional Public Health Laboratory. The results of numerous examinations in parallel by the two laboratories have been constantly identical.

1. *Stability of phage patterns*

Examinations to assess stability of phage patterns were: (i) repeated determination of phage pattern of strains maintained in the laboratory; (ii) determination of phage pattern of different colonies of the same strain and (iii) determination of phage pattern of strains supposedly originating from the same source. It was found that the phage patterns examined at the same time were fairly stable but varied when different colonies were examined or the typing was carried out at different times. Therefore it is advisable to perform phage typing of strains collected from a hospital outbreak or from a given geographical area simultaneously.

M. Epidemiological use of tests for colicinogeny and lysogeny

Nicolle *et al.* (1960) demonstrated that strains of the serogroup O111, phage type Sèvres, isolated in the North of France were more frequently colicinogenic than those obtained from other regions. Strains of phage type Tourcoing from the former region produced colicin E_1, those from other areas colicin I. Strains with phage type Tourcoing from Berlin (GDR) produced colicin I, those from Wernigerode colicin $E + I$. Nosocomial strains of phage type Sèvres var. lyonnaise produced two types of colicin(I, $I + E$) or were non-producers.

Strains of the serogroup O55 were often colicinogenic, producing colicin I. Strains of the phage type Weiler produced colicin I, E, $E + I$ or B or were non-colicinogenic. Strains of the phage type Béthune isolated in France, Belgium and Israel were non-colicinogenic, those from Berlin produced colicin J or B, others from Wernigerode colicin B or E.

Studies of lysogeny in *E. coli* serogroup O124 (Deák, 1965) showed that in Hungary type 1 was the most common, whereas in Roumania type 2 or 2b predominated. The method developed by Milch and Gyenes (1972) combined phage typing with tests of colicinogeny and lysogeny. For strains of the serogroups O124 and O78, primary importance is attached to colicin typing. Three indicator strains were used for the identification of colicins produced by strains of the serogroup O124. Colicin type of strains from the same geographical region proved identical. Colicins produced by strains of group O78 and isolated in various geographical regions were of different types. Therefore, additional colicin typing was performed using the method of Frédéricq. Strains collected from a hospital epidemic produced colicin V, those isolated from other hospitals and infants' communities produced colicin type S$_3$ or were non-colicinogenic.

VI. CORRELATION BETWEEN PHAGE SENSITIVITY AND OTHER CHARACTERISTICS OF *E. COLI* STRAINS

A. Correlation between pathogenicity, serogroup, phage type and colicinogenic properties

Clinical and epidemiological observations accompanied by serological identification, phage and colicin typing have shown' that certain sero-fermentative, phage or colicin types within the antigenic group O are involved in pathological processes causing diseases and outbreaks. Litera-ture data indicate that the individual serotypes may undergo changes within the organisms. This hypothesis is supported by *in vivo* transfer experiments (Guinée, 1968; Salzmann and Klemm, 1968; Jarolmen, 1971; Smith, 1971, 1974; Smith, 1975) and by studies on virulence using *Shigella* hybrids (Stenzel, 1962; Falkow *et al.*, 1963; Formal *et al.* 1965; Petrovskaya and Licheva, 1970; Radoutcheva *et al.*, 1975).

The relztion between certain plasmids (Ent, Col, R) and pathogenicity has been reported by a number of authors (Smith and Linggood, 1970, 1972; Skerman *et al.*, 1972; Smith, 1974; Evans *et al.*, 1975). However, no systematic investigations have been performed on the association between phage type, colicin type and pathogenicity. Smith and Crabb (1956) demonstrated, in a series of examinations, identical phage types in healthy calves and in animals suffering from white scours.

Information on the relationship between colicin type and pathogenicity are not conclusive. Hamon (1959) could distinguish four colicin types, I, E, B and G for the most common pathogenic serogroups (O111, O55, O26). Colicin V, often produced by non-pathogenic strains was not produced by the above strains.

Branche *et al.* (1963) reported that colicinogeny was related to a variety of serotypes in commensal *E. coli* of porcine origin. Vasenius (1967) could frequently demonstrate colicinogeny in strains isolated from diarrhoeal pigs, Papavassiliou (1961) in *E. coli* strains from enteritis of infants and calves, Likhoded and Kudlai (1963) in strains from infantile enteritis. Colicinogeny was found more frequently in the pathogenic porcine strains O141 than in the non-pathogenic ones (De Alwis and Thomlinson, 1973). Smith (1974) in studies on *E. coli* O78 reported a close correlation between colicin V production and pathogenicity for broiler chickens. Strains of the serogroup O78 isolated by Czirók *et al.* (1975) from outbreaks in neonatal wards and from healthy infants' collectives were of identical phage type. The nosocomial strains invariably produced colicin V, those isolated from healthy infants other colicin types or were non-colicinogenic (Milch *et al.*, 1975a).

B. Correlation between antibiotic resistance and phage type

Nicolle *et al.* (1960), and Viallier *et al.* (1963) when describing their phage-typing method for the O serogroups 111, 55 and 26 suggested that one of the main reasons for variability was the use of antibiotics. However, no experimental evidence has yet been produced to support the thesis.

Milch (1967) found that strains of the serotype O111: K58(B4): H⁻, phage type 111/1 were resistant to Streptomycin, those of other phage types sensitive.

Brown and Parisi (1966) failed to demonstrate any relationship between phage type and antibiotic resistance.

Marsik *et al.* (1975) examined O antigen, phage type and R factor of *E. coli* strains isolated from humans, animals and their rural environments. They rarely found strains with similar patterns of resistance and similar phage types or belonging to the same O antigen groups. This finding indicates that the presence of *E. coli* with multiple resistance in the gut of humans is not due to long-term colonisation by animal strains but possibly to the transfer of R factors by animal strains to the human resident flora.

Recently evidence accumulated on phage restriction by the R plasmid. A number of authors could demonstrate association between phage sensitivity and R plasmid in salmonellae and shigellae (Anderson, 1966, 1968; Guinée *et al.*, 1967; Bannister and Glover, 1968; Watanabe, 1968; Tschäpe and Rische, 1970; Tschäpe *et al.*, 1974; László and Rimanóczy, 1975).

Milch (1975b, c) transmitted R plasmids from strains of human and animal origin to an *E. coli* K12 Hfr lac⁻ recipient strain. Phages used for the restriction tests were: T1–T7, λ, Pl, *Salmonella* Vi I + IV and O, and 30 *E. coli* typing phages. The R plasmid of the O4, O6 and O55 strains isolated from infant enteritis cases having various antibiotic resistance patterns produced fairly similar phage restriction patterns. The R factors of O78 strains derived from the same hospital caused identical phage restriction patterns, but the R plasmid of some O78 strains originating from the same source did not restrict the phages examined. The phage patterns of the human and animal strains belonging to the O78 serogroup were uniform (4, 4a, 4b) except for one strain isolated from a calf suffering from enteritis. This strain was carrying an R factor with Tetracycline resistance determinant and the R factor caused restriction of typing phage 4. The enterotoxin producing strain belonging to serogroup O148 (Skerman *et al.*, 1972) showed Tetracycline resistance marker with an R plasmid not causing restriction. R factors of two strains of the serogroup O149 pathogenic for swine with identical antibiotic markers (Chloramphenicol, Tetracycline, Kanamycin) restricted different phages.

The 31R plasmid of strains belonging to serogroup O111 causing phage restriction could be divided into three groups on a basis of phage restriction patterns:

(i) T2 and typing phages 2, 3, 6
(ii) λ, T3, T4, T5 and typing phages 14, 18
(iii) T2, T7 and typing phages 2, 3, 4, 6.

Some association was found between restriction ability of the R plasmid of serogroup O111 strains and their phage pattern. A higher proportion of untypable strains carried R plasmids producing restriction than did the strains with identified phage patterns. For instance, 15 strains, with phage pattern 21, 28, isolated from the same source carried an R plasmid that did not produce restriction, whereas 14 untypable strains isolated from the same source carried R factors with an ability for restriction.

VII. INDICATIONS FOR USING PHAGE TYPING AND ADDITIONAL TYPING PROCEDURES OTHER THAN SEROTYPING

The serotyping of *E. coli* has technical limitations, since the procedure is time-consuming and only a few laboratories possess all known sera. Therefore, the use of phage typing and additional typing procedures has a two-fold indication: (i) identification and (ii) typing of a strain.

(i) Identification becomes necessary when there is no possibility for serotyping or the strain proves non-identifiable by the diagnostic sera available. Representative strains selected on the basis of identical lytic reactions and on the results of additional procedures can be subjected to serotyping, if necessary, at a later time. To follow this practice seems more expedient than performing serotyping without selection.

(ii) In the case of serologically typable strains belonging to known serogroups or serotypes, phage typing and additional typing procedures can reveal whether the strain in question is responsible for an outbreak, in which case also the common source can be traced.

Large scale investigations have suggested that the joint application of phage typing and additional procedures (tests on colicinogeny, lysogeny and PPR) serve the purpose best. Due to their simplicity, phage typing and additional procedures can be performed not only by National Reference Laboratories but also by a satisfactorily equipped and staffed regional laboratory.

Figure 1 illustrates the procedures used for identification and typing of *E. coli*. The methods can be used separately or in combination. For use and comparison of the procedures the reader is referred to the relevant Chapters in this Volume.

Figure 2 illustrates the methods approved in routine work.

In conclusion we may say that phage typing and additional typing procedures are suited to reveal the pathogenic role of serologically non-identified or non-identifiable *E. coli* strains. The application of these procedures allows a further subdivision of frequent serogroups and serotypes, thus serving as an efficient and valuable epidemiological tool.

REFERENCES

Abbott, J. D., and Shannon, R. (1958). *J. clin. Path.*, **11**, 71–77.
Ackermann, H. W., Nicolle, P., LeMinor, S., and Le Minor, L. (1962). *Ann. Inst. Pasteur*, **103**, 523–535.
Adam, A. (1923). *Jb. Kinderhk.*, **101**, 295–298.
Adzharov, M. (1966). *Epidemiologija, Mikrogiologija I Infekciozni bolesti*, 3, 160–164.
Adzharov, M. (1968). *Zh. Mikrobiol.* (*Mosk.*), **I.5**, 50–54.
Adzharov, M. (1969). *Epidemiologija, Mikrobiologija I Infekciozni bolesti*, 6, 311–315.
Anderson, E. S. (1966). *Nature* (*Lond.*), **212**, 795–799.
Anderson, E. S. (1968). *Ann. Rev. Mikrobiol.*, **22**, 131–180.
Atanassova, S., and Mladenova, L. (1975). *Arch. roum. Path. exp.*, **34**, 192–193.
Avdeeva, T. A., Polotsky, Yu. E., Smirnova, L. A., Dragunskaya, E. M., Poyasova, E. V., and Chalenko, V. G. (1973). *Zh. Mikrobiol.* (*Mosk.*), **II./11**, 9–12.
Bahr, L. (1912). *Zbl. Bakt. I. Abt. Orig.*, **66**, 335–365.
Bahr, L., and Thomsen, A. (1912). *Zbl. Bakt. I. Abt. Orig.*, **66**, 365–386.
Bailey, W. R., and Glynn, J. P. (1961). *Can. J. Microbiol.*, **7**, 901–905.
Bannister, D., and Glover, S. W. (1968). *Biochem. biophys. Res. Commun.*, **30**, 735–738.
Barbuti, S. (1961). *Boll. Ist. sieroterap. milan.*, **40**, 3–6.
Barry, G. T., Abbott, V., Everhart, D. L., Leffler, R. Y., and Mynant, E. (1962). *Bact. Proc.*, 51–52.
Bavastrelli, L. (1959). *Riv. Ist. sieroter. ital.*, **34**, 172–183.
Bayer, J. E. (1968). *J. Virol.*, **2**, 346–356.

Bercovici, C., Iosub, C., Besleaga, V., Greceanu, A., Trifan, G., and Brebenel, G. (1969). *Arch. roum. Path. exp.*, **28**, 964–974.
Bercovici, C., Iosub, C., Popovici, M., Constantinin, S., and Agafitei, V. (1975). *Arch. roum. Path. exp.*, **34**, 185.
Bernard, R. P. (1965a). *Ann. Inst. Pasteur*, **108**, 774–790.
Bernard, R. P. (1965b). *Ann. Inst. Pasteur*, **109**, 47–65.
Bettelheim, K. A., and Taylor, J. (1970). *J. med. Microbiol.*, **3**, 655–667.
Bettelheim, K. A., and Taylor, J. (1971). *Ann. N.Y. Acad. Sci.*, **176**, 301–313.
Borisov, L. B., Lents, E. K., and Klimashevskaia, V. F. (1970). *Zh. Microbiol. (Mosk.)*, **47**, 71–76.
Bradley, D. E. (1963). *J. gen. Microbiol.*, **31**, 435–445.
Branche, W. C., Young, V. M., Robinet, H. G., and Massey, E. D. (1963). *Proc. Soc. exp. Biol. (N.Y.)*, **114**, 198–201.
Braun, O. H. (1956). *In* "Säuglings-Enteritis" (Ed. Adam), Stuttgart.
Brinton, C. C., Jr, Gemski, P., Jr, and Carnahan, J. (1964). *Proc. natn. Acad. Sci. (Wash.)*, **52**, 776–783.
Brown, W. J., and Parisi, J. T. (1966). *Proc. Soc. exp. Biol. (N.Y.)*, **121**, 259–262.
Burnet, F. M. (1934). *J. Path. Bact.*, **38**, 285–299.
Buttiaux, R., Nicolle, P., Le Minor, L., Le Minor, S., and Gaudier, B. (1956). *Arch. Mal. Appar. dig.*, **45**, 225–247.
Buttiaux, R., Nicolle, P., Le Minor, S., and Gaudier, G. (1956). *Ann. Inst. Pasteur*, **91**, 799–809.
Carlquist, P. R. (1956). *J. Bact.*, **71**, 339–341.
Cavallo, G. (1951). *C.R. Soc. Biol., Paris*, **145**, 1885–1887.
Choy, Y., Fehmel, F., Frank, N., and Stirm, S. (1975). *J. Virol.*, **16**, 581–590.
Christiansen, M. (1917). *Zbl. Bakt. I. Abt. Orig.*, **79**, 196–248.
Chsistovich, G. N., and Matyko, N. A. (1967). *Zh. Mikrobiol. (Mosk.)*, **2**, 91–94.
Costin, J. D. (1966). *Path. Microbiol.*, **29**, 214–227.
Craigie, J., Felix, A., and (1947). *Lancet*, **i**, 823–827.
Czirók, É., Madár, J., Milch, H., and Gyengési, L. (1973). *Egészségtudomány*, **17**, 275–284.
Czirók, É., Milch, H., Madár, J., and Dombi, I. (1975), *Egészségtudomány*, **19**, 85–90.
Czirók, É., Milch, H., Sverteczky, Zs., Hérmán, G., and Losonczy, G. (1975). *Acta microbiol. Acad. Sci. hung.*, **22**, 299–304.
D'Allesandro, G. R., and Comes, R. (1952). *Boll. Ist. sieroterap. milan.*, **31**, 291–294.
De Alwis, M. C. L., and Thomlinson, J. R. (1973). *J. gen. Microbiol.*, **74**, 45–52.
Deák, S. (1965). *Acta microbiol. Acad. Sci. hung.*, **12**, 261–267.
Deák, S. (1966). *Egészségtudomány*, **10**, 228–233.
Deák, S., and Ádám, M. (1969). *Acta microbiol. Acad. Sci. hung.*, **16**, 97–105.
Donta, S. T., Moon, H. W., and Whipp, C. Sh. (1974). *Science*, **183**, 334–335.
Edwards, P. R., and Ewing, W. H. (1972). "Identification of Enterobacteriaceae", 3rd edn. Burgess, Minneapolis.
Edwards, P. R., Fife, M. A., and Ewing, H. W. (1956). *Am. J. med. Technol.*, **22**, 28–34.
Eörsi, M., Jablonszky, L., and Milch, H. (1953). *Népegészségügy*, **34**, 220–223.
Eörsi, M., Jablonszky, L., and Milch, H. (1954). *Acta microbiol. Acad. Sci. hung.*, 1–8.

Eörsi, M., Jablonszky, L., Milch, H., and Barsy, G. (1957). *Acta microbiol. Acad. Sci. hung.*, **5**, 201–215.

Evans, D. G., Silver, R. P., Evans, D. J., Chase, D. G., and Gorbach, S. L. (1975). *Infection and Immunity*, **12**, 656–667.

Ewing, W. H., Davis, B. R., and Edwards, P. R. (1960). *Publ. Hlth. Lab.*, **18**, 77–83.

Ewing, W. H. (1963). Isolation and identification of *Escherichia coli* serotypes associated with diarrhoeal diseases. U.S. Dept. of Health, Education and Welfare, Atlanta.

Falkow, S., Schneider, H., Baron, L. S., and Formal, S. B. (1963). *J. Bact.*, **86**, 1251–1258.

Fehmel, F., Feige, U., Niemann, H., and Stirm, S. (1975). *J. Virol.*, **16**, 591–601.

Formal, S. B., La Brec, E. H., Schneider, M., and Falkow, S. (1965). *J. Bact.*, **89**, 835–838.

Frédéricq, P. (1948). *Rev. belge Path.*, **19**, Suppl. 4, 1–5.

Frédéricq, P. (1950). *C. R. Soc. Biol.*, Paris, **144**, 295–297.

Frédéricq, P., Betz-Bareau, M., and Nicolle, P. (1956). *C. R. Soc. Biol.*, Paris, **150**, 2039–2042.

Gardner, J. F. (1950). *Brit. J. exp. Path.*, **31**, 102–111.

Glynn, A. A., and Howard, C. J. (1970). *Immunology*, **18**, 331–346.

Guinée, P. A. M. (1968). *Antonie v. Leeuwenhoek*, **34**, 93–98.

Guinée, P. A. M., Scholtens, R. Th., and Willems, H. M. C. C. (1967). *Antonie v. Leeuwenhoek*, **33**, 25–29.

Hamon, Y. (1958). *C.R. Acad. Sci.*, Paris, **247**, 1260–1261.

Hamon, Y. (1958). *Ann. Inst. Pasteur*, **95**, 117–121.

Hamon, Y. (1959). *Ann. Inst. Pasteur*, **96**, 614–629.

Hamon, Y. (1960). *Zbl. Bakt. I. Abt. Orig.*, **181**, 456–468.

Hamon, Y. (1960). 4. *Coll. Fragen der Lysotypie, Wernigerode.* (Tagungsbericht), S. 148–160.

Hamon, Y., and Brault, G. (1958). *C.R. Acad. Sci.* (Paris), **246**, 1779–1780.

Hamon, Y., Maresz, J., and Péron, Y. (1965). 5. *Coll. Fragen der Lysotypie, Wernigerode* (Tagungsbericht), S. 34–45.

Hart, J. M. (1965). *Zbl. Bakt. I. Abt. Orig.*, **196**, 364–369.

Jackson, L. E. et al. (1967). *Bact. Proc.*, p. 27., cit.: Bayer, M. E. (1968), *J. Virol.*, **2**, 246–356.

Jarolmen, H. (1971). *Ann. N.Y. Acad. Sci.*, **182**, 72–90.

Jensen, C. O. (1913). *In* "Handbuch der pathogenen Mikro-organismen" (Ed. Kolle-Wassermann), 2. Aufl. **6**, 121–144.

Kasatiya, S., cit.: Ackerman u. Mitarb. (1962). *Ann. Inst. Pasteur*, **103**, 523–535.

Kasatiya, S. (1963). Thèse Doctorat et sciences, Paris.

Kasatiya, S., and Hamon. Y. (1965). *Rev. Hyg. Med. soc.*, **13**, 35–48.

Kauffmann, F. (1948). *Acta path. microbiol. scand.*, **25**, 502–506.

Kauffmann, F. (1966). "The bacteriology of Enterobacteriaceae". 3rd edn. Munksgaard, Copenhagen.

Kauffmann, F., and Dupont, A. (1950). *Acta path. microbiol. scand.*, **27**, 552–564.

Kauffmann, F., and Ørskov, F. (1956). *In* "Säuglings-Enteritis" (Ed. Adam). Stuttgart.

Kauffmann, F., and Perch, B. (1948). *Acta path. microbiol. scand.*, **25**, 507–512.

Kauffmann, F., and Vahlne, G. (1945). *Acta path. microbiol. scand.*, **22**, 119–121.

Kayser, F. H. (1964). *Z. Hyg. Infekt.-Kr.*, **149**, 373–382.

Kolta, F., and Deák, S. (1962). *Egészségtudomány*, **6**, 363–371.

Krajlnovic, S. (1961). *Vojnosanit. Pregl.*, **18**, 857–862.

Krylova, M. D. (1962). *Zh. Mikrobiol.* (*Mosk.*), **33**, 65–69.

Kubinyi, L. (1965). *Egészségtudomány*, **9**, 84–87.

Kubinyi, L. (1971). *Egészségtudomány*, **15**, 125–130.

Kunin, C. M. (1966). *Ann. Rev. Med.*, **17**, 383–406.

Lachowicz, K. (1968). *In* "Enterobacteriacea-Infektionen" (Eds. J., Sedlák and H. Rische), pp. 445–512. VEB Georg Thieme, Leipzig.

La Brec, E. H., Schneider, H., Magnani, T. J., and Formal, S. B. (1964). *J. Bact.*, **88**, 1503–1518.

László, G. V., and Rimanóczy, I. (1976). *Acta Microbiol. Acad. Sci. hung.*, **23**, 251–257.

Le Minor, S., Le Minor, L., Nicolle, P., and Buttiaux, R. (1954). *Ann. Inst. Pasteur*, **86**, 204–226.

Le Minor, S., Le Minor, L., Nicolle, P., Drean, D., and Ackermann, H. W. (1962). *Ann. Inst. Pasteur*, **102**, 716–725.

Levine, P., and Frisch, A. W. (1934). *J. exp. Med.*, **59**, 213–228.

Levine, M., and Tanimoto, R. H. (1960). *Bact. Proc.*, 77–78.

Lieb, M. (1953). *J. Bact.*, **65**, 642–651.

Likhoded, V. G., and Kudlai, D. G. (1963). *Zh. Mikrobiol.* (*Mosk.*), **40**, 128–132.

Linton, K. B. (1960). *J. clin. Path.*, **13**, 168–172.

Linzenmeier, G. (1962). *Zbl. Bakt. I. Abt. Orig.*, **184**, 74–83.

Madár, J., Lakatos, M., Kiss Szabó, A., Milch, H., and Vánger, S. (1970). *Orv. Hetil.*, **111**, 1035–1038.

Marsik, F. J., and Parisi, J. T. (1971). *Appl. Microbiol.*, **22**, 26–31.

Marsik, F. J., Parisi, J. T., and Blenden, D. C. (1975) *J. Infect. Dis.*, **132**, 296–302.

McGeachie, J. (1965). *Zbl. Bakt. I. Abt. Orig.*, **196**, 377–384.

McGeachie, J., and McCormick, W. (1963). *J. Clin. Path.*, **16**, 278–280.

Medearis, D. N., Jr, and Kenny, J. F. (1968). *J. Immunol.*, **101**, 534–540.

Medearis, D. N., Jr, Camitta, B. M., and Heath, E. C. (1968). *J. exp. Med.*, **128**, 399–414.

Milch, H. (1967). "Epidemiological use of bacteriophage research" (in Hungarian). Budapest. 1967. Dissertation.

Milch, H., and Deák, S. (1961). *Acta microbiol. Acad. Sci. hung.*, **8**, 411–421.

Milch, H., Deák, S., and Lakatos, M. (1961). *Egészségtudomány*, **5**, 325–334.

Milch, H., and Gyenes, M. (1972). *Acta microbiol. Acad. Sci. hung.*, **19**, 213–244.

Milch, H. and Gyenes, M. (1975). VI International Colloquium on Phage Typing and Other Laboratory Methods for Epidemiological Surveillance, Wemigenode. Proceedings. Vol. 2, pp. 387–391.

Milch, H., Czirók, É., and Madár, J. (1975a). International Colloquium on Phage Typing and Other Laboratory Methods for Epidemiological Surveillance, Wemigenode. Proceedings. Vol. 2, pp. 350–357.

Milch, H., Gyenes, M. and Herman, G. (1975c). *In* "Drug-inactivating Enzymes and Antibiotic Resistance (Eds. J. Mitsuhashi, L. Rosival and V. Krěméry), pp. 391–396.

Moeller, V. (1955). *Acta path. microbiol. scand.*, **36**, 158–172.

Neish, A. S., Turner, P., Fleming, J., and Evans, N. (1975). *Lancet*, **ii**, 946–948.

Nhan, L. B., Jann, B., and Jann, K. (1971). *Europ. J. Biochem.*, **21**, 226–234.

Nicolle, P., Le Minor, L., Buttiaux, R., and Ducrest, P. (1952a). *Bull. Acad. nat. Méd.* (*Paris*), **136**, 480–483.

Nicolle, P., Le Minor, L., Buttiaux, R., and Ducrest, P. (1952b). *Bull. Acad. nat. Méd. (Paris)*, **136**, 483–485.
Nicolle, P., Le Minor, L., Le Minor, S., and Buttiaux, R. (1954). *C.R. Acad. Sci.*, **239**, 462–466.
Nicolle, P., Le Minor, L., Le Minor, S., and Buttiaux, R. (1957). *Zbl. Bakt. I. Abt. Orig.*, **168**, 512–528.
Nicolle, P., Le Minor, L., Le Minor, S., and Buttiaux, R. (1958). 7. *Internat. Congr. Microbiol.*, 320–322.
Nicolle, P., Le Minor, S., Hamon, Y., and Brault, G. (1960). *Rev. Hyg. Méd. soc.*, **8**, 523–562.
Nicolle, P., Brault, G., and Brault, J. (1963). *C.R. Acad. Sci.*, Paris, **257**, 2194–2197.
Nicolle, P., and Rische, H. (1968). *In* "Enterobacteriacea-Injektionen". (Eds. J. Sedlak and H. Rische). VEB Georg Thieme, Leipzig.
Novgorodskaya, E. M., Hazenson, L. B., Loseva, A. G., and Krivonosova, K. I. (1970). *In* "Dysenteria, Salmonella, Escherichia (colienteritis in infants)", **36**, 117–132., Leningrad.
Ørskov, F., Ørskov, I., Jann, B., and Jann, K. (1971). *Acta path. microbiol. scand.*, **B79**, 142–152.
Papavassiliou, J. (1961). *J. gen. Microbiol.*, **25**, 409–413.
Papavassiliou, J. (1962). *Path. et Microbiol. (Basel)*, **25**, 144–152.
Papavassiliou, J. (1963). *Z. Hyg. Infekt.-Kr.*, **149**, 164–169.
Parr, L. W., El Shavi, N. N., and Robbins, M. L. (1960). *J. Bact.*, **80**, 417–418.
Parisi, J. T., Russel, J. C., and Merlo, R. J. (1969). *Appl. Microbiol.*, **17**, 721–725.
Petrovskaya, V. G., and Licheva, T. A. (1970). *Ann. Inst. Pasteur*, **118**, 761–766.
Radoutcheva, T., Popov, Ch., Veljanov, D., and Bondarenko, V. M. (1975). *Zbl. Bakt. I. Abt. Orig.* **A231**, 116–121.
Randall-Hazelbauer, L., and Schwartz, M. (1973). *J. Bact.*, **116**, 1436–1446.
Rantz, L. A. (1962). *Arch. intern. Med.*, **109**, 37–42.
Rhodes, T. F., and Fung, H. C. (1970). *Bacteriol. Proc.*, 111–114.
Richard, C. (1966). *Ann. Inst. Pasteur*, **110**, 114–119.
Rische, H. (1968). *In* "Enterobacteriaceae-Infektionen" (Eds. J. Sedlák, an dH. Rische). VEB Georg Thieme, Leipzig.
Rogers, K. B., Benson, R. P., Foster, W. P., Jonas, L. F., Butler, E. B., and Williams, T. C. (1956). *Lancet*, **ii**, 599–604.
Rudnai, O., and Adamis, E. (1965). *Egészségtudomány*, **9**, 325–334.
Ryter, A., Shuman, H., and Schwartz, M. (1975). *J. Bact.*, **122**, 295–301.
Sakazaki, R., Murata, Y., and Iwanami, S. (1962). *Jap. J. Bacteriol.*, **17**, 973–975.
Salzmann, T. C., and Klemm, L. (1968). *Proc. Soc. exp. Biol. (N.Y.)*, **128**, 392–394.
Schell, J., and Glower, S. W. (1966). *J. gen. Microbiol.*, **45**, 61–72.
Schmidt, G. (1972). *Zbl. Bakt. I. Orig.* **A220**, 472–476.
Schmidt, G., Jann, B., and Jann, K. (1969). *Europ. J. Biochem.*, **10**, 501–510.
Serény, B. (1957). *Acta microbiol. Acad. Sci. hung.*, **4**, 367–376.
Shannon, R. (1957). *J. med. Lab. Technol.*, **14**, 199–214.
Sherman, F. J., Formal, S. B., and Falkow, S. (1972). *Infect. Immun.*, **5**, 622–624.
Smith, H. Williams (1963). *J. Path. Bact.*, **85**, 197–211.
Smith, H. Williams (1971). *Ann. N.Y. Acad. Sci.*, **182**, 80–90.
Smith, H. Williams (1974). *J. gen. Microbiol.*, **83**, 95–111.
Smith, H. Williams, and Crabb, W. E. (1956). *J. gen. Microbiol.*, **15**, 556–574.
Smith, H. Williams, and Halls, S. (1967). *J. gen. Microbiol.*, **47**, 153–161.
Smith, H. Williams, and Linggood, M. A. (1970). *J. gen. Microbiol.*, **62**, 287–299.

Smith, H. Williams, and Linggood, M. A. (1971). *J. Med. Microbiol.*, **4**, 467–485.
Smith, H. Williams, and Linggood, M. A. (1972). *J. Med. Microbiol.*, **5**, 243–250.
Smith, M. G. (1975). *J. Hyg. Camb.*, **75**, 363–370.
Sojka, W. J. (1972). "Enteropathogenic *E. coli* in man and farm animals". Symposium on Microbial Food-borne Infections and Intoxications. Ottawa. 5–17.
Stenzel, W. (1962). *Z. Hyg. Infekt.-Kr.*, **148**, 433–444.
Sterne, R. B., Wescott, R. B., and Parisi, J. T. (1970). *Am. J. Vet. Res.*, **31**, 2101–2103.
Stirm, S. (1968). *Nature, Lond.*, **219**, 637–639.
Stirm, S. (1968). *J. Virol.*, **2**, 1107–1114.
Stirm, S. (1971). *J. Virol.*, **8**, 330–342.
Stirm, S., Bessler, W., Fehmel, F., and Freund-Mölbert, E. (1972). *Zbl. Bakt. I. Abt. Orig.*, **A220**, 484.
Swanstrom, M., and Adams, M. H. (1951). *Proc. Soc. exp. Biol. (N.Y.)*, **78**, 372–375.
Tamaki, S., Sato, T., and Mitsuhashi, M. (1971). *J. Bact.*, **105**, 968–975.
Taylor, J. (1960). *Bull. Wld. Hlth. Org.*, **23**, 763–779.
Taylor, J., Maltby, M. P., and Payne, J. M. (1958). *J. Path. Bact.*, **76**, 491–499.
Taylor, K., and Taylor, A. (1963). *Acta microbiol. pol.*, **12**, 97–106.
Thibault, P., and Le Minor, L. (1957). *Ann. Inst. Pasteur*, **92**, 551–554.
Toft, G. (1947). *Acta path. microbiol. scand.*, **24**, 260–271.
Tschäpe, H., Dragaš, A., Rische, H., and Kühn, H. (1974). *Zbl. Bakt. I. Abt. Orig.*, **A226**, 184–193.
Tschäpe, H., and Rische, H. (1970). *Zbl. Bakt. I. Abt. Orig.*, **214**, 91–100.
Ujváry, G. (1957). *Zbl. Bakt. I. Abt. Orig.*, **170**, 394–405.
Vasenius, H. (1967). *Acta vet. scand.*, **8**, 195–200.
Viallier, J., Chassignol, S., Sédallian, A., and Sibille, S. (1963). *C.R. Soc. Biol. (Paris)*, **157**, 2004–2005.
Vosti, K. L. (1968). *J. Bact.*, **96**, 1947–1952.
Watanabe, T. (1968). *Proc. XII. Congr. Genetics*, **2**, 237–239.
Weidel, W. (1953). *Cold Spring Harbor Symp. Quant. Biol.*, **18**, 155–157.
Weidel, W. (1958). *Ann. Rev. Microbiol.*, **12**, 27–28.
Weidel, W., Koch, G., and Bobosch, K. (1954). *Z. Naturforsch.*, **9b**, 573–579.
Weidel, W., and Primosigh, J. (1958). *J. gen. Microbiol.*, **18**, 513–517.
Wilson, G. S., and Atkinson, J. D. (1945). *Lancet*, **i**, 647–648.
Wolberg, G., and De Witt, C. W. (1969). *J. Bact.*, **100**, 730–737.
Zichichi, M. L., and Kellenberger, G. (1963). *Virology*, **19**, 450–460.
Ziedt, C. H., Fox, F. A., and Norris, G. F. (1960). *Amer. J. clin. Path.*, **33**, 233–237.

CHAPTER IV

Phage Typing of *Salmonella*

P. A. M. Guinée and W. J. van Leeuwen

National Institute for Public Health, Bilthoven, The Netherlands

I: INTRODUCTION

A. General principles

One may distinguish several types of phage typing schemes for the phage typing of *Salmonella*, depending on the principles used for the development of the scheme as well as on how successfully they are applied.

(a) Phage typing systems consisting of one wild-type phage and a number of typing phages which were all derived from the wild-type phage by adaptation.

(b) Systems consisting of a combination of wild-type phages, their adaptations, and lysogenic phages.

(c) Systems consisting of wild-type phages only.

(d) Typing schemes consisting of lysogenic phages which were isolated from *Salmonella* strains usually belonging to the *Salmonella* serotype for which the typing scheme is evolved.

(e) Another principle for subdividing *Salmonella* strains of the same serotype, is the identification of the lysogenic phage carried by the bacterial strain. For this is needed a limited number of usually non-lysogenic indicator strains belonging to the same serotype.

(f) The lysogenic phages may moreover be identified in terms of cross-immunity against a standard set of lysogenic strains, e.g. one bacterial strain lysogenised by a series of standard lysogenic phages.

Ad a. This principle was successfully pursued by Craigie and Yen (1938a, b) when they developed the Vi-phage typing system for *S. typhi*. One of the characteristic features of Vi-phage II was its capacity to acquire specificity for the strain on which it had last been propagated. In this way, adaptations of the same phage having specificity for a particular collection of *S. typhi* strains were obtained. It was originally believed that the adapted phages were host-range mutants of the original Vi-phage II. It was demonstrated later that host-induced modification also played an important role (Anderson and Felix, 1952, 1953a). Host-induced modification (host-controlled modification) was first described in phage T2 by Luria and Human (1952) and later in phages P2 and lambda (Bertani and Weigle, 1953; Weigle and Bertani, 1953). A general description of host-induced modification was given by Luria (1953). It was termed phenotypic modification by Anderson and Fraser (1955). The phenomenon was further elucidated in terms of restriction and modification by Arber and Dussoix (1962) and Dussoix and Arber (1962). The essential feature is that a phage plates with an efficiency of plating (EOP) of 1 on a particular host A, but with an EOP of about 10^{-4} on host B. One of the rare plaques on B consists of particles which plate with an EOP of 1 on A as well as on B. As soon as the phage is again propagated on A, the EOP on host B is again reduced to 10^{-4} (see also Hayes, 1970). The terms restriction and modification will be used throughout this Chapter.

ad b. Although several phage typing systems attempted to copy the

Vi-II system, only a few were successful. The fact that a particular set of adapted typing phages did not sufficiently distinguish epidemiologically unrelated strains necessitated the inclusion of other wild-type phages into the phage typing system for *S. typhimurium* described by Scholtens and Rost (1972). During the process of adaptation, the phages may be contaminated, or even replaced, by lysogenic phages carried by the strains on which the phage was propagated before. An example of such a system is the Felix–Callow system for *S. typhimurium* (Felix, 1956).

ad c. Many systems make use only of wild-type phages isolated from sewage or similar sources, for instance the system for *S. typhimurium* described by Lilleengen (1948).

ad d. Several typing schemes have been described which use lysogenic phages mostly isolated from strains belonging to the serotype for which the typing scheme was evolved. Several terms are being used for such phages, such as prophages, natural phages, symbiotic phages, temperate phages or lysogenic phages. The latter term will be used throughout this Chapter. An example is the typing scheme for *S. dublin* described by Smith (1951c).

ad e. This principle is included for instance in the phage typing system for *S. blockley*, described by Sechter and Gerichter (1969).

ad f. This principle was extensively studied by Boyd (1950, 1954) and Boyd and Bidwell (1957).

B. Isolation and identification of typing phages

Phages occur abundantly in sewage, sewage-contaminated surface water and in faeces. Although phages isolated from such materials are usually called wild-type phages, they may equally well be lysogenic phages. For instance, phage 14-17 considered to be a wild-type and used for the *S. typhimurium* system of Scholtens (1962) was isolated from sewage-contaminated surface water. The phage is serologically identical with a number of lysogenic phages found in *S. typhimurium* strains. It is therefore possible that the phage 14-17 originated from a lysogenic *S. typhimurium* strain.

Bacteriophage, once isolated has to be purified by repeated single plaque isolation. Preliminary identification consists of establishing its host-range and plaque morphology. Neutralising antisera permit serological classification of the phages and are extremely useful tools to establish whether adapted phages have, in fact, been derived from wild-type phages or are contaminating lysogenic phages. Details of the above-mentioned procedures are given by Adams (1959).

C. Technique of phage typing

Although minor variations have been applied, the technique of phage typing in most typing systems, is basically that described by Craigie and Yen (1938a, b), Craigie and Felix (1947) and Anderson and Williams (1956).

1. Media

Media used in phage typing are usually those on which the bacterial strains concerned grow well, such as commercially available nutrient broth and nutrient agar.

Anderson and Williams (1956) recommended Bacto nutrient broth (Difco), pH 6·8, as liquid medium and the same medium with 1·3% David New Zealand agar (Difco) as solid medium.

Scholtens (1950a) described a solid nutrient agar medium for phage typing. The medium consists of beef extract (500 g fresh beef per 1000 ml of medium); NaCl, 0·85%; peptone (Difco) 1·25%; agar (Difco) 2%; pH: 6·8.

2. Production of typing phages

Prior to the preparation of phages on a large scale, they should be purified, or a pure state be confirmed by repeated single-plaque isolations. An equal concentration of phage particles and host cells generally yields the highest titre of phages.

Depending on the quantity of the phage preparation desired, a certain amount (for instance 100 ml) of prewarmed nutrient broth is inoculated with about 1×10^8 host cells in the logarithmic phase of growth together with the same number of phage particles. The mixture is incubated at 37°C (for Vi-phages 38·5°C is recommended by Anderson and Williams (1956)) for as long as lysis visibly continues and generally no longer than 7 h.

The mixture is then heated at 60°C for 30 min to kill the host cells. Most typing phages used for phage typing of *Salmonella* are thermostable and survive such heat treatment. If the phage is not thermostable, the host cells may be killed by the addition of a few drops of chloroform. In all cases, the lysates are centrifuged to remove the killed bacteria.

If one desires to avoid any heat damage to the phages, the following procedure is recommended. The mixture is centrifuged at 5000g for 15 min to sediment the majority of the host cells. The supernatant is then sterilised by filtering through a 0·45 μm membrane filter.

It is sometimes impossible to prepare a phage of sufficiently high titre in liquid medium. The agar-layer technique should then be recommended. It may be used for the production of small amounts of phage in 9 cm Petri dishes, or for the production of larger quantities of phage in Roux flasks.

An appropriate quantity of molten 0·4% nutrient agar is inoculated with bacteria and phage to give a final concentration of 1×10^8/ml of bacteria and 1×10^6/ml of phage particles. The mixture is poured on to a basal layer of 2% nutrient agar. After 16–20 h of incubation at 37°C, the top layer is scraped off. The phages may then be harvested using the techniques described elsewhere in this Chapter. The method used in the authors' laboratory is to freeze the agar-bacteria-phage mixture at -20°C and to centrifuge the frozen mixtures at 5000 g for 45 min at room temperature. The supernatant is then filtered through a membrane filter. The phage preparation is titrated on its host strain to establish its number of plaque forming units (pfu) per ml, as well as its critical test concentration (CTC). CTC is now usually called Routine Test Dilution (RTD). However RTD and CTC are not always identical in systems other than *Salmonella* (Rische, 1968).

The RTD of a phage is the highest dilution which in a spot test still produces confluent, or in some cases semiconfluent lysis on the strain on which the phage was propagated.

The advantage of using typing phages at their RTD rather than less or undiluted emerged from the Vi-phage typing system for *S. typhi*. It was found that any of the adapted Vi-phages adsorb to all Vi-types of *S. typhi* and this adsorption was shown to be lethal to the bacteria. In a spot test, such lethal adsorption may appear as confluent lysis and be indistinguishable from genuine lysis. Moreover, an undiluted phage preparation contains host-range mutants able to lyse other phage types of the same bacterial species (Anderson and Fraser, 1956). The principle of using typing phages at RTD has therefore become general practice. Most phages used in the phage typing of *Salmonella* remain stable for over a year, when kept in nutrient broth at 4°C. It is generally recommended that the RTD of each phage preparation be checked after storage, prior to using it for phage typing.

Most type strains of *Salmonella* remain stable for many years when maintained at room temperature on cork-stoppered Dorset egg slopes.

3. *Cultivation of strains to be phage typed*

Cultures to be phage typed are inoculated into 2–10 ml of nutrient broth to give a barely visible turbidity and incubated at 37°C, preferably with agitation in a waterbath. In this way, a concentration of 5×10^8 cells/ml is usually reached within approximately 3 h. The culture is then spread on the surface of a plate with solid medium either in discrete areas or as a complete lawn. After the inoculated agar plates have been allowed to dry, the phages are spotted in small amounts on to the lawn or discrete areas.

Laborious manual spotting has been replaced by a number of mechanical devices (Parker, 1972), of which the Lidwell apparatus or contrivances similar to it are the most frequently used (Lidwell, 1959)*. More recently, another mechanical device was described by Guinée et al. (1973). Farmer et al. (1975) described a bacteriophage applicator which dispenses 59 drops of phage simultaneously†. The phage typing plates are usually read after over-night incubation at 37°C.

Methods of recording degrees of lysis may vary from one system to another, but the most generally adopted procedure is that presented by Anderson and Williams (1956) (Table 1).

TABLE I

Method of recording degrees of lysis on enteric phage typing plates according to Anderson and Williams (1956)

Plaque sizes	Plaque numbers	Lysis
L = large	0–5 plaques	SCL = semi-confluent lysis
N = normal	± = 6–20 plaques	CL = confluent lysis
S = small	+ = 21–40 plaques	< SCL = ∫ intermediate degrees
		< CL = { of lysis
m = minute	+ ± = 41–60 plaques	OL = confluent "opaque"
Visible only	+ + = 61–80 plaques	lysis (opacity due to
with hand	+ + ± = 81–120 plaques	heavy secondary
lens	+ + + = >120 plaques	growth)
μ = micro		

Depending on the system, phage types are recognised either by one or more specific phage reactions, particularly so in type a or b systems, or by recognition of specific phage patterns, particularly so in type c or d systems.

II. INTERNATIONALLY EMPLOYED PHAGE TYPING SYSTEMS

A. Phage typing of S. typhi

In 1934, Felix and Pitt announced their discovery of the Vi-antigen of S. typhi. This antigen is present in the great majority of freshly isolated strains of the typhoid bacillus. Its presence renders the organism in-

Commercial sources, e.g.

* Lidwell phage typing: Biddulph and Company Ltd, Commercial Street, Knott Mill, Manchester 15, England. Nichron coiled wire for phage typing loops $2\frac{1}{2} \times 22$ s.w.g.: Mallinson Greenhount Works Ltd, Halifax, England.

† The phage applicator is available commercially from Johnny Brown Machine Shop, P.O. Box 239, Route 3, Tuscaloosa, Alabama 35401, USA.

agglutinable by sera containing only antibodies against the somatic O-antigen of *S. typhi*. Phages specific for the Vi-form of *S. typhi* were isolated independently by a number of workers (Craigie, 1936; Craigie and Brandon, 1936; Scholtens, 1936; Sertic and Boulgakov, 1936). These Vi-phages adsorb only to organisms possessing the Vi-antigen.

Craigie and Yen (1938a, b) described four virulent Vi-phages which they designated I, II, III and IV. These phages differed in physical properties and were differentiated serologically. Vi-phage II proved to have the unusual property of adapting itself to strains of *S. typhi* in such a way that it became highly specific for the last strain on which it had been propagated. By means of a series of adaptations of the Vi-phage II, the authors were able to subdivide a collection of *S. typhi* strains into 11 phage types.

The value of this scheme was recognised by Felix (1943) and it was largely due to his work that it was accepted for international use. As it was employed by an increasing number of workers, it was improved and extended.

1. *International organisation*

Craigie and Felix (1947) published suggestions for the standardisation of typing methods, and the International Committee for Enteric Phage Typing (I.C.E.P.T.) was formed. The Enteric Reference Laboratory in London acts as the International Reference Laboratory for Enteric Phage Typing (I.R.L.).* The typing system in current international use consists of 87 phages, distinguishing 87 types of *S. typhi*. (Rische and Ziesché, 1973a). For the typing of *S. typhi* strains, only phage preparations and type strains distributed by the I.R.L. should be used.

2. *Typing phages and type strains*

All the Vi-typing phages were adapted from the original Vi-phage II of Craigie and Yen (1938a, b) and are neutralised by an antiserum against phage A, which is the form in which the phage was first isolated. In the typing scheme, the phage type of a particular organism and the corresponding adaptation of the Vi-phage II carry the same symbol. All typing phages, type strains with year of isolation and references have been listed by Rische and Ziesché (1973a).

3. *Type designations*

Recognised Vi-phage types have either lettered or numerical designations, for instance E₁, F₁, 40 etc. The designation "I+IV" indicates

* Address: Central Public Health Laboratory Service, Colindale Avenue, London NW9, England.

cultures resistant to all specific typing adaptations of Vi-phage II of Craigie and Yen (1938a, b), but sensitive to phages I and/or IV of Craigie and Yen. The designation "degraded Vi-strains" indicates strains which cross-react widely with the Vi-typing phages, but do not belong to a specific type. The designation "Vi-negative" means that such cultures are devoid of Vi-antigen, and cannot be typed with Vi-phages. Cultures that do not react with the Vi-typing phages are designated as "untypable Vi-strains", whether their lack of lysis is due to inherent resistance to the Vi-phages or to the fact that they belong to new Vi-types.

4. Technical methods

Full details of the media, conditions of growth, propagation and testing of the phages, typing methods and reading and reporting of results have been described by Anderson and Williams (1956) and Rische and Ziesché (1973a), and are outlined on pages 159–162.

The typing phages are used at RTD. To determine the RTD, titration of the phage is carried out on type A, which is sensitive to all the typing phages and on the homologous Vi-type strain on which the phage has been propagated.

Since S. typhi cultures tend to lose their Vi-antigen quite rapidly in vitro, it is preferable to subculture S. typhi cultures intended for phage typing from primary platings on selective media such as desoxycholate-citrate agar or Wilson–Blair agar, and to prepare a pooled subculture of a number of colonies. For routine purposes, some phage preparations corresponding to rare Vi-types can be pooled so that the test may be completed on one plate (Anderson and Williams, 1956; Rische and Ziesché, 1973a). When one of the pooled phage suspensions gives a positive reaction, the particular phages of the pool must be tested separately.

5. Auxiliary typing method

The method has its limitations. There are untypable strains because they are completely devoid of Vi-antigen. Some strains are basically resistant to the Vi-phage and there are strains for which an adapted phage has not yet been evolved. Most untypable strains, however, belong to the group of so-called "degraded" strains which are sensitive to a large number of the adapted phages. A more serious difficulty is that, in many countries, one Vi-phage type—usually either type A or type E_1—is so common as to limit the epidemiological information obtainable from typing. Attempts have been made to further subdivide the common types.

(1) Nicolle et al. (1954) and Desranleau and Martin (1950) subdivided strains of type A by means of unadapted Vi-phages and lysogenic phages.

(2) Brandis (1955) subdivided type E_1 strains by means of Vi-phage VII into types E_{1a} and E_{1b}.

(3) A subdivision of "I + IV" strains was made by Nicolle and Diverneau (1961). By means of seven phages, isolated from lysogenic "I + IV" strains, and phage Taunton of the *S. schottmuelleri* (*S. paratyphi B*) scheme, "I + IV" strains can be subdivided into 13 subtypes.

(4) An auxiliary system consisting of unadapted Vi-phages was developed by Scholtens (1950a). By means of this system some groups of untypable Vi-strains gave characteristic phage patterns.

6. *Frequency distribution of the Vi-types of* S. typhi

As a result of the formation of the I.C.E.P.T., much information has been gained concerning the distribution of the phage types of *S. typhi*. If a type which is not indigenous in a given country is isolated from a case of typhoid fever, it is sometimes possible to trace its origin if the countries in which the type is prevalent, are known.

The Vi-phage types of *S. typhi* can be divided into groups according to their geographical distribution (Report of the I.C.E.P.T., 1973):

(1) Cosmopolitan phage types generally the commonest in Europe: A, B_2, C_1, D_1, F_1, N, T, 28, 46 and the three groups of untypable strains: "I + IV", degraded Vi-strain, "Vi-negative".

(2) Phage types frequently found in North Africa (Morocco, Algeria, Tunisia, Egypt) but less common or even non-existent in other parts of the world: L_1, L_2, 40, 42.

(3) Phage types common in Central Africa and rare, often non-existent in Europe and most other parts of the world: C_3, C_4, C_6, Central Africa variant of type C_1 (Nicolle *et al.*, 1955).

(4) Phage type especially common in countries bordered by the Indian Ocean: G.

(5) Phage types common or present only in the Far East and rare or non-existent in most other parts of the world: C_3, C_5, D_2, D_4, E_3, E_4, E_7, E_9, E_{10}, G_3, G_5, J_3, J_5, M_2, M_3, M_4 and, above all, type 37 and some "I + IV" groups.

(6) Phage type M, rare or non-existent in Europe and Africa, but in contrast, common in the Far East, in the countries of the West coast of the Pacific Ocean and also found in Australia.

(7) Phage types present almost exclusively in America especially Latin America: 26, 35, 38.

7. *Vi-type specificity in* S. typhi

Vi-type specificity in *S. typhi*, that is the pattern of reactions with the typing phages (adaptations of Vi-phage II) is in many instances partly dependent on lysogenicity of the bacterial strains concerned (Craigie, 1946; Felix and Anderson, 1951a; Anderson, 1951; Anderson and Felix, 1953b). Phages decisive for phage sensitivity are called type-determining phages.

Such phages were isolated from a number of strains and lysogenisation of type A with them produced specific types that seemed similar in all respects to types found in nature. For example, the phage already shown by Craigie (1942, 1946) to be carried by type D_1, converted type A into type D_1; that isolated from type D_6 changed type A into D_6; that from 25 converted A into 25; and so on. Type-determining phages were given small letter symbols when those of the corresponding Vi-types were capitals, and a number with a superscript prime sign when the type carrying them was designated numerically. For example, the type-determining phage carried by type D_1 strains is designated phage d1, and that carried by type 25 strains as phage 25′. The type-determining phages differ from Vi-phage II in not being specific for the Vi-form of S. typhi; some of them will even lyse other Salmonella serotypes. They also differ from the Vi-phage in serological, physiological and physical properties.

Anderson and Felix (1953b) found that the specificity of a lysogenic Vi-type was determined by two factors:

(a) The identity of the non-lysogenic precursor type and,
(b) that of the type-determining phage with which it was lysogenised.

In this context, the terms lysogenic and non-lysogenic are used only in relation to the presence or absence of type-determining phages. Like most Salmonella species, S. typhi carries many lysogenic phages, but relatively few have type-determining properties.

Anderson and Fraser (1955) composed structural formulae consisting of the symbol of the non-lysogenic precursor type followed by that of the type-determining phage in parenthesis; thus type 29, which is type A carrying phage f2, is structurally A (f2); type F_2 is type F_1 carrying the same phage, and is structurally F_1 (f2); type D_6 is type A carrying phage d6 and is therefore A (d6), and so on. The structural formulae are useful in the study of this typing system, because they show relationships between types that may not be indicated by the symbols in routine use and also give information on the probable genetic nature of the corresponding typing adaptations of Vi-phage II. The structural formulae of the typing strains have been summarised by Rische and Ziesché (1973a).

Vi-types which fall into groups in the typing scheme are so classified because they react as though they possessed a common precursor. Thus, types C_8, E_8 and 26, although they carry phage 26′, fall into three different groups in the typing scheme because they have the different precursors C_1, E_1 and A respectively.

8. *Genetic categories of the typing phages*

It has been shown by Anderson and Fraser (1955, 1956) that the Vi-typing preparations fall into four genetic categories:

(1) The wild type of the Vi-phage II (phage A).

(2) Preparations that have undergone restriction and modification. The adapted phage changes to phage A when grown on type A. It has undergone restriction and modification and it possesses the wild genotype of Vi-phage II (e.g. phages E_1, C_1, F_1, etc.).

(3) Preparations that are pure host-range mutants, e.g. the phages F_2, C_3 and E_1 which indicate Vi-types having the formulae F_1 (f2), C_1 (f2) and E_7 (f2) respectively. When these phages are grown on type A, they all change to phage 29, indicating the Vi-type with the formula A (f2). This phage represents the host-range mutant which is capable of overcoming the block in the multiplication of the wild genotype of phage A produced by the presence of the type-determining phage f2.

(4) Phages that represent a combination of restriction, modification and mutation, e.g. phage E^9, indicating the Vi-type with the formula E_1 (d6). Firstly there is the mutation which enables the Vi-phage to overcome the barrier to its multiplication erected by the determining phage d6. Secondly the mutant must undergo modification to enable it to multiply in the E_1 component of the E_1 (d6) complex. The selected mutant has the genotype of phage D_6 and the phenotype of phage E_1. All information about structural formulae of the type strains, and phenotype and genotype of the typing phages has been summarised by Rische and Ziesché (1973a).

9. *Adaptation of Vi-phage II to new types of* S. typhi

The most satisfactory phage to adapt to new types is phage A, as it represents the wild phenotype and wild genotype of Vi-phage II. But if difficulty is experienced in adapting phage A to a new type, attempts should be made to isolate and identify the type-determining phage from the new type. If, for instance, the type-determining phage is identified as phage d6, it will be evident that the satisfactory method of preparing a typing phage for this new type, is to start with phage D_6, thus by-passing the step which involves the uncertainty of selecting this host-range mutant from phage A (Anderson, 1956).

B. Phage typing of S. schottmuelleri (S. paratyphi B)

1. System of Felix and Callow

Felix and Callow (1943, 1951) evolved a series of adaptations of a phage which was specific for *S. schottmuelleri* (Type 1 phage, Aberdeen). They were able to define several epidemiologically valid phage types.

The typing phages of Felix and Callow: 3a and 3aI which were found to be identical (Scholtens, 1959), 2, 3b, Jersey, Taunton and Dundee were obtained by adaptation of the type 1 phage termed Aberdeen (Felix and Callow, 1943). However, the phages 3b, Taunton and Dundee were found to be serologically different from the parent type 1 phage. It was concluded that they were unlikely to have been adaptations, but had been derived from lysogenic phages carried by the strains on which the adaptations had been propagated (Nicolle and Hamon, 1951; Hamon and Nicolle, 1951).

Phage Beccles was an adaptation of another type 1 phage termed Exeter. The adapted phage, however, was not neutralised by an antiserum prepared against its parent type 1 phage Exeter. Phage B.A.O.R. was a lysogenic phage carried by the type 1 strain termed Epping.

Thus the typing scheme of Felix and Callow consists of nine phages: phage 1 (Aberdeen), three phage preparations adapted from phage 1 (2, 3a and Jersey), and 5 lysogenic phages (3b, Taunton, Dundee, Beccles, B.A.O.R.) (See Table II).

TABLE II

Typing system for the phage typing of S. schottmuelleri (S. paratyphi B) according to Felix and Callow

Typing strains	Typing phages								
	1	2	3a	3b	Jersey	Beccles	Taunton	BAOR	Dundee
1	CL	CL	+ +	—	CL	—	—	—	SCL
2	—	CL	—	—	—	—	—	—	SCL
3a	—	—	CL	CL	—	CL	CL	CL	CL
3aI	—	—	CL	—	—	—	—	—	CL
3b	—	—	—	CL	—	CL	CL	CL	CL
Jersey	—	—	—	—	CL	CL	CL	CL	CL
Beccles	—	—	—	—	—	CL	CL	—	CL
Taunton	—	—	—	—	—	—	CL	—	CL
BAOR	—	—	—	—	—	—	—	CL	—
Dundee	—	—	—	—	—	—	—	—	CL

CL = confluent lysis.
SCL = semiconfluent lysis.
+ + = 20 to 50 plaques.
— = no lysis.

The phage preparations and type strains are distributed by the International Reference Laboratory, Colindale, London. The technique has been internationally standardised in a manner similar to that used for Vi-phage typing of *S. typhi* (Anderson and Williams, 1956). The frequency distribution of phage types during 1962–1965 was given by Rische (1968). Phage type Taunton was found to be the predominant type in Europe. Strains with this phage type were not found during these periods in Japan, Ceylon or Australia. The frequency distribution of *S. schottmuelleri* phage types in a number of countries during 1966–1969 has been reported by the I.C.E.P.T. (Report, 1973).

2. *System of Scholtens*

Scholtens (1950b, 1952, 1955, 1956, 1959) extended the Felix and Callow scheme with a number of phages. He divided *S. schottmuelleri* strains into groups which he further subdivided into types.

The groups are indicated by: Phage 1 of Felix and Callow, those phages of the Felix and Callow system which were found to be adaptations of type 1 phage (phages 2, 3a and Jersey), phage Beccles–Meppel (BM) which was adapted by Scholtens from phage 1 of Felix and Callow, and a group of six lysogenic phages (phages e, d, c, f, k and b), which can only be isolated by the mixed culture technique (Scholtens, 1952). Phage d appeared to be identical to phage Dundee of Felix and Callow. With these eleven phages, Scholtens divided *S. schottmuelleri* strains into eleven groups (groups A, M, S, J, BM, B, P, Q, BT, 1 and 2).

Each group could be further subdivided into types by means of a set of lysogenic phages (I, IVa, IVb and VI) isolated from pure broth cultures of lysogenic strains (Table III). These phages were isolated from the supernatants of lysogenic strains, in contrast to the lysogenic phages used for grouping which could only be isolated by mixed culture techniques. Phage I was found to be identical with the typing phage 3b of Felix and Callow. Phages IVa and IVb, were found to be identical to the typing phages Beccles and Taunton of the Felix and Callow system. Phages I, IVa, IVb and VI differ from those found in mixed cultures with respect to their host-range, plaque morphology and thermolability.

It is generally agreed that the phage types according to Scholtens are of practical value (Rische and Schneider, 1960; Polanetzki and Brandis, 1971; Rische and Ziesché, 1973b). For the final designation of the *S. schottmuelleri* phage types, Polanetzki and Brandis (1971) proposed to use the nomenclature of Felix and Callow followed by the designation according to Scholtens placed in brackets, e.g. Taunton (B7), Beccles (A3), and so on.

TABLE III

Typing system for the phage typing of *S. schottmuelleri* (*S. paratyphi B*) according to Scholtens

Phage type according to Felix and Callow	Phage type and group according to Scholtens	Adapt. 1	2	3a	J.B.M.	Jersey 1	2	3a	Dundee e	d	c	f	h	b	I	IVb	IVa	VI	Beccles 3b	Taunton	BAOR	Type-determining phages found in the types
		Adaptation of phage 1 of Felix–Callow				*indication acc. to system of Felix–Callow*			*phages from mixed cultures*						*type-determining phages*							
A																						
3b	A1	–	–	–	–	–	–	–	–	CL	CL	CL	CL	CL	CL	CL	CL	CL	–	–	CL	–
3b var.2	A2	–	–	–	–	–	–	–	–	CL	CL	CL	CL	CL	CL	–	CL	CL	–	–	–	–
Beccles	A3 (22)	–	–	–	–	–	–	–	–	CL	CL	CL	CL	CL	–	CL	CL	CL	–	–	–	I
Dundee	A4	–	–	–	–	–	–	–	–	CL	CL	CL	CL	CL	–	CL	CL	CL	–	–	–	I
3b var.2	A5	–	–	–	–	–	–	–	–	CL	CL	CL	CL	CL	+ +	–	–	–	++	–	CL	I VI
Dundee	A6	–	–	–	–	–	–	–	–	CL	CL	CL	CL	CL	–	CL	–	–	–	–	–	I VI
Taunton	A7 (Hague)	–	–	–	–	–	–	–	–	CL	CL	CL	CL	CL	–	–	–	CL	–	CL	–	I IVb
M																						
3b var.1	M1	–	–	–	–	–	–	–	–	CL	CL	CL	CL	CL	CL	CL	CL	CL	–	–	CL	–
Beccles	M3 (Midwoud)	–	–	–	–	–	–	–	–	CL	CL	CL	CL	CL	–	CL	CL	CL	–	–	(CL)	I
BAOR	M5	–	–	–	–	–	–	–	–	CL	CL	CL	CL	CL	+ +	–	–	–	++	–	CL	I VI
untypable	M6	–	–	–	–	–	–	–	–	CL	CL	CL	CL	CL	–	–	–	–	–	–	–	I VI
untypable	M7	–	–	–	–	–	–	–	–	CL	CL	CL	CL	CL	–	CL	–	–	–	–	–	I IVb

Type	Group	Specimen																		
3a	S	S1 ("54")	—	—	CL	—	CL	CL	CL	CL	CL	—	CL	CL	CL	—	—	CL	—	—	—
Beccles or 3aI		S3 (Sittard)	—	CL	—	—	CL	CL	CL	CL	CL	—	—	CL	CL	CL	+	CL	I	—	—
Dundee or 3aI		S4	—	—	—	CL	CL	—	CL	CL	CL	—	CL	—	—	—	++	CL	I	—	—
3aI or Dundee		S5 ("18")	(CL)	—	CL	CL	CL	CL	CL	CL	CL	—	—	—	—	—	+++	CL	I	VI	—
		S6 ("87")	(CL)	—	CL	CL	CL	CL	CL	CL	CL	—	—	—	—	—	—	CL	I	VI	—
3aI or Dundee		S7	—	—	—	CL	CL	CL	—	CL	CL	—	—	CL	—	CL	·	—	I	—	IVb
3a or Jersey	J	J1 ("60")	—	CL	CL	CL	CL	CL	CL	CL	CL	—	CL	CL	CL	—	CL	CL	—	—	—
Jersey		J3	—	CL	CL	CL	CL	CL	CL	—	CL	—	CL	CL	CL	—	CL	—	I	—	—
3a var.3		J5	—	CL	CL	CL	CL	CL	—	CL	CL	—	—	—	—	CL	CL	++	—	VI	—
new type		J6	—	CL	CL	CL	CL	CL	—	—	CL	—	—	—	—	—	·	—	I	VI	—
new type		J7	—	CL	CL	CL	CL	CL	—	—	CL	—	—	—	—	CL	—	—	I	—	IVb
Beccles	B.M.	B.M.3 (Meppel)	—	CL	—	CL	—	CL	—	—	CL	—	—	CL	CL	CL	·	—	I	—	—
3a	B	B1	—	CL	CL	CL	CL	CL	CL	CL	CL	—	CL	CL	CL	CL	CL	CL	—	—	—
3a var.2		B2	—	CL	CL	CL	CL	CL	CL	—	CL	—	—	CL	CL	—	—	CL	—	—	—
3aI var.1.2		B3	—	CL	CL	CL	CL	CL	CL	CL	CL	—	CL	CL	CL	CL	CL	CL	—	—	—
3aI		B4 (Schiedam)	—	CL	CL	CL	CL	CL	CL	—	CL	—	CL	CL	CL	—	—	CL	—	—	—
3a var.4		B5	—	CL	CL	CL	CL	CL	CL	—	CL	—	++	—	CL	—	CL	CL	I	VI	—
3aI		B6 (Leeuwarden)	—	(CL)	CL	CL	CL	CL	CL	—	CL	—	—	—	—	—	—	CL	I	VI	—
Taunton		B7 (Kampen)	—	—	CL	CL	CL	CL	CL	CL	CL	—	—	—	—	—	—	—	I	—	IVb
3b	P	P1	—	—	—	CL	CL	CL	CL	CL	CL	—	CL	CL	CL	CL	CL	CL	—	—	—
Beccles		P3	—	—	—	CL	CL	CL	—	—	CL	—	—	CL	CL	CL	—	CL	I	—	—
Dundee		P4	—	—	—	CL	CL	CL	—	—	CL	—	—	CL	CL	CL	—	CL	I	VI	—
Dundee		P6	—	—	—	CL	CL	CL	—	—	CL	—	—	—	—	—	·	—	I	VI	—

TABLE III (continued)

Phage type according to Felix and Callow	Phage type and group according to Scholtens		Group reactions										type reactions					Type-determining phages found in the types
			Adaptation of phage 1 of Felix–Callow				phages from mixed cultures						type-determining phages				BAOR	
			1	2	3a	J. B.M.	e	d	c	f	h	b	I	IVb	IVa	VI		
			indication according to the system of Felix–Callow										Bec-cles	Taun-ton				
			1	2	3a	Jersey				Dundee			3b					
3a	Q1	Q	–	–	CL	–	CL	CL	CL	CL	CL	CL	CL	CL	CL	CL	·	–
	Q2		–	–	CL	–	CL	CL	CL	CL	CL	CL	CL	–	CL	CL	–	–
3aI var.1.2	Q3		–	–	CL	–	CL	CL	CL	CL	CL	CL	–	CL	CL	CL	·	I
	Q4		–	–	CL	–	CL	CL	CL	CL	CL	CL	–	–	–	CL	–	I
3aI	Q6		–	–	CL	–	CL	CL	CL	CL	CL	CL	–	–	–	–	–	I, VI
Taunton	Q7		–	–	–	–	CL	CL	CL	CL	CL	CL	–	–	CL	–	–	I, IVb
2	2	2	–	CL	–	–	CL	CL	–	CL	CL	–	–	–	–	CL	–	–
3b	BT1	BT	–	–	–	–	CL	CL	–	CL	CL	CL	CL	CL	CL	CL	CL	–
Beccles	BT3		–	–	–	–	–	CL	CL	CL	CL	CL	–	CL	CL	CL	·	I
Dundee	BT4		–	–	–	–	CL	CL	CL	CL	CL	CL	–	–	–	CL	·	–
Dundee	BT6		–	–	–	–	–	CL	CL	CL	CL	CL	–	–	–	–	·	I, VI
1	1	1	CL	CL	CL	CL	CL	CL	CL	CL	CL	CL	V	V	V	CL	V	·

CL = confluent lysis; (CL) = unstable reaction; V = variable reaction; + to + + + = increasing number of plaques; – = no lysis; · = without explanation.

C. Phage typing of *S. paratyphi A*

Felix and Banker divided *S. paratyphi A* into two types, using a wild type phage which had been isolated from sewage in Bombay by Dhayagude and one adaptation (type 1 and 2) (See Felix, 1955). Banker reported in 1955 that two more adaptations had been obtained (type 3 and 4) (Banker, 1955). Four further types have been added to the scheme (Anderson, 1964). The phage types were found to be stable after passage through guinea pigs (Rische, 1958). Buczowski (1960) found that different phage types may occur simultaneously in one patient. This may make the practical value of phage typing questionable. A report on the geographical distribution of *S. paratyphi A* phage types was published by the International Committee for Enteric Phage Typing (Report, 1973).

D. Phage typing of *S. typhimurium*

Soon after the introduction of the phage typing schemes for *S. typhi* and *S. schottmuelleri* (*S. paratyphi B*), attempts were made to develop a typing scheme for *S. typhimurium*. In contrast to *S. typhi* and *S. schottmuelleri*, there is not one internationally employed phage typing system for *S. typhimurium*, but various systems have been developed and have come into use in several countries.

1. Colindale systems

The first attempts were made by Felix and Callow (Felix and Callow, 1943; Felix, 1951) during the Second World War at the Enteric Reference Laboratory in Colindale, London. The scheme evolved was published in some detail in 1956 after it had been used for over ten years in Colindale (Felix, 1956). It consists of 12 typing phages. Phages 1, 1a and 1b were derived from phage 3b of *S. schottmuelleri* (Felix and Callow, 1951). Another phage (phage 2) was isolated from faeces containing a *S. typhimurium* strain and yielded two adapted phages, 2a and 2c, one of which (2a) yielded 3 other adapted phages (3, 3a and 3b = 1a var.1). Two other phages (2b and 2d) were found in a *S. enteritidis* strain and the 12th phage (phage 4) was a lysogenic phage from *S. typhimurium*.

Later, the two phages 5 and 2e and associated types 5 and 2e were added (Anderson, 1964). Phage 2e is apparently also known as phage 35 (Kühn *et al.*, 1973). In this form, but with minor modifications and extensions, the scheme is being used in a number of countries (Table IV) (Kühn *et al.*, 1973; Beumer and van Oye, 1965; Sechter and Gerichter, 1967; Rische, 1968). In addition, a number of variations of the original 14 types were recognised by Anderson (1964) and enabled the extension of the scheme to 33 patterns.

TABLE IV

Schematic presentation of the typing scheme of Felix–Callow for *S. typhimurium*

Type strain	| Typing phages													
	1	1a	1a var.1 (= 3b)	1b	2	2a	2b	2c	2d	3	3a	4	5	35 (= 2e)
1	cl	cl	cl	cl	cl	cl	cl	cl	cl	cl	3	–	cl	cl
1a	–	cl	cl	cl	cl	cl	cl	cl	cl	cl	cl	–	cl	cl
1a var.1 (= 3b)	–	cl	–	cl	–	–	cl	–	cl	–	cl	–	3	cl
1b	–	–	cl	–	scl	scl	cl	scl	cl	–	scl	–	–	cl
2	–	–	–	–	cl	cl	cl	–	cl	–	–	–	–	–
2a	–	–	–	–	–	cl	cl	cl	cl	–	–	–	–	cl
2b	–	–	–	–	–	–	cl	cl	cl	–	–	–	–	scl
2c	–	–	–	–	cl	cl	cl	cl	cl	–	–	–	–	cl
2d	–	–	–	–	–	–	–	cl	cl	cl	–	–	–	–
3	–	–	–	3	–	–	–	–	–	cl	scl	–	–	–
3a	–	–	3	–	–	–	cl	–	scl	–	cl	–	cl	cl
4	–	–	–	–	–	–	–	–	–	–	–	cl	–	–
5	–	–	–	–	–	–	–	–	–	–	–	–	cl	cl
35 (= 2e)	–	–	–	–	–	–	–	–	–	–	–	–	–	cl

cl = confluent lysis, sometimes opaque lysis; scl = semi-confluent lysis; 3 = nearly scl; – = no plaques.

The original scheme of 12 phages, termed scheme 1, was further developed into a phage typing system by Callow. The new scheme, termed scheme 2, comprised in addition to the 12 original scheme 1 phages, another 17 typing phages enabling the recognition of 34 types (Callow, 1959).

Seventeen of the 29 scheme 2 phages were adaptations of the *S. schottmuelleri* phage 3b. However, some of them were found to be serologically different from phage 3b which was due to replacement of the starting phage by lysogenic phages carried by the bacterial strain on which the phage was propagated. The other typing phages were derivates of lysogenic phages of lysogenic *S. typhimurium* strains.

The scheme was further extended by Anderson and Wilson (1961) to 31 phages enabling the recognition of more than 90 types. This number increased later to 124 (Wuthe *et al.*, 1975). Details of these types have not been published.

A number of workers, using the original 12 scheme 1 phages extended with the phages 5 and 35 (Felix–Callow system), felt that too many of the strains originating from their country were untypable with this system and that the system had to be extended with "local phages". Brandis *et al.* (1973) obtained 31 typing phages from Anderson and phage typed *S. typhimurium* strains isolated in 1970 and 1971 in Northern Germany. More than half of all strains tested were found to be untypable. Harris and Khakhria (1969) found that 33% of strains isolated in Canada were untypable.

The percentage of strains isolated in Belgium found to be untypable with the Felix–Callow system was 26% in 1963 and 33% in 1964 (Beumer and van Oye, 1965). Denis (1966) isolated therefore three wild-type phages from Belgian waters distinguishing 8 patterns and lysing 90% of the strains untypable with the Felix–Callow phages. Sechter and Gerichter (1967) in Israel encountered in 1964–1966 34·0% of *S. typhimurium* strains untypable with the Felix–Callow system. A supplementary typing scheme consisting of three lysogenic phages from untypable *S. typhimurium* strains and three wild-type phages from sewage distinguishing 11 types reduced the percentage of untypable strains from 34·0% to 2·2%. Chirakadze and Chanishvili (1974) encountered 37·5% atypically reacting or untypable strains among 2708 *S. typhimurium* strains from USSR and Poland. A typing scheme of 14 phages selected from 950 bacteriophages (915 from lysogenic *S. typhimurium* strains and 35 from sewage) distinguishing 36 phage types, enabled the authors to type 98·8% of all strains.

2. Lilleengen system

This phage typing system, published in 1948 (Lilleengen, 1948), was developed independently of that of Felix–Callow. Initially, the system contained 44 phages isolated from sewage, faecal filtrates or lysogenic *S. typhimurium* strains. The system finally adopted by Lilleengen contained only 12 phages; 2 were isolated from sewage (phages 2 and 31), 3 from lysogenic *S. typhimurium* strains (phages 4, 8 and 32) and 7 were adaptations (phages 28B, 30, 33, 34, 36, 37 and 39) derived from phages which had been omitted from the system. These 12 typing phages permitted the identification of 24 types. Three more types (types 24, 25 and 26) were added to the scheme by Kallings and Laurell (1957). The system has been extensively used in the Scandinavian countries (Kallings *et al.*, 1959) as well as in the German Federal Republic, albeit in a slightly modified form (Kühn *et al.*, 1973).

3. *Method of identification according to Boyd*

Boyd (1950, 1954) discovered that the great majority of *S. typhimurium* strains are lysogenic. A group of heat-resistant lysogenic phages was termed type A phages. The identification of these phages was based on antigenic structure, host-range and plaque characters, until Boyd and Bidwell (1957) discovered an apparently phage-free strain of *S. typhimurium* (strain Q1) which was sensitive to all type A phages. This strain was lysogenised with the lysogenic phages and the cross-immunity of the lysogenised strains was tested. Thus, the immunity-pattern of a collection of type A phages could be established and the original types further subdivided. In this way, a subdivision of *S. typhimurium* strains was obtained which was based on the identification of the phage, carried by the lysogenic strain.

The conviction of Boyd and Bidwell (1957) that this method should provide better results in the differentiation of bacteria was confirmed by Kallings *et al.* (1959), Milch *et al.* (1963) and Rische (1968). According to Anderson and Felix (1953b), this method has only limited possibilities, because it involves a lot of work and because it is impossible to use it to type non-lysogenic strains.

4. *Bilthoven system*

Scholtens (1962, 1969) developed a system using 3 wild-type phages; 14-17, 726 and E1311. He obtained 48 adapted phages from phage 14-17. With 11 adaptations of wild-type phage E1311 and 2 from phage 726, he was able to subdivide a group of otherwise undistinguishable organisms. At a later stage, 5 additional wild-type phages were added to enable the subdivision of a group of related strains (Scholtens and Rost, 1972). The system was extensively used in the Netherlands. After Scholtens' retirement, the system and its nomenclature were modified (Guinée *et al.*, 1974).

In its modified form, the system consists of 54 routinely used typing phages. In this form the system permitted the identification of 54 phage types recognised by means of phage-indicated type reactions and 36 phage types recognised by specific pattern reactions. However, the system is continuously extended when new patterns, clearly recognisable and reproducible, occur with sufficient frequency to accept them as new types.

This system permitted the identification of about 95% of strains isolated in the Netherlands in 1971–1973 (van Leeuwen *et al.*, 1974). The system was also employed to type strains isolated in 13 other countries. The frequency distribution of types as found in the German Federal Republic, Belgium and Switzerland was somewhat similar to that in the Netherlands. The strains originating from other countries often showed phage patterns not observed in the Netherlands. If such patterns were accepted as new types, it was generally possible to identify about 90% of all strains (van Leeuwen and Guinée, 1975).

5. Atlanta system

Wilson *et al.* (1971) developed a phage typing system for strains isolated in the U.S.A. The system consisted of 12 typing phages, all of them adaptations from Scholtens' wild-type phage 14-17 which was also one of the primary phages in the Bilthoven system. Nevertheless, Wilson's type strains behaved atypically when typed with Scholtens' system and vice versa (Guinée and van Leeuwen, unpublished observations).

Some authors use a combination of two systems. Kühn *et al.* (1973) use the Felix–Callow system in combination with the Lilleengen system. Some Felix–Callow types can be further subdivided by means of the Lilleengen system and vice versa. Similar results were obtained by Lalko and Buczowski (1970).

III. PHAGE TYPING SYSTEMS FOR OTHER *SALMONELLA* SEROTYPES

A. S. adelaide

Atkinson *et al.* (1952) found 12 out of 48 *S. adelaide* strains to be lysogenic when cross-tested on each other. With respect to their host-range, plaque morphology and serological properties, the lysogenic phages obtained from the 12 strains could be divided into 4 types: 1a, 1b, 2a and 2b (Atkinson and Geytenbeek, 1953). Atkinson and Klauss (1954) reported on two more lysogenic phages, encountered in an additional collection of *S. adelaide* strains (types 2c and 3). The six lysogenic phages were used as typing phages. Six phage patterns could be distinguished. A correlation

7

was observed between the phage pattern and the type of lysogenic phages carried by a particular strain (Atkinson, 1955). Of 423 strains of *S. adelaide* tested, 392 were found to be typable with this phage typing scheme (Atkinson and Klauss, 1955a).

B. *S. anatum*

A phage typing system for this serotype was described by Gershman (1974b). He isolated 6 wild-type phages from sewage and used them at RTD to subdivide a collection of *S. anatum* strains from avian, animal and human sources. With the 6 phages, he was able to distinguish 12 different phage patterns among 328 cultures of *S. anatum*. The number of untypable strains is not mentioned.

Popovici *et al.* (1972) found 80·4% of 418 *S. anatum* strains to be lysogenic, when testing the strains by means of cross-testing. Twelve lysogenic phages were selected for further study and also used for phage typing of 2464 strains of *S. anatum*. Of these, 177 strains (7·2%) were untypable and 60 strains (2·4%) reacted atypically (Popovici *et al.*, 1975).

C. *S. bareilly*

Majumdar and Singh (1973) described a phage typing system for *S. bareilly*. Three wild-type phages isolated from sewage and three lysogenic phages isolated from strains of *S. bareilly* were used at RTD. Twenty-one strains tested could be subdivided into 4 distinct phage types. Only the presence or absence of lysis, not the degree of lysis was taken into account.

D. *S. blockley*

Sechter and Gerichter (1969) presented a phage typing system for *S. blockley*. Two-hundred and forty strains were tested for lysogenicity by means of cross-testing using the same strains as indicator strains. Three distinct lysogenic phages were found and termed bl.A, bl.B and bl.C. The phages bl.B and bl.C were found to be serologically related but distinct in host-range. These 3 lysogenic phages were used as typing phages. The use of three different phages (bl.AV bl.B and bl.C) would theoretically lead to 8 combinations of reactions, i.e. 8 types (A through H).

For further characterisation of the strains, they were tested for lysogenicity, using three different strains of *S. blockley* as indicator. For this purpose the strain to be tested was grown together with one of the indicator strains, which was non-lysogenic itself. The 20 h mixed culture was chloroform-killed and tested on the three indicator strains. The use of three distinct indicator strains would be expected, according to Sechter and Gerichter (1969), to lead to the distinction of 8 types (1 through 8). The combination of the two techniques would theoretically enable the distinction of 64 types. Fourteen of these types were observed among 1256 strains of *S. blockley*. They were found to be stable and to have epidemiological significance in the investigation of food poisoning outbreaks. The most frequently encountered types were A_1, E_5, E_2 and H_5. None of the strains was untypable.

E. *S. bovis-morbificans*

Atkinson *et al.* (1952) found all of 60 *S. bovis-morbificans* strains to be lysogenic, often carrying more than one type of lysogenic phage. On the basis of lysis and lysogenicity patterns, four phage types could be distinguished (1 through 4). Two new provisional types (5 and 6) were found in another collection of strains (Atkinson and Klauss, 1955b). The system finally adopted, using 5 lysogenic phages as typing phages and 5 indicator strains of *S. bovis-morbificans* for lysogenicity patterns, was able to distinguish 5 distinctive patterns (Atkinson, 1956). Of 153 strains, 120 could be typed (Atkinson and Bullas, 1956a).

F. *S. braenderup*

Sechter and Gerichter (1968) presented a phage typing system for *S. braenderup* using the same techniques as those described for *S. blockley* (Sechter and Gerichter, 1969). Three lysogenic phages (brA, brB and brC) isolated from *S. braenderup* strains were used as typing phages. The strains were further tested for lysogenicity using 3 *S. braenderup* strains (brl, br6 and br7) as indicator. Fifteen of the 64 theoretically possible types were encountered among 424 strains tested. None of the strains was untypable (see also: *S. blockley*).

G. *S. dublin*

Lilleengen (1950) was able to subdivide *S. dublin* strains into 6 types and subtypes by means of 5 wild-type phages isolated from sewage. All strains examined were typable. Strains related epizootiologically always showed the same phage type.

Smith (1951c) evolved a phage typing system for *S. dublin*, independent of that described by Lilleengen. Six lysogenic phages (A through F) isolated from *S. dublin* strains were used at RTD to subdivide 306 strains of *S. dublin*. Of these, 12 strains (4%) were not lysed by any of the phages. The remaining 294 strains were typable and could be subdivided into 11 types. Unfortunately, from the point of view of analysing epidemics, 205 (66·9%) of the strains belonged to one type: phage type 1. Attempts to further subdivide phage type 1 strains failed. The phage types of strains kept under laboratory conditions were found to be stable. Some of the lysogenic phages, also used as typing phages, had phage type-determining properties.

H. S. enteritidis

Lilleengen (1950) was the first to describe a phage typing method for *S. enteritidis*. By means of 4 wild-type phages isolated from sewage, he distinguished 8 types and subtypes among 116 strains of *S. enteritidis*. All strains were typable.

Anderson (1964) reported that a phage typing system had been developed at Colindale. Details have not been published.

Since Lilleengen's preparations as well as those used by Anderson were not available, Macierewicz *et al.* (1968) developed an alternative phage typing scheme for *S. enteritidis*, consisting of 7 typing phages. They were isolated from lysogenic cultures of *S. schottmuelleri* (*S. paratyphi B*), *S. typhimurium* and *S. enteritidis* and propagated on one or more strains of *S. enteritidis*. When used at RTD, the seven phages distinguished 11 phage patterns among 613 strains of *S. enteritidis*. Only 4 strains (0·1%) reacted atypically. Stability of the phage pattern could be established for all types except types 6 and 8.

Sechter (personal communication, 1974) reported that a phage typing system had been developed in his laboratory. Details have not yet been published.

I. S. gallinarum–pullorum

Lilleengen (1952) obtained a number of wild-type phages from sewage and faecal filtrates of various animals. With 4 of these phages, he was able to subdivide strains belonging to *S. gallinarum* and *S. pullorum* into 6 types. When, in addition, six selected *S. dublin* and *S. enteritidis* phages were used (Lilleengen, 1950), the same strains could be subdivided into 7 types. Seventeen of 149 *S. pullorum* strains belonged to the same phage type as the *S. gallinarum* strains examined. When using the *S. dublin* and *S. enteritidis* phages, the 17 *S. pullorum* strains were recognised as a group distinct from *S. gallinarum*. Two types of *S. gallinarum* and four types of *S. pullorum* could be identified with the *S. gallinarum* and *S. pullorum* phages. All strains were typable.

Anderson (1964) reported that a phage typing system for *S. pullorum* had been evolved at Colindale. Details have not been published.

J. S. newport

Gershman (1974a) described a typing system for this organism. Seven wild-type phages were isolated from sewage with strains of *S. newport*. Using these phages, Gershman was able to establish 16 distinct phage types among 228 *S. newport* strains which had been isolated mainly from animal sources. The number of untypable strains was not mentioned.

Petrow *et al.* (1974) evolved a phage typing system independent of that described by Gershman. The frequency of lysogeny in *S. newport* was established using 4 strains of *S. newport* as indicator and using the mixed culture technique or induction with ultra-violet light or Mitomycin (Otsuji *et al.*, 1959) to show lysogeny. Of 588 strains, 560 were found to be lysogenic. Four lysogenic phages were finally selected as typing phages for the phage typing scheme. In addition, 5 phages isolated from sewage were incorporated in the phage typing system. The morphology of the 9 typing phages was extensively studied. Of 854 strains tested, 97·9% were typable and could be divided into 35 types. Type 9 was the most common, representing 69·1% of all isolates from human and other sources. There was agreement between the expected phage types and the epidemiological data in 69 of 81 outbreaks representing 315 cultures.

K. *S. panama*

Guinée and Scholtens (1967) described a phage typing system for *S. panama*. The parental phage (phage A) was isolated from sewage with *S. panama* strain 47. Seven typing phages were obtained by adaptation of phage A and were found to be serologically identical with the parental phage A. These 8 phages (termed A through H) when used in RTD, permitted the identification of 8 stable phage types termed A through H. Untypable strains were termed Z.

All adapted phages were derived directly from the parental phage A except phage H which was derived from the adapted phage G. Phages B, D and F were found to be restricted and modified variants of the phages A, C and E respectively. The restriction and modification of the three phages B, D and F was found to be due to a resistance factor (Guinée *et al.*, 1967). Phage H was a restricted and modified variant of phage G.

Phages C and E were host-range mutants of the parental phage A. Phage G was also a host-range mutant which was not restricted by the resistance factor in contrast to the host-range mutants C and E. Most type strains were lysogenic using a strain of *S. gallinarum* as indicator. However, none of the lysogenic phages lysed any of the *S. panama* strains. Type-determining properties of the lysogenic phages could not be established.

Approximately 95% of all *S. panama* strains isolated in the Netherlands are typable with this phage typing system. The majority of strains isolated in other European countries also showed characteristic phage patterns (Guinée, 1969).

The system was employed on a large scale in Rumania by Nestorescu *et al.* (1969). Of 1588 strains tested, 81·5% were found to be typable.

Type A predominated in this material in contrast to the situation in the Netherlands where type A became relatively rare after 1965.

Further studies in Rumania by Szegli *et al.* (1975) on 7530 strains revealed that 90·3% of the strains were typable; 9·7% were untypable (group Z); however 55·5% of the strains belonged to type A. Szegli *et al.* (1975) isolated 11 lysogenic phages from group Z strains which enabled them to type 84·3% of the group Z strains, as well as to subdivide all type A strains into 8 subtypes.

L. *S. thompson*

Smith (1951a) observed that 42 of 50 *S. thompson* strains were lysogenic, when tested on each other. A preliminary examination of these lysogenic phages showed many of the phages to differ from one another in terms of host-range. Six of the phages were selected for further work and propagated on *S. thompson* strain 19 which itself was very unlikely to be lysogenic (Smith, 1951b). The six phages, used at RTD, and one of them also used at 100 × RTD, were employed to subdivide the 50 *S. thompson* strains. Forty-one were typable and 9 strains (18%) were not lysed by any of the phages. The 41 typable strains could be divided into 11 types. One type comprised 28% of the strains, the number of the other types was not large. The phage types of strains maintained under laboratory conditions appeared to be relatively stable. Strains from animal origin could not be differentiated from strains of human origin. There was no opportunity to determine whether phage typing would be of any value in studying the epidemiology of *S. thompson* infections. Some of the lysogenic phages also used as typing phages were found to have phage type-determining properties.

Anderson (1964) reported that a phage typing system for *S. thompson* has been evolved independently. Details of this scheme have not been published.

Gershman (1972) developed an alternative system for *S. thompson*. Eight wild-type phages were isolated independently from sewage and used at RTD. Thirteen distinct phage types could be distinguished among an unknown number of cultures tested. The number of untypable strains is not mentioned.

M. *S. virchow*

Velaudapillai (1959) described a phage typing system for *S. virchow*, one of the commonest serotypes of *Salmonella* in Ceylon at that time. He found that nearly all strains of *S. virchow* were lysogenic (50 out of 52), when using the cross plating method of Fisk (1942). With a selection of 5 phages obtained from lysogenic *S. virchow* cultures, 7 phage types could be distinguished among 55 strains tested. Three strains (6%) were untypable. Strains from Denmark and Sweden fell, with one exception in each case, into characteristic and distinct phage types. Most strains of English origin belonged to another type.

N. *S. weltevreden*

Phage typing systems for this organism were evolved by Garg and Singh (1971, 1973) and by Kaliannan and Mallick (1973).

Garg and Singh (1971) isolated wild-type phages from sewage as well as lysogenic phages by means of UV induction from strains of *S. weltevreden*, using strain nr. 30 of *S. weltevreden* as indicator. Twelve wild-type phages selected from 110 phages isolated and 6 lysogenic phages selected from 46 phages isolated, were finally compared and the lysogenic phages were found to be slightly superior to the wild-type phages with regard to their specificity. The six lysogenic phages selected for the typing scheme were serologically indistinguishable. They were all DNA phages and were completely inactivated by heating at 80°C. The phages were different with regard to their plaque morphology and their host-range. Amongst 64 theoretically possible types, 17 were identified among 149 *S. weltevreden* strains. Some types were frequently encountered, such as type 1 (48%), type 11 (11%) and type 64 (28%). No relation of phage types with respect to host or geographic distribution was observed. No untypable strains were encountered. Garg and Singh (1973) attempted also to subdivide *S. weltevreden* strains by testing their lysogenicity. They tested 245 strains for lysogenic phages using 6 strains of *S. weltevreden* including strain nr. 30 as indicator. Of the 245 strains tested, 207 (84·5%) were found to be lysogenic. Using 6 different indicator strains, 64 possible phage patterns might be expected. Of these, 15 were detected among 245 strains tested. Garg and Singh considered the latter method to have distinct advantages over the phage typing system earlier described (Garg and Singh, 1971), because of the permanent heritable character of lysogeny and because the maintenance of indicator strains is easier than that of the typing phages. They realised however, that the latter method is more time-consuming than phage typing.

Kaliannan and Mallick (1973) evolved a phage typing system for *S. weltevreden* independently. Wild-type phages were obtained from sewage. Lysogenic phages were isolated from strains of *S. weltevreden* with the cross-culture technique of Fisk (1942). The phages were characterised in terms of host-range, plaque characteristics, heat resistance, sodium citrate and urea sensitivity, and antigenic properties. Six lysogenic phages and 2 wild-type phages were selected for the phage typing scheme. Twenty-five phage types were distinguished among 180 strains. Only 3·9% of the strains were untypable. The stability of the phage types was established by retyping all the cultures after subculturing them twice. A high degree of correlation between the phage type and the epidemic source was observed.

O. *S. abortus-ovis, S. cholerae-suis, S. good, S. minnesota, S, montevideo, S. oranienburg, S. potsdam and S. waycross*

Keyhani (1969) found that all cultures of *S. abortus-ovis* isolated in Iran were lysed by one particular phage whereas biochemically and serologically identical cultures isolated in England and Canada were not lysed by the same phage.

Lalko (1974) described a provisional differentiation of *S. cholerae-suis* strains into different phage types.

Kawanishi (1972) isolated 10 lysogenic phages from *S. good*. He could distinguish 24 phage types among 159 strains of *S. good* when these 10 lysogenic phages were used as typing phages. Thirty-one (19·4%) of the strains were untypable. The same phages also lysed strains of *S. minnesota* having the same O antigen as *S. good*. *S. minnesota* could thus be divided into 5 types. Thirty-eight per cent of the strains were untypable.

Thal (1957) could distinguish four phage types among *S. montevideo* strains isolated from South American meat-meal.

Bordini (1970) reported on the phage typing of *S. oranienburg*. Lysogenic phages were isolated from 28·5% of 561 cultures tested. More than 99% of the cultures were lysed by one or more of these phages. Fourteen selected lysogenic phages distinguished a number of phage types, not mentioned in Bordini's paper.

Atkinson and Bullas (1956b) provisionally defined six groups of *S. potsdam* strains in terms of lysis by lysogenic phages and lysogenicity determined on 6 indicator strains of *S. potsdam*. Two strains out of 15 failed to produce any phage. The other 13 strains were all lysogenic.

Atkinson *et al.* (1952) reported the occurrence of two lysogenic strains of *S. waycross* among 31 tested. The two lysogenic phages were found to be identical. The 31 strains of *S. waycross* could be divided into two groups. Group 1 carries the lysogenic phage and is immune to its lytic action, group 2 does not carry the phage and is sensitive to its lytic action (Atkinson and Carter, 1953).

IV. EFFECT OF PLASMIDS ON THE PHAGE PATTERN OF *SALMONELLA*

Phage typing is primarily used as a method to establish the epidemiological relation of bacterial strains which cause outbreaks of disease. It has repeatedly been found in recent years that the phage pattern is influenced by extrachromosomal elements (plasmids) such as resistance factors (R-factors) or transfer factors.

Restriction of bacteriophages by R-factors was first demonstrated by Yoshikawa and Akiba (1962), when fi⁻ R-factors* introduced into *Escherichia coli* K12 were shown to inhibit the multiplication of externally infecting λ and T phages. Anderson and Lewis (1965) and Anderson (1966) found that some fi⁻ R-factors restricted *S. typhimurium* typing phages, reducing the sensitivity of the host strains in a characteristic fashion. The introduction of the transfer factor Δ of *S. typhimurium* phage type 29 reduced the sensitivity of *S. typhimurium* phage type 36 which is sensitive to all typing phages, to type 6 which is sensitive to only six of the typing phages. When only the Ampicillin resistance determinant of the same strain was introduced into type 36, only two phages were restricted. Type 36 was then altered into type 125 (Anderson, 1966; Anderson *et al.*, 1968). The introduction of the transfer factor (Δ) into *S. schottmuelleri* phage type 1 var. 2, converted the phage pattern of the recipient strain into type Beccles var. 2. When introduced into *S. typhi* phage type A, this transfer factor inhibited lysis by all the Vi-II typing phages (Anderson, 1966).

Anderson *et al.* (1973) tested a total of 2716 R-factors and transfer factors isolated from *E. coli* and salmonellae of human and animal origin for their phage-restricting effects in *S. typhimurium* phage type 36. All of 1402 fi⁺ factors tested were non-restrictive. In contrast, eleven distinct changes in the phage type 36 of the recipient were produced by fi⁻, I-like R- and transfer factors.

Scholtens (1967) also observed phage restricting effects of R-factors in *S. typhimurium* strains. When the R-factor of *S. typhimurium* phage type 10 (=phage type 221 in the modified nomenclature (Guinée *et al.*, 1974)) was transferred to strain *S. typhimurium* A (= 1), the phage pattern of this strain was altered into phage type 10 (= 221).

Phage types 500, 501, 502, 503, 504 and 505 (Guinée *et al.*, 1974) have very close relationships with regard to the presence of plasmids and lysogenic phages. Originally, these phage types were considered as identical and termed type 661 (Scholtens and Rost, 1972). They noted that not all strains of group 661 were entirely identical. By transferring their R-factors to strains of *S. typhi* and *S. typhimurium*, they observed changes of the phage pattern of the recipient strain. Referring to these observations they established new subgroups within type 661, which could be more easily distinguished by means of an additional set of 5 wild-type phages, isolated with these strains. There were indications that other restriction deter-

---

* fi⁻ = fertility inhibition negative.

minants were present in addition to the restriction determinant 10 (Scholtens and Rost, 1972). By means of transfer experiments as well as molecular studies, it was established that strains of phage type 505 harbour two lysogenic phages, a transfer factor (= restriction determinant 10 already indicated by Scholtens and Rost, 1972) and 3 other, non self-transmissable plasmids, each coding for a specific pattern of phage restriction. Phage types 501, 502 and 504 were found to harbour another combination of the same extrachromosomal elements. When all lysogenic phages and plasmids from each of these donor strains were introduced into strain *S. typhimurium* 1, its phage type 1 was converted into the phage type of the donor strain (van Embden *et al.*, 1976).

Guinée *et al.* (1967) and Guinée and Willems (1967) observed phage restriction by R-factors in *S. panama*. The majority of R-factors, encountered in natural isolates of *S. panama* with various phage types, were found to be identical with respect to restriction properties. This type of R-factor was only sporadically found in the other members of the *Enterobacteriaceae* (*Salmonella*, *E. coli*, *Citrobacter*, *Enterobacter*, *Klebsiella* and *Proteus*). They concluded from their observations that transfer of R-factors in nature e.g. from *E. coli* to *S. panama* was not as common as originally thought.

Scholtens and Rost (1972) observed that the R-factor of *S. panama* strains converted the phage type 1 of *S. typhimurium* strain 1 in the same way as did some R-factors from *S. typhimurium* strains.

Similar observations were described by Toucas and Vieu (1973) for *S. typhi*.

It may appear from the foregoing that a particular phage type can be determined by the presence of one or more plasmids with restricting properties. Introduction or segregation of such elements may alter the phage pattern of the host. It will be clear that the tool of phage typing may contribute to the classification of R-factors and other plasmids as well as to the understanding of their epidemiological behaviour.

V. THE STABILITY OF PHAGE TYPING AND THE REPRODUCIBILITY OF PHAGE PATTERNS

It should be clear that any phage typing method has practical value only, if its results are reproducible. This means that established phage types should be stable.

The majority of authors describing phage typing systems claim that their system meets this standard. The reproducibility of their phage patterns has been shown either by retyping cultures and obtaining the same results (e.g. Smith, 1951a; 1951c; Macierewica *et al.*, 1968; Kaliannan and

Mallick, 1973) or by showing that epidemiologically related strains belong to the same phage types (Anderson and Williams, 1956; Rische, 1958; Sechter and Gerichter, 1969; Petrow *et al.*, 1974).

The reproducibility strongly depends on rigorous standardisation of the methods (Anderson and Williams, 1956). Felix and Anderson (1951b) observed a variation in phage types of *S. typhi* strains isolated from an outbreak in Oswestry (phage type A and degraded strains). Cultures, isolated at the same time from different patients, or even from blood and faeces of the same patient, did not react identically with the standard Vi-typing phages. These variations were found to be caused by the failure of several strains to develop their Vi-antigen on certain batches of agar.

The same variation as reported by Felix and Anderson (1951b) was also observed by the present authors in chronic carriers of *S. typhi* in The Netherlands, although they believed that they had standardised their technical procedures. Similar observations were made by Buczowsky (1960) in carriers of *S. paratyphi A*.

It will be obvious that variation in phage reaction is more likely to occur in strains which are lysed by many of the typing phages than in strains which are lysed only by the homologous adapted phage. But even then reproducible results may be obtained as shown in the studies of van Leeuwen and Guinée (1975). They studied the frequency distribution of *S. typhimurium* phage types using strains from 13 countries. Many strains possessed a phage pattern which in one or more major reactions was aberrant from any of the accepted types. Such atypically reacting strains were retyped several times for confirmation. The reproducibility of the patterns was found to be very high.

In a number of cases, lack of reproducibility can be explained in terms of segregation or acquisition of plasmids or lysogenic phages.

Tschäpe *et al.* (1974) found in an outbreak of *S. typhimurium* in a duck breeding farm, that two or three epidemic strains of *S. typhimurium* were involved. The analysis of plasmids indicated that the variation of phage types was due to plasmids with phage-restricting properties.

Kallings and Laurell (1959) in several faecal specimens noted the presence of *S. typhimurium* strains, showing different phage patterns. The typing of single colony lines and retyping of cultures stored for half a year revealed a specific change by steps in the sensitivity to the typing phages. These steps corresponded to the different phage patterns observed in the outbreak. They found that the variations of these phage types were due to lysogenic phages.

Three different phage types (Bilthoven system 250, 251 and an atypical pattern) were observed among *S. typhimurium* strains isolated from the faeces of a cat. Analysis of the R-factor and lysogenic phages indicated that strains with phage type 250 carried neither lysogenic phages nor an R-factor. Strains with phage type 251 carried only an R-factor, while strains with atypically reacting patterns carried the R-factor as well as the lysogenic phage (van Leeuwen, unpublished observations).

The examples presented indicate that strains with different phage types may be epidemiologically related.

REFERENCES

Adams, M. H. (1959). "Bacteriophages". Interscience Publishers Inc., New York.
Anderson, E. S. (1951). *J. Hyg. (Camb.)*, **49**, 458–470.
Anderson, E. S. (1956). *J. gen. Microbiol.*, **14**, 676–683.
Anderson, E. S. (1964). In "The world problem of salmonellosis" (Ed. E. van Oye), pp. 89–110. Junk, The Hague.
Anderson, E. S. (1966). *Nature (Lond.)*, **212**, 795–799.
Anderson, E. S., and Felix, A. (1952). *Nature, Lond.*, **170**, 492–494.
Anderson, E. S., and Felix, A. (1953a). *J. gen. Microbiol.*, **8**, 408–420.
Anderson, E. S., and Felix, A. (1953b). *J. gen. Microbiol.*, **9**, 65–88.
Anderson, E. S., and Fraser, A. (1955). *J. gen. Microbiol.*, **13**, 519–532.
Anderson, E. S., and Fraser, A. (1956). *J. gen. Microbiol.*, **15**, 225–239.
Anderson, E. S., and Lewis, M. J. (1965). *Nature (Lond.)*, **208**, 843–849.
Anderson, E. S., and Williams, R. E. O. (1956). *J. clin. Path.*, **9**, 94–114.
Anderson, E. S., and Wilson, E. M. J. (1961). *Zbl. Bakt. Hyg. I. Abt. Orig.*, **224**, 368–373.
Anderson, E. S., Kelemen, M. V., Jones, C. M., and Pitton, J. S. (1968). *Genet. Res. (Camb.)*, **11**, 119–124.
Anderson, E. S., Threlfall, E. J., Carr, J. M., and Savoy, L. G. (1973). *J. Hyg. (Camb.)*, **71**, 619–631.
Arber, W., and Dussoix, D. (1962). *J. mol. Biol.*, **5**, 18–36.
Atkinson, N. (1955). *Aust. J. exp. Biol. med. Sci.*, **33**, 371–374.
Atkinson, N. (1956). *Aust. J. exp. Biol. med. Sci.*, **34**, 231–234.
Atkinson, N., and Bullas, L. R. (1956a). *Aust. J. exp. Biol. med. Sci.*, **34**, 349–360.
Atkinson, N., and Bullas, L. R. (1956b). *Aust. J. exp. Biol. med. Sci.*, **34**, 461–470.
Atkinson, N., and Carter, M. C. (1953). *Aust. J. exp. Biol. med. Sci.*, **31**, 591–594.
Atkinson, N., and Geytenbeek, H. (1953). *Aust. J. exp. Biol. med. Sci.*, **31**, 441–451.
Atkinson, N., and Klauss, C. (1954). *Aust. J. exp. Biol. med. Sci.*, **32**, 221–228.
Atkinson, N., and Klauss, C. (1955a). *Aust. J. exp. Biol. med. Sci.*, **33**, 375–380.
Atkinson, N., and Klauss, C. (1955b). *Aust. J. exp. Biol. med. Sci.*, **33**, 421–428.
Atkinson, N., Geytenbeek, H., Swann, M. C., and Wollaston, J. M. (1952). *Aust. J. exp. Biol. med. Sci.*, **30**, 333–340.
Banker, D. D. (1955). *Nature, Lond.*, **175**, 309–310.
Bertani, G., and Weigle, J. J. (1953). *J. Bacteriol.*, **65**, 113–121.
Beumer, J., and Oye, E. van (1965). In "Tagungsbericht 5. Coll. Fragen der Lysotypie Wernigerode", pp. 81–83. Institut für Experimentelle Epidemiologie, Wernigerode, SDR.
Bordini, A. (1970). *C.R. Acad. Sci. (Paris)*, **270D**, 567–570.
Boyd, J. S. K. (1950). *J. Path. Bact.*, **62**, 501–517.
Boyd, J. S. K. (1954). *J. Path. Bact.*, **68**, 311–314.
Boyd, J. S. K., and Bidwell, D. E. (1957). *J. gen. Microbiol.*, **16**, 217–228.

Brandis, H. (1955). Zbl. Bakt. I. Abt. Orig., 162, 223–224.
Brandis, H., Posch, J., and Andries, L. (1973). Off. Gesundh. Wesen, 35, 99–108.
Buczowski, Z. (1960). Bull. Inst. mar. trop. Med. Gdansk, 11, 101–108.
Callow, B. R. (1959). J. Hyg. (Camb.), 57, 346–559.
Chirakadze, I. G., and Chanishvili, T. G. (1974). Zh. Mikrobiol. (Mosk.), 51, 72–78.
Craigie, J. (1936). J. Bacteriol., 31, 56.
Craigie, J. (1942). Canad. publ. Hlth. J., 33, 41–42.
Craigie, J. (1946). Bact. Rev., 10, 73–88.
Craigie, J., and Brandon, K. F. (1936). Can. publ. Hlth. J., 27, 165–170.
Craigie, J., and Felix, A. (1947). Lancet, i, 823–827.
Craigie, J., and Yen, C. E. (1938a). Can. publ. Hlth. J., 29, 448–483.
Craigie, J., and Yen, C. E. (1938b). Can. publ. Hlth. J., 29, 484–496.
Denis, R. (1966). Arch. Belg. Med. Soc. Hyg., 24, 451–460.
Desranleau, J. M., and Martin, J. I. (1950). Can. J. publ. Hlth., 41, 128–132.
Dussoix, D., and Arber, W. (1962). J. molec. Biol., 5, 37–49.
Embden, J. D. A. van, Leeuwen, W. J. van, and Guinée, P. A. M. (1976). J. Bact. 127, 1414–1426.
Farmer, J. J., III, Hickman, F. W., and Sikes, J. V. (1975). Lancet, ii, 787–790.
Felix, A. (1943). Br. med. J., 1, 435–438.
Felix, A. (1951). Br. med. Bull., 7, 153–162.
Felix, A. (1955). Bull. Wld. Hlth. Org., 13, 109–170.
Felix, A. (1956). J. gen. Microbiol., 14, 208–222.
Felix, A., and Anderson, E. S. (1951a). Nature, Lond., 167, 603.
Felix, A., and Anderson, E. S. (1951b). J. Hyg., 49, 349–364.
Felix, A., and Callow, B. R. (1943). Br. med. J., 2, 127–130.
Felix, A., and Callow, B. R. (1951). Lancet, ii, 10–14.
Felix, A., and Pitt, R. M. (1934). Lancet, ii, 186–191.
Fisk, R. T. (1942). J. infect. Dis., 71, 153–160.
Garg, D. N., and Singh, I. P. (1972). Ann. Inst. Pasteur, 121, 751–762.
Garg, D. N., and Singh, I. P. (1973). Antonie van Leeuwenhoek, 39, 41–50.
Gershman, M. (1972). Appl. Microbiol., 23, 831–832.
Gershman, M. (1974a). Can. J. Microbiol., 20, 769–771.
Gershman, M. (1974b). Avian Diseases, 18, 565–568.
Guinée, P. A. M. (1969). Zbl. Bakt. I. Abt. Orig., 209, 331–336.
Guinée, P. A. M., and Scholtens, R. Th. (1967). Antonie van Leeuwenhoek, 33, 25–29.
Guinée, P. A. M., and Willems, H. M. C. C. (1967). Antonie van Leeuwenhoek, 33, 397–412.
Guinée, P. A. M., Scholtens, R. Th., and Willems, H. M. C. C. (1967). Antonie van Leeuwenhoek, 33, 30–40.
Guinée, P. A. M., Jansen, W. H., Schuylenburg, A. van, and Leeuwen, W. J. van (1973). Appl. Microbiol., 26, 474–477.
Guinée, P. A. M., Leeuwen, W. J. van, and Pruys, D. (1974). Zbl. Bakt. Hyg., I. Abt. Orig., A,226, 194–200.
Hamon, Y., and Nicolle, P. (1951). Ann. Inst. Pasteur, 80, 496–507.
Harris, W., and Khakhria, R. (1969). Can. J. publ. Hlth., 60, 35.
Hayes, W. (1970). "The genetics of bacteria and their viruses". Blackwell Scientific Publications, Oxford and Edinburgh.
Kaliannan, K., and Mallick, B. B. (1973). Indian J. Anim. Sci., 43, 24–26.

Kallings, L. O., and Laurell, A. B. (1957). *Acta path. microbiol. scand.*, **40**, 328–342.
Kallings, L. O., Laurell, A. B., and Zetterberg, B. (1959). *Acta path. microbiol. scand.*, **45**, 347–356.
Kawanishi, T. (1972). *Bull. natn. Inst. Hyg. Sci. Tokyo*, **47**, 75–80.
Keyhani, M. (1969). *Br. vet. J.*, **125**, 568–572.
Kühn, H., Falta, R., and Rische, H. (1973). *In* "Infektions-krankheiten und ihre Erreger. Eine Sammlung von Monographien. Band 14. Lysotypie" (Ed. H. Rische), pp. 101–140. VEB Gustav Fischer, Jena.
Lalko, J. (1974). *In* "Salmonella and Salmonellosis in Bulgaria. Second National Conference, Sofia 1971", pp. 175–178.
Lalko, J., and Buczowski, Z. (1970). *Bull. Inst. mar. trop. Med. Gdansk*, **21**, 47–65.
Leeuwen, W. J. van, and Guinée, P. A. M. (1975). *Zbl. Bakt. Hyg., I. Abt. Orig.*, **A,230**, 320–335.
Leeuwen, W. J. van, Pruys, D., and Guinée, P. A. M. (1974). *Zbl. Bakt. Hyg., I. Abt. Orig.*, **A,226**, 201–206.
Lidwell, O. M. (1959). *Monthly Bull. Minist. Hlth.*, **18**, 49–52.
Lilleengen, K. (1948). *Acta path. microbiol. scand.*, Suppl., 77.
Lilleengen, K. (1950). *Acta path. microbiol. scand.*, **27**, 625–640.
Lilleengen, K. (1952). *Acta path. microbiol. scand.*, **30**, 194–202.
Luria, S. E. (1953). *Cold Spring Harb. Symp. quant. Biol.*, **18**, 237–244.
Luria, S. E., and Human, M. L. (1952). *J. Bacteriol.*, **64**, 557–569.
Macierewicz, M., Kaluzewski, S., and Lalko, J. (1968). *Exp. med. Microbiol.*, **20**, 138–146.
Majumdar, A. K., and Singh, S. P. (1973). *Indian vet. J.*, **50**, 1161–1162.
Milch, H., Lásló, V. G., and Biró, G. (1963). *Acta microbiol. Acad. Sci. hung.*, **10**, 41–52.
Nestorescu, N., Popovici, M., Szégli, L., Negut, M., and Chiriac, Fl. (1969). *Arch. Roum. Path. Exp. Microbiol.*, **28**, 997.
Nicolle, P., and Diverneau, G. (1961). *Zbl. Bakt. I. Orig.*, **181**, 385–387.
Nicolle, P., and Hamon, Y. (1951). *Ann. Inst. Pasteur*, **81**, 614–630.
Nicolle, P., Pavlatou, M., and Diverneau, G. (1954). *Ann. Inst. Pasteur*, **87**, 493–509.
Nicolle, P., Oye, E. van, Crocker, C. G., and Brault, I. (1955). *Bull. Soc. Path. exot.*, **48**, 492–510.
Otsuji, N., Sekiguchi, M., Iijima, T., and Takagi, Y. (1959). *Nature, Lond.*, **184**, 1079–1080.
Parker, M. T. (1972). *In* "Methods in Microbiology", vol. 7b, pp. 1–27. Academic Press, London and New York.
Petrow, S., Kasatiya, S. S., Pelletier, J., Ackermann, H.-W., and Peloquin, J. (1974). *Ann. Microbiol. Pasteur*, **125**, 433–445.
Polanetzki, U., and Brandis, H. (1971). *Zbl. Bakt., I. Abt. Orig.*, **216**, 314–322.
Popovici, M., Szegli, L., Soare, L., Negut, M., and Florescu, E. (1972). *Arch. roum. Path. exp. Microbiol.*, **31**, 664–665.
Popovici, M., Szégli, L., Soare, L., Negut, M., Florescu, E., Calin, C., and Manolache, D. (1975). *Arch. Roum. Path. Exp. Microbiol.*, **34**, 223–230.
Report of the International Committee for Enteric Phage Typing (ICEPT) (1973). *J. Hyg. (Camb.)*, **71**, 59–84.
Rische, H. (1958). *Zbl. Bakt. I. Orig.*, **171**, 568–572.
Rische, H. (1968). *In* "*Enterobacteriaceae*-Infektionen" (Eds I. Sedlak and H. Rische), pp. 204–211. VEB Georg Thieme, Leipzig.

Rische, H., and Schneider, H. (1960). *J. Hyg. (Praha)*, **IV-32**, 32–53.
Rische, H., and Ziesché, K. (1973a). *In* "Infektionskrankheiten und ihre Erreger. Eine Sammlung von Monographien. Band 14. Lysotypie" (Ed. H. Rische), pp. 23–64. VEB Gustav Fischer Verlag, Jena.
Rische, H., and Ziesché, K. (1973b). *In* "Infektionskrankheiten und ihre Erreger. Eine Sammlung von Monographien. Band 14. Lysotypie" (Ed. H. Rische), pp. 65–86. VEB Gustav Fischer Verlag, Jena.
Scholtens, R. Th. (1936). *J. Hyg.*, **36**, 452–455.
Scholtens, R. Th. (1950a). *Antonie van Leeuwenhoek*, **16**, 245–255.
Scholtens, R. Th. (1950b). *Antonie van Leeuwenhoek*, **16**, 256–268.
Scholtens, R. Th. (1952). *Antonie van Leeuwenhoek*, **18**, 257–274.
Scholtens, R. Th. (1955). *J. Hyg. (Camb.)*, **53**, 1–11.
Scholtens, R. Th. (1956). *Antonie van Leeuwenhoek*, **22**, 65–88.
Scholtens, R. Th. (1959). *Antonie van Leeuwenhoek*, **25**, 403–421.
Scholtens, R. Th. (1962). *Antonie van Leeuwenhoek*, **28**, 373–381.
Scholtens, R. Th. (1967). *C.R. Soc. Biol.*, **161**, 1050–1054.
Scholtens, R. Th. (1969). *Arch. roum. exp. Path. Microbiol.*, **28**. 984–990.
Scholtens, R. Th., and Rost, J. A. (1972). *Ann. Sclavo*, **14**, 309–344.
Sechter, I., and Gerichter, Ch. B. (1967). *Ann. Inst. Pasteur*, **113**, 399–409.
Sechter, I., and Gerichter, Ch. B. (1968). *Appl. Microbiol.*, **16**, 1708–1712.
Sechter, I., and Gerichter, Ch. B. (1969). *Ann. Inst. Pasteur*, **116**, 190–199.
Sertic, V., and Boulgakov, N. A. (1936). *C.R. Soc. Biol.*, **122**, 35–37.
Smith, H. W. (1951a). *J. gen. Microbiol.*, **5**, 458–471.
Smith, H. W. (1951b). *J. gen. Microbiol.*, **5**, 472–479.
Smith, H. W. (1951c). *J. gen. Microbiol.*, **5**, 919–925.
Szégli, L., Popovici, M., Soare, L., Negut, M., Calin, C., and Florescu, E. (1975). *Arch. Roum. Path. Exp. Microbiol.*, **34**, 217–222.
Thal, E. (1957). *Nord. vet.-Med.*, **9**, 831–838.
Tschäpe, H., Dragas, A. Z., Rische, H., and Kühn, H. (1974). *Zbl. Bakt. Hyg., I. Abt. Orig.*, **A,226**, 184–193.
Toucas, M., and Vieu, J. F. (1973). *Ann. Microbiol. Pasteur*, **124A**, 477–487.
Velaudapillai, T. (1959). *Z. Hyg.*, **146**, 84–88.
Weigle, J. J., and Bertani, G. (1953). *Ann. inst. Pasteur*, **84**, 175–179.
Wilson, V. R., Hermann, G. J., and Balows, A. (1971). *Appl. Microbiol.*, **21**, 774–776.
Wuthe, H. H., Brandis, H., Andries, L., and Wuthe, S. (1975). *Zbl. Bakt. Hyg., I. Abt. Orig.*, **A,230**, 172–185.
Yoshikawa, M., and Akiba, T. (1962). *Jap. J. Microbiol.*, **6**, 121–132.

Note added in proof

After finishing the manuscript, following papers were published which provide information on already described or new phage typing systems.

Anderson, E. S., Ward, L. R., de Saxe, M. J. and de Sa, J. D. H. (1977). *J. hyg. Camb.*, **78**, 297–300.
Gershman, M. (1976). *Appl. environment. Microbiol.*, **32**, 190–191.
Gershman, M. (1976). *J. clin. Microbiol.*, **3**, 214–217.
Gershman, M. (1976). *J. Milk Food Technol.*, **39**, 682–683.
Gershman, M. (1977). *J. clin. Microbiol.*, **5**, 302–314.
Gershman, M. (1977). *J. Food Prot.*, **40**, 43–44.
Slavkov, I. (1975). *Veterinarnomed. naouki.*, **12**, 55–60.

CHAPTER V

Phage Typing of *Klebsiella*

STEFAN ŚLOPEK

*Institute of Immunology and Experimental Therapy,
Polish Academy of Sciences, Wrocław*

I. INTRODUCTION

Strains of *Klebsiella* are responsible for a wide variety of diseases of man, perhaps a greater variety than any other kind of bacteria. In addition to being the primary cause of infections of the respiratory tract (rhinoscleroma, ozaena, sinusitis, pneumonia, otitis), infections of the alimentary tract (enteritis, appendicitis, cholecystitis), infections of the urinary tract (pyelitis, cystitis), of the genital tract (metritis), and of the eye (conjunctivitis, dacryocystitis, ulcus corneae), they may also induce sepsis. They have a marked tendency to occur in mixed and secondary infections with other pathogenic bacteria (Ślopek, 1961, 1968). The increasing number of *Klebsiella* infections occurring recently, particularly those of the respiratory, urinary, and alimentary tracts, besides specific diseases like rhinoscleroma and ozaena justify attempts at typing *Klebsiella* by bacteriophages (Ślopek *et al.*, 1967; Ślopek, 1973).

II. TAXONOMY OF *KLEBSIELLA*

Characteristics of the genus *Klebsiella* according to the Subcommittee on Taxonomy of the *Enterobacteriaceae* (Subcommittee Rep., 1963) are shown in Table I.

Cowan *et al.* (1960) proposed a classification of *Klebsiella* into six subgroups (categories), which were given the names *K. aerogenes*, *K. pneumoniae*, *K. edwardsii* var. *edwardsii*, *K. edwardsii* var. *atlantae*, *K. ozaenae* and *K. rhinoscleromatis*. The classification was based on morphologic, and biochemical characteristics, and resistance to exogenous factors. The

TABLE I

Characteristics of the genus *Klebsiella*
(*Enterobacteriaceae* Subcommittee
Report, 1963)

Characteristic biochemical properties[a]	
Gas from glucose	+
Lactose	+
Sucrose	+
Mannitol	+
Dulcitol	d
Salicin	+
Adonitol	+
Inositol (gas)	+
Glycerol (gas)	+
Cellobiose (gas)	+
Insoluble starch (gas)	+
Indole	−
Methyl red	−
Voges Proskauer	+
Ammonium citrate	+
Hydrogen sulfide	−
Urease	(+)
Gelatin liquefaction	−
Growth in KCN medium	+
Phenylalanine deaminase	−
Sodium malonate	+
Lysine decarboxylase	+
Arginine dihydrolase	−
Ornithine decarboxylase	−

[a] + positive in one or two days
　− negative
　(+) delayed positive
　d different biochemical types.

TABLE II

Differentiation of *Klebsiella* (Ślopek and Durlakowa, 1967)

Test	*Klebsiella* species[a]					
	1	2	3	4	5	6
Fimbriae	+	−	+	−	−	−
Glucose (gas)	+	+	+	+	d	−
Lactose	+	+	+	+	+	−
Indole	−	−	−	−	−	−
Methyl red	−	−	+	+	+	+
Voges Proskauer	+	+	−	−	−	−
Citrate	+	+	+	+	+	−
Urease	+	+	+	+	d	−
Malonate	+	+	+	+	−	+
Lysine decarboxylase	+	+	+	+	−	−

[a] 1 – *K. aerogenes*
2 – *K. aerogenes* var. *edwardsii*
3 – *K. pneumoniae*
4 – *K. pneumoniae* var. *atlantae*
5 – *K. ozaenae*
6 – *K. rhinoscleromatis*
+ = above 75% of positive results
− = below 25% of positive results
d = 25–75% of positive or negative results.

number of strains studied by these authors was 176, the majority of which belonged to the subgroup (category) *K. aerogenes*, while the remaining subgroups, were each represented by several strains. Other writers (Darrell and Hurdle, 1964; Foster and Bragg, 1962) have confirmed the value of this classification.

Based on 851 strains of *Klebsiella*, Ślopek and Durlakowa (1967) showed the suitability of the Cowan *et al.* classification of *Klebsiella*. Categories 1 and 2 (*K. aerogenes* and *K. edwardsii* var. *edwardsii*) differ only with respect to fimbriation. Categories 3 and 4 (*K. pneumoniae* and *K. edwardsii* var. *atlantae*) also differ only with regard to fimbriation. Categories 5 (*K. ozaenae*) and 6 (*K. rhinoscleromatis*) differ distinctly from the foregoing categories, but the two share a number of properties. *K. rhinoscleromatis* differs most from the other strains of *Klebsiella*.

In view of the fact that category pairs 1 and 2, and 3 and 4, differ only in respect to fimbriation, it might be more appropriate to distinguish only four species with two subspecies, which is the terminology to be used in this Chapter:

1. *K. aerogenes,*
2. *K. aerogenes* var. *edwardsii* (*K. edwardsii* var. *edwardsii,* Cowan *et al.*),
3. *K. pneumoniae,*
4. *K. pneumoniae* var. *atlantae* (*K. edwardsii* var. *atlantae,* Cowan *et al.*),
5. *K. ozaenae,*
6. *K. rhinoscleromatis.*

Differential diagnosis of the various subgroups (categories) of *Klebsiella* is based on the criteria shown in Table II. Essentially, the criteria applied correspond to those used by Cowan *et al.* (1960). Some tests which proved to be of little differential value were omitted (gluconate decomposition, the 44°C test, sensitivity to KCN). The adhesive properties of *Klebsiella* are also a useful characteristic for their classification (Durlakowa *et al.*, 1967a).

The classification of Ślopek and Durlakowa (1967) was confirmed by Szulga (1971) using the numerical taxonomic method of "centrifugal correlation". This allows the selection of taxonomically valuable characteristics according to their ranking order. Figure 1 shows the relationships of the *Klebsiella* subgroups.

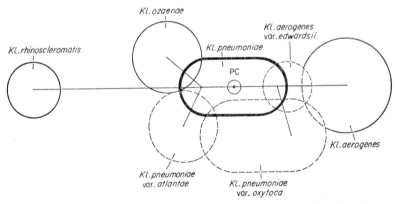

FIG. 1. Representation of numerical taxonomy of *Klebsiella.*.

III. MORPHOLOGICAL AND PHYSIOLOGICAL CHARACTERISTICS OF *KLEBSIELLA*

Klebsiella are non-motile rods with well developed capsules, some with fimbriae. The properties of *Klebsiella* are shown in Table III. With the criteria in Table III, 16 biochemical types (biotypes) can be established. Table IV shows the characteristic properties of the different biotypes and the frequency of their occurrence in the various subgroups of *Klebsiella* according to Durlakowa *et al.* (1967b).

TABLE III

Biochemical characteristics of the genus *Klebsiella*
(Durlakowa *et al.*, 1967b)

Test	\#1	2	3	4	5	6	Characteristic of the genus
			Klebsiella species[a]				
Gas from glucose	+	+	+	+	d	−	+
Lactose	+	+	+	+	+	−	+
Maltose	+	+	+	+	+	+	+
Sucrose	+	+	+	d	d	+	+
Mannitol	+	+	+	+	+	+	+
Dulcitol	d	d	d	d	−	−	d
Inositol	+	+	+	+	+	+	+
Glycerol	+	+	+	+	+	+	+
Sorbitol	+	+	+	+	+	+	+
Adonitol	+	+	+	+	+	+	+
Salicin	+	+	+	+	+	+	+
Arabinose	+	+	+	+	+	+	+
Xylose	+	+	+	+	+	+	+
Rhamnose	+	+	+	+	d	+	+
Raffinose	+	+	+	+	+	+	+
Hydrogen sulfide	−	−	−	−	−	−	−
Gelatin liquefaction	−	−	−	−	−	−	−
Indole	−	−	−	−	−	−	−
Methyl red	−	−	+	+	+	+	d
Voges-Proskauer	+	+	−	−	−	−	d
Citrate/Simmons	+	+	+	+	+	−	+
Citrate/Christensen	+	+	+	+	+	+	+
Urease	+	+	+	+	d	−	d
Growth in KCN medium	+	+	+	+	+	+	+
MacConkey 44°C	−	−	−	−	−	−	−
Phenylalanine deaminase	−	−	−	−	−	−	−
Lysine decarboxylate	+	+	+	+	−	−	+
Malonate	+	+	+	+	−	+	+
Gluconate	+	+	d	d	d	−	d
Number of investigated strains	310	63	94	141	171	72	851

[a] See Table II. The compilation of Table III is based on the criteria of Cowan *et al.* (1960)
+ means that more than 75% of the strains gave positive tests
− means that less than 25% of the strains gave positive tests
d means that the number of strains giving positive tests ranged between 25% and 75%.

TABLE IV

Biochemical types of *Klebsiella* (Durlakowa et al., 1967b)

Bio-chemical type	Glucose (gas)	Lactose	Indole	Methyl red	Voges-Proskauer	Dulcitol	Sucrose	*Klebsiella* species[a]						Total number of strain tested
								1	2	3	4	5	6	
1	+	+	−	−	+	+	+	119	35					154 (18·1%)
2	+	+	−	−	+	+	−	0	0					0
3	+	+	−	−	+	−	+	189	56					245 (28·8%)
4	+	+	−	−	+	−	−	2	3					5 (0·6%)
5	+	+	−	+	−	+	+			25	77			102 (11·9%)
6	+	+	−	+	−	+	−			1	5			6 (0·7%)
7	+	+	−	+	−	−	+			26	26			52 (6·1%)
8	+	+	−	+	−	−	−			11	33			44 (5·2%)
9	d	+	−	+	−	+	+					10		10 (1·2%)
10	d	+	−	+	−	+	−					1		1 (1·0%)
11	d	+	−	+	−	−	+					47		47 (5·5%)
12	d	+	−	+	−	−	−					113		113 (13·3%)
13	−	−	−	+	−	+	+						0	0
14	−	−	−	+	−	+	−						0	0
15	−	−	−	+	−	−	+						71	71 (8·3%)
16	−	−	−	+	−	−	−						1	1 (0·1%)

[a] See Table II.

IV. ANTIGENIC STRUCTURE OF *KLEBSIELLA*

One of the characteristic features of *Klebsiella* is the production of well developed capsules. These are antigenic and constitute the basis for serotyping of *Klebsiella* according to K-antigens. Currently, 80 K-serotypes of *Klebsiella* are known. These were discovered in the following order:

serotypes 1–3   Julianelle (1926)
         4–6   Goslings and Snijders (1936)
        7–14   Kauffmann (1949)
         15   Worfel and Ferguson (1951)
     16–25   Edwards and Fife (1952)
     26–49   Brooke (1951)
     50–57   Edwards and Fife (1952)
     58–63   Edwards and Fife (1955)
     64–69   Edmunds (1954)
     69–72   I. Ørskov (1955)
     73–76   Maresz-Babczyszyn (1962)
     77–80   Durlakowa *et al.* (1963).

The reader is referred to the original papers for details of the methods used.

The high percentage of untypable strains (10·2% according to Durlakowa *et al.* (1967c)) suggests the existence of further K-serotypes.

Nearly all known types are represented in *K. aerogenes*. The most frequent being: K1, K3, K4, K7, K9, K10, K13, K14, K15, K16, K17, K20, K30, K31, K33, K35, K37, K40, K43, K47, K55, K60, K61, K62, K64, K72, K74, K75, and K76.

A somewhat smaller number of K-serotypes was found in the subspecies *K. aerogenes* var. *edwardsii* where the most frequent types were: K1, K3, K4, K31, K35, K62, and K73.

In *K. pneumoniae* and *K. pneumoniae* var. *atlante*, few K-serotypes were found: K1, K3, K4, K30, K37, and K62.

In *K. ozaenae*, serotype K4 predominates. Less frequently, serotypes K3 and K5 occur.

In *K. rhinoscleromatis*, only serotype K3 was observed.

Several authors (Brooke, 1951; Durlakowa *et al.*, 1963; Edmunds, 1954; Kauffmann, 1949) have demonstrated that the capsular antigens of *Klebsiella* frequently exhibit cross-reactions indicating their serological relationships. The capsular-swelling type reactions observed are shown in Table V.

The frequent concurrence of different types of antigens points to the necessity of using absorbed sera and of checking every new lot of anti-K immune sera.

TABLE V

Capsular (quellung) reactions of *Klebsiella* antigens and antisera
(Durlakowa *et al.*, 1967c)

Antigens reacting with antisera		Antisera reacting with antigens	
Antigens	Antisera	Antisera	Antigens
1	5 6 7 30 72	1	6
2	30 69	2	13 14 55 69
3	4 31 64 68 75	3	68
4	6	4	
5	6 7	5	
6	1 5 7 72	6	4 5 7 58
7	6 10 58 61	7	5 6 58 61
8	20 25 30 59	8	9 25 71
9	8	9	
10	61	10	7 61
11	21 69	11	21
12	29 41 42 72	12	29 41
13	2 7 27 30 69 76	13	26 30
14	2 15 30 64 68	14	64
15	14 20	15	
16		16	
17	19	17	
18	44 61	18	44 61
19		19	17
20	13 22 27	20	8 15 23 27 45
21	11 69	21	11 26 33 69
22	23 37	22	23 37 41
23	8 22	23	22
24	43 59 76	24	
25	8 36	25	8
26	13 21	26	
27	20 46 65	27	71
28	64 71	28	71
29	12 65 70 72	29	12 70 72
30	13 33 53 55	30	2
31	60 64	31	60
32		32	
33	4 21 35	33	56
34	70 72	34	72
35		35	33
36	64 68	36	25
37	22 40 45	37	
38	71	38	71
39	65 71	39	65 71
40	65	40	

TABLE V (*Cont.*)

Antigens reacting with antisera		Antisera reacting with antigens	
Antigens	Antisera	Antisera	Antigens
41	12 22 23 26 43 61	41	12 61
42	65	42	12
43	24 76	43	41
44	18 72	44	18 56
45	20 27 71 72	45	37 71 72
46	71	46	27
47	3 64	47	
48	10 44 49 61	48	49
49	63 64	49	
50		50	
51	35 70	51	
52	53	52	53 70
53	30 52 55 70	53	52
54		54	
55	13 30 53	55	
56	33 44 64	56	
57	3	57	
58	6 7 58	58	7
59	24	59	
60	18 31	60	31
61	7 10 18 41 58	61	7 10 18 41
62		62	
63		63	49
64	14 36	64	14 31 36 47 49 56 59 68
65	39	65	27 28 29 39 40 42
66		66	
67		67	
68	3	68	3 14 36
69	2 21	69	2 11 13 21
70	29 52 53 72	70	29 34 51 72
71	8 27 28 38 39 45 72	71	28 38 39 45 46 72
72	29 34 45 55 70 71	72	1 6 12 29 34 44 45 70 71
73		73	
74	10	74	
75	3 8	75	3 8
76	24 43	76	24 43
77		77	
78		78	
79		79	41 44
80		80	70 71

V. RESISTANCE TO ANTIBIOTICS

Resistance of *Klebsiella* strains isolated from patient specimens during the years 1950–1965 to Streptomycin, Chlortetracycline, Oxytetracycline and Chloramphenicol indicated annual fluctuations in resistance. The fluctuations were considerable in some cases; for instance, Streptomycin between 29·1% and 67·2%, Chlortetracycline between 7·6% and 24·9%, Oxytetracycline between 7·2% and 55·0%, and Chloramphenicol between

TABLE VI

Types of antibiotic resistance of *Klebsiella* (Tkaczowa *et al.*, 1967)

Type of resistance	Antibiotics[a]						Number of strains
	Sm	Au	Tr	Cm	Ne	Po	
1	+	+	+	+	−	+	2
2	+	+	+	+	−	−	10 (3·6%)
3	+	+	+	−	+	+	2
4	+	+	+	−	+	−	2
5	+	+	+	−	−	+	2
6	+	+	+	−	−	−	1
7	+	−	+	+	−	+	2
8	+	−	+	+	−	−	2
9	+	−	+	−	+	−	1
10	+	−	−	+	−	+	7
11	+	−	−	+	−	−	13 (4·7%)
12	+	−	−	−	+	−	2
13	+	−	−	−	−	+	13 (4·7%)
14	+	−	−	−	−	−	117 (42·4%)
15	−	+	+	+	−	−	4
16	−	+	+	−	+	−	1
17	−	+	+	−	−	+	3
18	−	+	+	−	−	−	1
19	−	−	+	+	−	−	1
20	−	−	+	−	−	+	4
21	−	−	+	−	−	−	15 (5·4%)
22	−	−	−	+	−	+	1
23	−	−	−	+	−	−	34 (12·3%)
24	−	−	−	−	−	+	36 (13·0%)

[a] Sm = Streptomycin
Au = Chlortetracycline (Aureomycin)
Tr = Oxytetracycline (Terramycin)
Cm = Chloramphenicol
Ne = Neomycin
Po = Polymyxin.

17·0% and 60·0%. A steady increase in the proportion of resistant strains was also observed, especially for antibiotics widely used in medical practice. *Klebsiella* strains showed 24 patterns of resistance towards antibiotics (Tkaczowa *et al.*, 1967) (Table VI).

VI. BACTERIOCIN TYPING OF *KLEBSIELLA*

Attempts to differentiate *Klebsiella* on the basis of their sensitivity to the known colicins of the Fredericq and Abbott-Shannon types demonstrated either too few sensitive strains (Fredericq's set), or insufficient differentiation of sensitive strains (Abbott-Shannon set of colicins). The practical value of colicins for differentiation of *Klebsiella* is therefore small (Maresz-Babczyszyn *et al.*, 1967a, 1967b).

Results of preliminary studies by Durlakowa *et al.* (1964a, 1964b) demonstrated that bacteriocinogeny is fairly widespread in *Klebsiella*

TABLE VII

Origins of the pneumocins (Ślopek and Maresz-Babczyszyn, 1967)

Strain number	Species	Type of pneumocin[a]	Serotype
114	*K. pneumoniae* var. *atlantae*	A	K-3
902	*K. aerogenes*	A	K-16
L-4	*K. ozaenae*	A	K-4
F-5052	*K. ozaenae*	A	K-6
L-34	*K. ozaenae*	A	K-4
E-5051	*K. ozaenae*	A	K-5
D-33	*K. rhinoscleromatis*	A	K-3
D-35	*K. rhinoscleromatis*	A	K-3
D-36	*K. rhinoscleromatis*	A	K-3
D-40	*K. rhinoscleromatis*	A	K-3
1205	*K. aerogenes*	B	K-72
823	*K. aerogenes*	C	K-10
525	*K. pneumoniae*	D	K-62
212	*K. aerogenes*	E	K-1
2813	*K. aerogenes*	F	K-23

[a] In the first phase of the study, all 15 pneumocins listed in Table VII were used. Analysis of the material, however, showed that the use of only eight of the pneumocins is sufficient for typing 99% of the strains typable by means of 15 pneumocins. Since the use of all the colicins is not clearly superior to the use of only one-half, while differentiating 124 pneumocin types, it was decided to determine the sensitivity of the studied *Klebsiella* strains to 8 selected pneumocins, embracing nearly the whole spectrum of typable strains (see Table VIII).

(about 35%). *Klebsiella* mainly produce bacteriocins (also known as pneumocins) against isolates of homologous groups. This suggested the use of bacteriocins (pneumocins) to differentiate *Klebsiella* (Maresz-Babczyszyn *et al.*, 1968; Ślopek and Maresz-Babczyszyn, 1967). A list of the *Klebsiella* strains producing bacteriocins (pneumocins) and types of pneumocins produced is given in Table VII.

Pneumocins were obtained by the method described by Jacob *et al.* (1952). Cultures of the pneumocinogenic strain in plain broth were exposed to UV rays (W. Philips 57413 p/UV lamp) for one minute at a distance of 68 cm.

The pneumocins obtained were titrated by the method described by Fredericq (1957). The strains tested for sensitivity were typed using undiluted pneumocins whose titres were 10^{-2} to 10^{-5}.

Four-hour plain broth cultures of the studied strains were diluted 1 : 10 (density about 10^{-7}) and inoculated on the surface of agar medium in Petri plates. After drying the plates (30 min, 37°C), a drop of undiluted bacteriocin, sensitivity to which was being tested, was placed on the surface of the agar. The plates were incubated for 8 h at 37°C before reading the results. Sensitivity of the strains to the bacteriocin was manifested by the presence of a growth inhibition zone around the bacteriocin drop.

Typing with pneumocins was performed on 504 *Klebsiella* strains belonging to the most frequently encountered serotypes, namely: 49 strains belonging to serotype K1, 180 serotype K3, 215 serotype K4, 29 serotype K13, and 31 serotype K37 strains. The strains also represented all the species and subspecies of the genus *Klebsiella*. The numbers of strains of different species were as follows:

K. aerogenes	100 strains
K. aerogenes var. *edwardsii*	30 strains
K. pneumoniae	34 strains
K. pneumoniae var. *atlantae*	113 strains
K. ozaenae	155 strains
K. rhinoscleromatis	72 strains

Sensitivity of the most frequently encountered *Klebsiella* serotypes to the eight pneumocins is presented in Table VIII.

The data in Table VIII show that out of 504 strains of serotypes K1, K3, K4, K13 and K37, an average of 58·5% were sensitive to one or more of the pneumocins. On the basis of the different sensitivities of the strains, 65 pneumocin types were distinguished, represented by at least one strain, usually, however, by a larger number. The most numerously represented pneumocin types were: 1, 15, 34, 41, 43, 44, 46, 49, 56, and 65.

Compared with previous studies (Maresz-Babczyszyn *et al.*, 1967a,

TABLE VIII

Pneumocin types of *Klebsiella* serotypes K-1, K-3, K-4, K-13 and K-37
(Ślopek and Maresz-Babczyszyn, 1967)

Pneumocin types	Pneumocins								Number of strains
	114	902	L-4	K-6	D-33	D-40	823	525	
1	+	+	+	+	+	+	+	+	13
2	+	+	+	+	+	+	+		6
3	+	+	+	+	+		+	+	1
4	+	+	+	+		+	+	+	1
5	+	+	+	+			+	+	7
6	+	+	+	+			+		3
7	+	+	+		+	+	+	+	3
8	+	+	+		+	+	+		2
9	+	+	+		+	+			1
10	+	+	+			+			1
11	+	+	+				+	+	8
12	+	+	+					+	1
13	+	+	+				+		2
14	+	+		+			+	+	1
15	+	+					+	+	15
16	+	+						+	2
17	+		+	+					1
18	+		+						1
19	+								2
20		+	+	+	+	+	+		2
21		+	+	+	+	+			2
22		+	+	+	+		+		1
23		+	+	+	+				1
24		+	+			+	+		2
25		+	+				+		7
26		+	+				+	+	1
27		+	+						3
28		+			+	+	+		3
29		+			+	+		+	1
30		+			+	+			1
31		+			+		+	+	2
32		+			+			+	1
33		+					+		6
34		+					+	+	10
35		+						+	1
36		+							5
37		+			+	+	+		1
38		+			+	+		+	1
39		+			+	+			3
40		+			+				3
41		+					+		21

TABLE VIII (*Cont.*)

Pneumocin types	Pneumocins								Number of strains
	114	902	L-4	K-6	D-33	D-40	823	525	
42	+							+	7
43	+						+	+	11
44	+								17
45			+	+	+	+		+	1
46			+	+	+	+			10
47			+	+	+				1
48			+	+		+			4
49			+	+					10
50			+		+	+	+		1
51			+		+	+			5
52			+		+			+	1
53			+		+				1
54			+			+			2
55			+					+	2
56			+						16
57				+		+			2
58				+					3
59					+	+		+	1
60					+	+			16
61					+				7
62						+			4
63							+	+	1
64							+		3
65								+	20
Typable strains									295
Non typable									209
Total number of strains examined									504

1967b) on the differentiation of *Klebsiella* on the basis of sensitivity to known colicins of the Fredericq and Abbott-Shannon types, use of pneumocins produced by *Klebsiella* allowed differentiation of 53·5% of the studied strains with variation between 13·9% and 75·7%. Among the sensitive strains, 32 pneumocin types were distinguished.

A fairly high percentage of *K. ozaenae* strains (42·7%) could be typed. Since these are difficult to type by serological or biochemical methods or by using bacteriophages, the possibility of pneumocin typing is of practical interest.

The method may prove particularly useful for differentiating *K. aerogenes*

(64·2% sensitive strains) and *K. aerogenes* var. *edwardsii* (75·7% sensitive strains).

VII. PHAGE TYPING OF *KLEBSIELLA*

A. Isolation of *Klebsiella* Bacteriophages

Bacteriophages were isolated from sewage, stools and nasopharyngeal secretions of ozaena patients (Przondo-Hessek, 1966).

Mucosal secretions were collected by means of sterile cotton swabs, which were immersed in nutrient broth (beef-extract peptone broth (beef extract 3 g; peptone 5·0 g; distilled water 1000 ml; pH 7·0–7·2)) inoculated previously with a homologous bacterial strain (isolated from the patients) or some other strain against which bacteriophages were sought. After 12 or 24 h incubation at 37°C, one part of the cultures was Seitz-filtered, and the other was centrifuged. Presence of phages in the filtrate and supernatant was checked by the plate method.

Human stool samples were suspended in physiological saline solution, shaken, and centrifuged twice for 30 min at 4000 rev/min. The supernatant was examined by the plate method, using standard strains representing 72 types of encapsulated *Klebsiella* as detector strains.

Sewage material was treated in the same manner; after two centrifugations, several drops of the fluid were placed on the surface of an agar plate previously inoculated with a 4-h culture of the detector strain. Plaques were counted after 6 h incubation at 37°C.

Several plaques usually appeared on plates with sensitive strains. If the plaques were identical, one was transferred to 5 ml of broth inoculated with the sensitive strain. The cultures were then incubated at 37°C for 4–6 h, centrifuged, and diluted logarithmically. One drop of each dilution was placed on an inoculated agar plate and, after further incubation at 37°C, a single, separate plaque was transferred to broth. Each phage was cloned by this method five times. If different types of plaques occurred, each was isolated separately.

In some cases, complete lysis (CL) was observed at the site of application of sewage samples. An agar square was cut out from the area of lysis and diluted in broth. After shaking, the broth was diluted 1 : 5, 1 : 10, and 1 : 100, and one drop of each dilution was placed on an agar plate inoculated with the homologous strain. In this way, single plaques suitable for isolation were obtained.

The purified bacteriophages were passaged on homologous strains growing in broth medium containing 1% glucose. For this purpose, 0·1 ml of phage lysate and 0·1 ml of bacterial suspensions with a density about 10^8 cells per ml were added to 10 ml of broth. After 4 h incubation at 37°C, the lysate was stored in the refrigerator overnight before passage on the following day.

In order to determine the length of the incubation period needed to obtain maximal concentration of bacteriophages, 10 ml of broth medium was inoculated with 4-h cultures of the corresponding host propagating strains (with an inoculum of 10^8 cells per ml medium), and 0·1 ml of the phage preparation with a titre of 10^8 was added. Turbidity of cultures incubated for 12 h was checked nephelometrically at intervals of 2 h. Cultures to which no phage preparation was added served as controls. The lysates were centrifuged for 30 min at 4000 rev/min, filtered through membrane filters (Coli 5 sartorius membrane filter), titrated, and stored in the refrigerator.

The sewage samples yielded 42 phages. Of these, 33 phages were easy to propagate, exhibited different ranges of lytic activity and produced distinct plaques. From 50 samples of human stools, only 4 phages were isolated. Only one of these was propagated and used further. Isolation of phages from the nasopharyngeal secretions of ozaena patients yielded only one phage although 200 samples were examined. This phage was propagated and used in the further experiments. A total of 35 phage preparations was obtained for further studies.

B. Propagation of *Klebsiella* Bacteriophages

The optimal time of propagation of *Klebsiella* phages is 4–5 h. After 2–3 h, a majority of the cultures contained mixtures of homologous bacteria and phages producing light turbidity. After 4–5 h clearing of the cultures occurred, which either persisted to the end of the incubation period (12 h), or, more often, showed secondary bacterial growth after 5 h. Highest concentrations of phages were obtained in lysates harvested 4 h after inoculating the media; the titres did not increase during continued incubation. In cases where secondary growth occurred, titres often dropped slightly.

C. Determination of Lytic Activity of *Klebsiella* Bacteriophages

Preliminary titration of lytic activity was performed by the method of Craigie and Yen (1938a, 1938b). On Petri-plates containing 20 ml of 1·5% agar medium, 0·2 ml of 4-h broth cultures of the homologous strains were pipetted. After drying the plates at 37°C for 10 min, one loopful of the phage suspension was applied to segments equal to 1/12 of the plate. A logarithmic series, from 10^{-1} to 10^{-12}, of each phage preparation was studied, the capacity of the loop being about 0·01 ml. With this method, the Routine Test Dilution (RTD) of the phages was determined. The titres of the phage preparations were determined by means of a technique based on the method described by Gratia (1936, 1942), Makower *et al.* (1953) and Nicolle *et al.* (1951).

For this purpose, dilutions in broth of the phage lysates were prepared in logarithmic series from 10^{-1} to 10^{-10}, and from the last three dilutions 0·5 ml of phage suspension was transferred to 2 ml of 1% agar medium (1% Difco agar, see page 214) at 50°C. After thorough mixing, 0·5 ml of a 1:10 dilution of a 24-h broth (nutrient broth) culture of the homologous bacterial strain was added, mixed with the agar, and poured out on Petriplates already containing 30 ml of 1·2% Difco agar (1·2% Difco agar, see

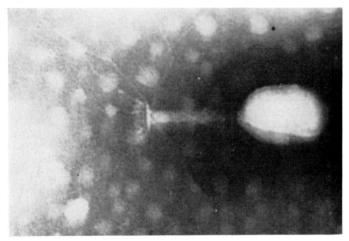

FIG. 2. Electron-microscopic morphology of Kl-1 bacteriophage ($\times$ 240,000).

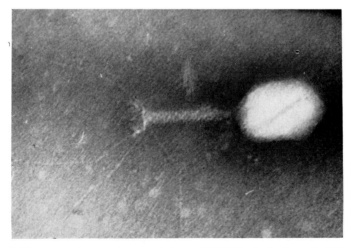

FIG. 3. Electron-microscopic morphology of Kl-2 bacteriophage ($\times$ 240,000).

8

page 214). After the second layer solidified, the cultures were incubated for 6 h at 37°C, followed by 6 h at room temperature when plaque counts were determined.

D. Morphology of *Klebsiella* Bacteriophages

Electron-microscopical morphology of *Klebsiella* bacteriophages used for typing of *Klebsiella* (all species except *K. ozaenae*) are shown in Figs. 2–15, uranyl acetate negative stained preparations × 240,000 magnification (Krzywy *et al.*, 1971; Krzywy and Ślopek, 1974).

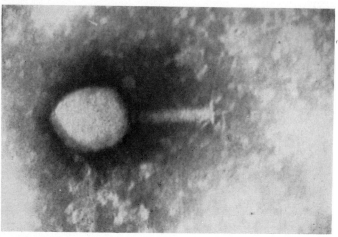

Fig. 4. Electron-microscopic morphology of Kl-6 bacteriophage (× 240,000).

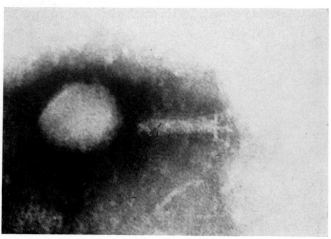

Fig. 5. Electron-microscopic morphology of Kl-7 bacteriophage (× 240,000).

Morphological control of the test bacteriophages is helpful in order to ensure homogeneity of the bacteriophages in the lysate. Irreproducible results of phage typing are usually due to non-homogeneity of the phage lysate or changes in antigenic structure of the tested strains.

E. Determinations of Sensitivity of *Klebsiella* to Bacteriophages

Klebsiella strains were inoculated on solid medium of the following

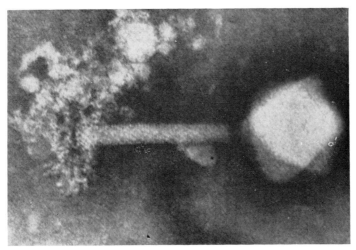

FIG. 6. Electron-microscopic morphology of Kl-9 bacteriophage (× 240,000).

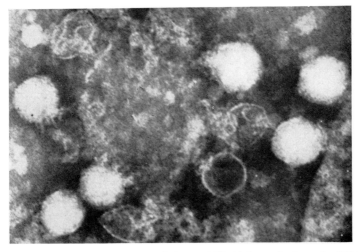

FIG. 7. Electron-microscopic morphology of Kl-11 bacteriophage (× 240,000).

composition:

Peptone (Difco)	0.5%
Meat extract (Difco)	0.3%
Lactose	1.0%
Bromocresol purple	0.0025%
Agar (Difco)	2.0%

pH of the medium 6·8–7·0

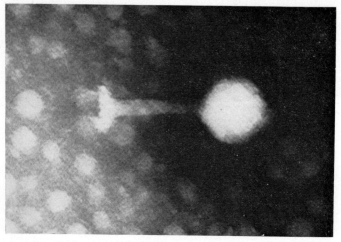

FIG. 8. Electron-microscopic morphology of Kl-14 bacteriophage (× 240,000).

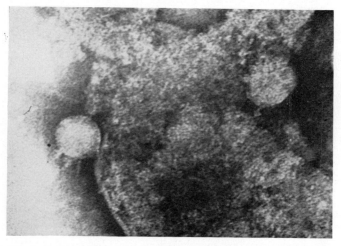

FIG. 9. Electron-microscopic morphology of Kl-16 bacteriophage (× 240,000).

After incubation for 24 h at room temperature, single colonies were transferred to broth and incubated for 24 h at 37°C. A few slowly growing strains were incubated for 6 h. The number of viable bacterial cells in the 4- (or 6-) h cultures was about 10^8 cells per 1 ml; the absolute number of cells was approximately equal to the number of cells capable of producing colonies. Two ml of the broth culture were poured out on a Petri-plate

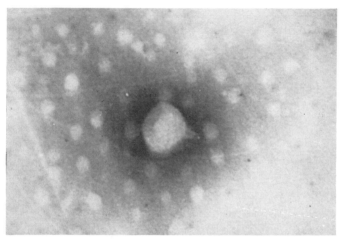

Fig. 10. Electron-microscopic morphology of Kl-17 bacteriophage ($\times$ 240,000).

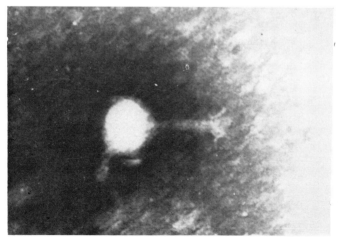

Fig. 11. Electron-microscopic morphology of Kl-21 bacteriophage ($\times$ 240,000).

containing about 20 ml of agar medium of the following composition:

Peptone (Difco)	1·0%
NaCl	0·3%
Na_2HPO_4	0·2%
Agar (Difco)	1·2%

pH of the medium 7·2

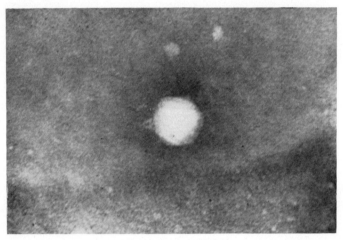

FIG. 12. Electron-microscopic morphology of Kl-23 bacteriophage (× 240,000).

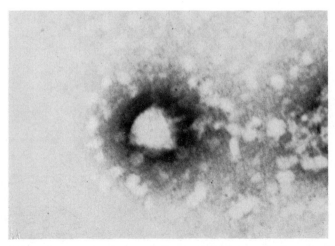

FIG. 13. Electron-microscopic morphology of Kl-24 bacteriophage (× 240,000).

The inoculum was spread evenly on the surface of the agar. Excess broth was removed by means of a Pasteur pipette. The inoculated plates were dried for 30 min in the incubator. Immediately after drying, 12 drops of undiluted phage lysates of concentration given in Table IX, or diluted 1 : 5 to 1 : 25, were placed on each plate. The phage suspensions were applied to the plates by means of Pasteur pipettes selected to give drops of 0·01 ml volume. The plates were dried for 20 min at room temperature and then

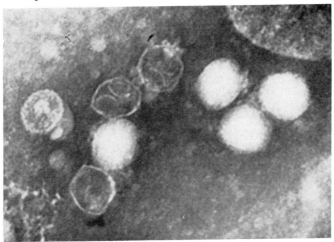

Fig. 14. Electron-microscopic morphology of Kl-26 bacteriophage (× 240,000).

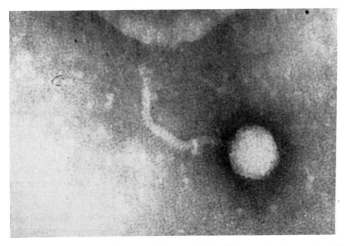

Fig. 15. Electron-microscopic morphology of Kl-42 bacteriophage (× 240,000).

TABLE IX

List of *Klebsiella* bacteriophages (Przondo-Hessek *et al.*, 1966)

Phage designation	*Klebsiella* propagation strain	Bacteriophage titre	Phage origin
1	K-67 264/1	2×10^8	City sewage
2	K-59 2212	2×10^8	City sewage
6	K-53 1756	5×10^8	City sewage
7	K-41 6177	1×10^8	City sewage
9	K-14 119	1×10^9	City sewage
11	K-35 7444	2×10^8	City sewage
14	K-3 L-174	3×10^9	Nasal secretion
16	K-30 7824	5×10^9	City sewage
17	K-18 1754	2×10^8	City sewage
21	K-2 B-5055	2×10^9	Stool
23	K-11 390	5×10^8	City sewage
24	K-24 1680	2×10^8	City sewage
26	K-42 1702	5×10^8	City sewage
42	K-71	1×10^8	Stool

incubated at 37°C for 6 h. Results were recorded according to the following scheme:

CL	Confluent lysis
OL	Opaque lysis
SCL	Semiconfluent lysis
+ + +	80 isolated plaques
+ +	40–80 isolated plaques
+	20–40 isolated plaques
−	Negative reaction

All the *Klebsiella* strains were examined three times at intervals of six months.

F. Working Set of *Klebsiella* Bacteriophages Used for Typing of *Klebsiella*

The working set is composed of representative phages from the two largest collections of *Klebsiella* phages, those of Milch and Deák (1964/65) and Przondo-Hessek (1966). Out of 46 phages from Hungarian and Polish collections, 14 phages were selected, 10 of them for the typing of *K. aerogenes*, *K. aerogenes* var. *edwardsii*, *K. pneumoniae* and *K. pneumoniae* var. *atlantae*, and 6 for the typing of *K. rhinoscleromatis*. The bacteriophages selected are shown in Table IX. The lytic spectrum of the selected

9

TABLE X

The lytic spectrum of *Klebsiella* bacteriophages from Hungarian (Milch and Deák) and Polish (Przondo-Hessek) collections (Ślopek et al., 1967)

Test strains	Bacteriophages (Collection of Przondo-Hessek)												Bacteriophages (Collection of Milch and Deák)	
	1	2	6	7	9	11	14	16	17	21	23	24	26	42
K-67 264/1	Cl													
K-59 2212	Cl	Cl		Cl										
K-53 1756	Cl	Cl	Cl	Cl		Cl								
K-41 6177	Cl	Cl	Cl	Cl										
K-14 1193	Cl	Cl	Cl	Cl	Cl									
K-35 7444	Cl	Cl	Cl	Cl		Cl								
K-3 L-174	Cl	Cl	Cl	Cl		Cl	Cl							
K-30 L-107							Cl	Cl						
K-18 1754					Cl				Cl					
K-2 B-5055	Cl				Cl					Cl				
K-11 390		Cl	Cl								Cl			
K-24 1680												Cl		
K-42 1702													Cl	
K-71	Cl	Cl	Cl	Cl										

Cl — Confluent lysis.

TABLE XI

Serological relationship of *Klebsiella* bacteriophages (Przondo-Hessek *et al.*, 1967)

Antiphage serum	Bacteriophages													
	1	2	6	7	9	11	14	16	17	21	23	24	26	42
1	100[a]	70	70	50	0	0	0	0	0	0	0	0	0	0
2	70	100	90	90	0	0	0	0	0	50	0	0	0	0
6	50	90	100	90	0	0	0	0	0	0	0	0	0	0
7	70	90	100	100	0	0	0	0	0	0	0	0	0	0
9	0	0	0	0	100	0	0	0	0	0	0	0	0	0
11	0	0	0	0	0	100	0	0	0	0	0	0	0	0
14	0	0	0	0	0	0	100	0	0	0	0	0	0	0
16	p	0	0	0	0	0	0	100	0	0	0	0	0	50
17	0	50	0	0	0	0	0	0	100	0	0	0	0	0
21	0	0	0	0	0	0	0	0	0	100	0	80	0	0
23	0	0	0	0	0	0	0	0	0	0	100	0	0	50
24	0	0	0	0	0	0	0	0	0	80	0	100	0	0
26	0	0	0	0	0	0	0	90	90	0	0	0	100	0
42	0	0	0	0	0	0	0	50	0	0	50	0	0	100

[a] Percentages of neutralisation.

Antiphage sera were diluted with nutrient broth. Phage preparations containing 10^6 plaque forming units per ml were used in the test. To 0·1 ml of diluted phage, 0·9 ml of diluted serum was added, and the mixture was placed in a water bath at 37°C for 30 min. After that time, a 0·1 ml aliquot of the mixture was transferred to 0·9 ml of nutrient broth, and 0·1 ml of this dilution was added to 2·0 ml of 0·7% Difco agar at 50°C. The detector strain was added in the amount of 0·5 ml of a 4-h nutrient broth culture. The mixture was poured out on agar plates and incubated for 8 h at 37°C. A mixture of the phage with nutrient broth served as a control. About 100 plaques appeared on the control plates and in the cases in which no neutralisation occurred.

TABLE XII
Phage types of *Klebsiella aerogenes* and *K. pneumoniae*

Phage types	Bacteriophages										Number of strains
	1	2	9	14	16	21	23	24	26	42	
1	Cl										50
2	Cl	Cl									2
3	Cl	Cl			Cl						2
4	Cl		Cl								2
5	Cl		Cl			Cl					1
6	Cl				Cl						1
7	Cl								Cl		2
8	Cl	Cl									80
9	Cl	Cl	Cl								27
10	Cl	Cl	Cl						Cl		3
11	Cl	Cl		Cl							2
12	Cl	Cl				Cl			Cl		1
13	Cl	Cl						Cl			1
14	Cl	Cl							Cl		6
15	Cl	Cl	Cl								18
16	Cl	Cl	Cl	Cl							3
17	Cl	Cl	Cl				Cl				2
18	Cl	Cl	Cl				Cl		Cl		2
19	Cl	Cl	Cl						Cl		4
20		Cl									31
21		Cl	Cl			Cl					2
22		Cl		Cl							1
23		Cl				Cl					1
24		Cl						Cl			1
25		Cl						Cl	Cl		1
26			Cl								8
27			Cl	Cl		Cl					7
28			Cl			Cl					4
29			Cl							Cl	1
30				Cl							67
31				Cl	Cl						4
32				Cl		Cl					2
33				Cl				Cl			1
34				Cl				Cl	Cl		1
35				Cl						Cl	4
36					Cl						9
37						Cl					10
38						Cl		Cl			2
39							Cl				13
40								Cl			6
41								Cl	Cl		2
42								Cl		Cl	1
43									Cl		6
44										Cl	7

bacteriophages toward the host propagating strains used is illustrated in Table X. In Table XI are assembled data pertaining to the serological relationships of the bacteriophages.

1. *Phage Typing of* K. aerogenes *and* K. pneumoniae

For typing of *K. aerogenes, K. aerogenes* var. *edwardsii, K. pneumoniae* and *K. pneumoniae* var. *atlantae* the following bacteriophages were chosen: 1, 2, 9, 14, 16, 21, 23, 24, 26, and 42.

Out of 404 *K. aerogenes* strains 252 (61·9%) were typable and out of 204 *K. pneumoniae* strains 149 (72·5%) were typable. On the basis of typing of 252 strains of *K. aerogenes, K. aerogenes* var. *edwardsii*, and 149 strains of *K. pneumoniae* and *K. pneumoniae* var. *atlantae*, 44 phage types were found (Table XII). The most frequent phage types in both subgroups were: 1, 8, 9, 14, 15, 19, 20, 26, 27, 28, 30, 31, 35, 36, 37, 39, 40, 43, and 44.

2. *Phage Typing of* K. ozaenae

At present it is impossible to use bacteriophages for the typing of *K. ozaenae* because the percent of strains sensitive to the bacteriophages known so far is too low.

3. *Phage Typing of* K. rhinoscleromatis

For typing of *K. rhinoscleromatis* the following bacteriophages were chosen: K1 6, K1 7, K1 11, K1 14, K1 17, and K1 21 (Przondo-Hessek *et al.*, 1968). On the basis of typing of 155 strains of *K. rhinoscleromatis* isolated in Poland and Belorussian Soviet Socialistic Republic, 8 phage

TABLE XIII

Phage types of *Klebsiella rhinoscleromatis* (Przondo-Hessek *et al.*, 1968)

Phage types	Bacteriophages						Number of strains
	7	6	17	14	11	21	
1	Cl	Cl	SCl	Cl	Cl	—	7
2	—	Cl	SCl	Cl	Cl	—	11
3	—	—	SCl	Cl	Cl	—	46
4	—	—	SCl	Cl	—	—	2
5	—	—	—	Cl	Cl	—	82
6	—	—	—	Cl	Cl	Cl	1
7	—	—	—	Cl	—	—	5
8	—	—	—	—	Cl	—	1

Cl = Confluent lysis.
SCl = Semi-confluent lysis.

types were found (Table XIII). The most frequent phage types were: 1, 2, 3, 5, and 7.

VIII. CONCLUSION

The materials presented above indicate that *Klebsiella* can be subdivided on the basis of their morphological, physiological and antigenic characteristics, their resistance to antibiotics, and sensitivity to bacteriocins (pneumocins) and bacteriophages.

The practical value of these different methods of typing should be the subject of further clinical and epidemiological evaluations.

REFERENCES

Brooke, M. S. (1951). *Acta Path. et Microbiol. Scand.*, **28**, 313–327.
Cowan, S. T., Steel, K. J., Shaw, C. and Duguid, J. P. (1960). *J. gen. Microbiol.*, **23**, 601–612.
Craigie, J., and Yen, Ch. H. (1938a). *Can. Publ. Health J.*, **29**, 448–463.
Craigie, J., and Yen, Ch. H. (1938b). *Can. Publ. Health J.*, **29**, 484–496.
Darrell, J. H., and Hurdle, A. D. (1964). *J. Clin. Pathol.*, **17**, 617–621.
Durlakowa, I., Lachowicz, Z., and Ślopek, S. (1967a). *Arch. Immunol. et Ther. Exper.*, **15**, 488–489.
Durlakowa, I., Lachowicz, Z., and Ślopek, S. (1967b). *Arch. Immunol. et Ther. Exper.*, **15**, 490–496.
Durlakowa, I., Lachowicz, Z., and Ślopek, S. (1967c). *Arch. Immunol. et Ther. Exper.*, **15**, 497–504.
Durlakowa, I., Maresz-Babczyszyn, J., Przondo-Hessek, A., and Lusar, Z. (1963). *Arch. Immunol. et Ther. Exper.*, **11**, 549–562.
Durlakowa, I., Maresz-Babczyszyn, J., Przondo-Hessek, A., Lusar, Z., and Mróz-Kurpiela, E. (1964a). *Arch. Immunol. et Ther. Exper.*, **12**, 308–318.
Durlakowa, I., Maresz-Babczyszyn, J., Przondo-Hessek, A., Lusar, Z., and Mróz-Kurpiela, E. (1964b). *Arch. Immunol. et Ther. Exper.*, **12**, 319–331.
Edmunds, P. N. (1954). *J. infect. Dis.*, **94**, 65–71.
Edwards, P. R., and Fife, M. A. (1952). *J. Infect. Dis.*, **91**, 92–104.
Edwards, P. R., and Fife, M. A. (1955). *J. Bact.*, **70**, 382–390.
Enterobacteriaceae Subcommittee Report (1963). *Interntl. Bull. Bact. Nomen. Tax.*, **13**, 69–93.
Foster, W. D., and Bragg, J. (1962) *J. Clin. Pathol.*, **15**, 478–481.
Fredericq, P. (1957). *Ann. Rev. Microbiol.*, **11**, 7–22.
Goslings, W. R. O., and Snijders, E. P. (1936). *Zbl. Bakt. Abt. I. Orig.*, **136**, 1–24.
Gratia, A. (1936). *Ann. Inst. Pasteur*, **57**, 652–676.
Gratia, A. (1942). *Arch. ges. Virusforsch.*, **2**, 325–344.
Jacob, F., Siminovich, L., and Wollman, E. L. (1952). *Ann. Inst. Pasteur*, **83**, 295–315.
Julianelle, L. A. (1926). *J. Exptl. Med.*, **44**, 113–126.
Kauffmann, F. (1949). *Acta Path. et Microbiol. Scand.*, **26**, 381–406.
Krzywy, T., Przondo-Hessek, A., and Ślopek, S. (1971). *Arch. Immunol. Ther. Exper.*, **19**, 1–14.

Krzywy, T., and Ślopek, S. (1974). *In* "Morphology and Ultrastructure of *Shigella* and *Klebsiella* Bacteriophages", pp. 1–166. Polish Medical Publ. Warszawa.

Makower, H., Skurska, Z., Stankiewicz, C., and Sypula, A. (1953). *Arch. Immunol. et Ther. Exper.*, **1**, 353–385.

Maresz-Babczyszyn, J. (1962). *Arch. Immunol. et Ther. Exper.*, **10**, 589–617.

Maresz-Babczyszyn, J., Mróz-Kurpiela, E., and Ślopek, S. (1967a). *Arch. Immunol. et Ther. Exper.*, **15**, 512–516.

Maresz-Babczyszyn, J., Mróz-Kurpiela, E., and Ślopek, S. (1967b). *Arch. Immunol. et Ther. Exper.*, **15**, 517–520.

Maresz-Babczyszyn, J., Mróz-Kurpiela, E., and Ślopek, S. (1968). *Arch. Immunol. et Ther. Exper.*, **16**, 406–409.

Milch, H., and Deák, S. (1964/65). *Acta Microbiol. Acad. Hung.*, **11**, 251–261.

Nicolle, P., Hamon, Y., and Edlinger, E. (1951). *Ann. Inst. Pasteur*, **80**, 479–495.

Ørskov, I. (1955). *Acta Path. et Microbiol. Scand.*, **36**, 449–453.

Przondo-Hessek, A. (1966). *Arch. Immunol. et Ther. Exper.*, **14**, 413–435.

Przondo-Hessek, A., and Ślopek, S. (1967). *Arch. Immunol. et Ther. Exper.*, **15**, 563–577.

Przondo-Hessek, A., Ślopek, S., and Miodońska, J. (1967). *Arch. Immunol. et Ther. Exper.*, **15**, 557–562.

Przondo-Hessek, A., Ślopek, S., and Miodońska, J. (1968). *Arch. Immunol. et Ther. Exper.*, **16**, 402–505.

Szulga, T. (1971). *Arch. Immunol. et Ther. Exper.*, **19**, 67–122.

Ślopek, S. (1961). *In* "*Enterobacteriaceae*-Infektionen" (Eds. J. Sedlak and H. Rische), pp. 313–337. Georg Thieme, Leipzig.

Ślopek, S. (1968). *In* "*Enterobacteriaceae*-Infektionen" (Eds. J. Sedlak and H. Rische), second ed., pp. 531–557. Georg Thieme, Leipzig.

Ślopek, S. (1973). *In* "Lysotypie" (Ed. H. Rische), pp. 343–368. Gustav Fischer, Jena.

Ślopek, S., and Durlakowa, I. (1967). *Arch. Immunol. et Ther. Exper.*, **15**, 481–487.

Ślopek, S., and Maresz-Babczyszyn, J. (1967). *Arch. Immunol. et Ther. Exper.*, **15**, 525–529.

Ślopek, S., Przondo-Hessek, A., Milch, H., and Deák, S. (1967). *Arch. Immunol. et Ther. Exper.*, **15**, 589–599.

Tkaczowa, A., Durlakowa, I., and Ślopek, S. (1967). *Arch. Immunol. et Ther. Exper.*, **15**, 505–511.

Worfel, M. T., and Ferguson, W. W. (1951). *Am. J. Clin. Pathol.*, **21**, 1097–1100.

CHAPTER VI

Bacteriocin Typing of Clinical Isolates of *Serratia marcescens*

WALTER H. TRAUB*

Institut für klinische Mikrobiologie und Infektionshygiene,
Universität Erlangen-Nürnberg, D-8520 Erlangen, Wasserturmstr. 3,
B.R.D. (F.R.G.)

I. INTRODUCTION

A. Current clinical significance of *Serratia marcescens*

For decades, *Serratia marcescens*, the type species of genus IV *Serratia* Bizio, Tribe IV *Klebsielleae* Trevisan, of the family *Enterobacteriaceae* (Edwards and Ewing, 1972), was considered a harmless, ubiquitous micro-organism, which only very rarely caused infections in mammals and in man (Ewing *et al.*, 1962). Since the early 1960s, however, an increasing number of investigators have reported the isolation of *S. marcescens* from cases of respiratory tract, urinary tract, soft tissue, wound,

---

* Present address: Clinical Microbiology Laboratory, V.A. Hospital, 4150 Clement Street, San Francisco, CA. 94121, U.S.A.

and central nervous system infections, as well as from cases of septicaemia, bacteraemia, and endocarditis; the majority of patients revealed severe, underlying illnesses. Furthermore, this opportunistic-pathogenic micro-organism proved responsible for a number of outbreaks of nosocomial cross-infection (for additional references see Clayton and von Graevenitz, 1966; Davis *et al.*, 1970; Ewing *et al.*, 1962; Traub and Raymond, 1971; Wilfert *et al.*, 1970; Wilkowske *et al.*, 1970).

B. Previous attempts to type isolates of *S. marcescens*

In the late 1950's, Ewing and co-workers (see Edwards and Ewing, 1972) successively defined the somatic (O) and flagellar (H) antigens of *S. marcescens*. Currently, 15 serogroup and 13 flagellar antigens are recognised; with appropriate monospecific anti-O and anti-H rabbit hyperimmune sera, it was possible to serotype approximately 95% of clinical isolates of *S. marcescens* from a variety of sources and geographic localities (Edwards and Ewing, 1972). In a subsequent study, it was possible to differentiate 97 serotypes among 652 clinical isolates of this micro-organism (B. R. Davis, 1967; unpublished results). Sedlak *et al.* (1965) examined 91 isolates of *S. marcescens*; 89 of the isolates (97·8%) were typable and assigned to a provisional scheme of 11 O-antigen groups (see also Sedlak, 1968).

Another approach was taken by Pillich *et al.* (1964); these authors examined the susceptibility of 100 isolates of *S. marcescens* and 7 isolates of *S. marcescens* var. *kiliensis* to 10 selected bacteriophages. Only 13 isolates proved resistant to all 10 phages. On the basis of increasing susceptibility to 6 of the 10 phages, 77 of the 100 strains of *S. marcescens* could be allocated to 1 of 7 provisional groups.

C. Bacteriocins of *S. marcescens*

The first report of a bacteriocin in *S. marcescens*, designated "marcescin", was that of Fuller and Horton (1950). Recovered from a pigmented isolate of *S. marcescens* (strain 82B) from soil the bacteriocin was a heat-stable (100°C) polypeptide. The substance was active against *Corynebacterium diphtheriae*, *Staphylococcus aureus*, *Pasteurella multocida* (*P. septica*), and *Clostridium perfringens*, and was very toxic for mice. Sublethal concentrations of the substance failed to protect mice against "marcescin"-sensitive bacteria. In addition, "marcescin" was haemolytic for horse red cells, soluble in aqueous organic solvents, dialysable, and resistant to trypsin.

Bacteriocinogeny among 73 of 85 isolates of *Serratia* (*S. marcescens*, 80; *S. plymuthicum*, 2; *S. indica*, 2; *S. kiliensis*, 1) was first observed by Hamon and Peron (1961) after induction with ultraviolet light (UV). However, 12 of the 73 bacteriocinogenic strains elaborated several bacteriocins that differed in their host range and thus were assigned to two fractions— fraction 1 and fraction 2 (Table I). The bacteriocins of fraction 1 were "colicin-like", in that they were active against strains of *Escherichia coli*, but not against *S. marcescens*, and were susceptible to trypsin. The property of "colicinogeny" of *S. marcescens* was subsequently transferred *in vitro* through conjugation to recipient strains of *E. coli* (Mandel and Mohn, 1962a, b); likewise, colicin plasmids E1 and E2 proved transferable by conjugation to *S. marcescens* strain HY. Foulds worked extensively with a colicin-like bacteriocin from *S. marcescens* strain no. JF 246 (Foulds, 1971, 1972; Foulds and Shemin, 1969a, b). This trypsin-sensitive bacteriocin

TABLE I

Bacteriocins of *Serratia marcescens*

Property	Fraction 1* Group B†	Fraction 2 Group A
Resistance against chloroform‡	—	+
Resistance against trypsin‡	—	+
Electrophoretically mobile§	—	+
Biological activity against:‡		
S. marcescens	—	+
Escherichia coli	+	—

* Nomenclature of Hamon and Peron (1961).
† Nomenclature of Prinsloo (1966).
† Determined by Hamon and Peron (1961) and verified by Prinsloo (1966).
§ Prinsloo *et al.* (1965).

inhibited the growth of 11 of 20 *Escherichia coli* strains, but none of 7 *S. marcescens* strains tested. The purified bacteriocin had a molecular weight of 64,000 Daltons and inhibited the incorporation of leucine into protein, thymidine into deoxyribonucleic acid (DNA), as well as the active transport of leucine into the indicator strain *E. coli* JF 135.

The bacteriocins of fraction 2, on the other hand, were active against strains of *S. marcescens*, but not against *E. coli*, and were trypsin resistant. These observations were confirmed by Prinsloo (1966), who employed a different nomenclature. Her "group B" bacteriocins correspond to fraction 1 of Hamon and Peron (1961), those of "group A" to fraction 2 (Table I). A total of 71 of 139 strains of *S. marcescens* produced group A bacteriocins that were active against strains of *S. marcescens*; some were active against strains of *Salmonella*, *E. coli*, and *Aerobacter* as well. The group A bacteriocins were divided into 8 subgroups on a basis of their host range against indicator strains and resistant mutants derived therefrom.

Furthermore, group A bacteriocins were resistant to chloroform, trypsin and proteolytic enzymes from certain micro-organisms and were non-dialysable, and heat-stable (60°C, 30 min). In contrast, group B bacteriocins were inactive against *S. marcescens*, but inhibited strains of *E. coli, Hafnia,* and *Aerobacter* (19 *S. marcescens* strains produced only these latter bacteriocins, while an additional 35 strains elaborated group A and group B bacteriocins after induction with UV). Group B bacteriocins proved heat-labile, non-dialysable, and susceptible to chloroform and proteolytic enzymes. Earlier, Prinsloo *et al.* (1965) had demonstrated that group A bacteriocins were electrophoretically mobile, migrating towards the cathode, whereas group B bacteriocins remained stationary (Table I).

In 1970, after the encounter of an impressive upsurge in the number of isolates of *S. marcescens* from various clinical sources at the North Carolina Baptist Hospital, Winston-Salem, North Carolina, U.S.A., a simple and rapid method for bacteriocin typing isolates of this micro-organism was developed (Traub *et al.*, 1971). Of 50 strains of *S. marcescens* examined after induction with Mitomycin C (MC), 37 strains (74%) regularly produced group A bacteriocins, which were tested in a checkerboard-like fashion with respect to their host range against all 50 isolates employed. A significant number of isolates produced bacteriocins with identical host ranges. Computer analysis of the data allowed one to limit the number of group A bacteriocinogenic (bA⁺) strains to 10 isolates (bA⁺ isolates no. 5, 10, 12, 16, 17, 18, 31, 33, 43 and 46), yet maintain the extent of typability of the *S. marcescens* isolates at the same level as before (46/50 = 92% typable), as based on the susceptibility or resistance/tolerance of the isolates against the 10 selected group A bacteriocins.

TABLE II

Summary of properties shared by *S. marcescens* group A bacteriocins of both subgroups

Property	References
Inducible with	
Ultraviolet light	Hamon & Peron (1961); Prinsloo (1966)
Mitomycin C	Traub *et al.* (1971); Farmer (1972a, b)
Nalidixic acid	Traub *et al.* (1973)
Trimethoprim	Traub & Kleber (1974b)
Inhibition of induction by Rifampin	Traub (1972d)
Resistance against trypsin, pronase, pepsin, carboxypeptidase A, beta-glucosidase	Traub *et al.* (1974)
Variable sensitivity to papain	Traub *et al.* (1974)
Resistance against 0·01 M HCl, pH 2·08	Traub *et al.* (1974)
Ultrastructure: phage tails	Traub (1972c), Traub *et al.* (1974)

TABLE III

Properties of subgroup I and II group A bacteriocins of *S. marcescens**

Property	Sub-group I	Subgroup II
Freeze-stability ($-65\,^\circ$C)	+	—
Precipitation by ammonium sulphate without loss of biological activity	+	—
Concentration by polyethylene glycol (PEG 6000)	+	—
Resistance to 0·05 M sodium acetate, pH 4·0	+	—
Biological (lethal) activity on S medium with added Casamino Acids, Technical (Difco)	+	—
Electrophoretic mobility (2 mA, 2 h)†	+	—
Neutralisation of biological activity by rabbit hyper-immune sera against:		
Subgroup I bacteriocins†	+	—
Subgroup II bacteriocins†	—	+
Ultrastructure of phage tails	core	core, con-tractile sheath, basal plate, tail fibres

* Modified from Traub *et al.* (1974).
† Unpublished observations.

Subsequently, these 10 group A bacteriocins were examined more closely (Tables II and III). Most important, all 10 bacteriocins were shown to be phage tails. Based on their ultrastructural morphology and several biological and physico-chemical properties, these 10 phage tail bacteriocins were assigned to two subgroups: subgroup I = bA$^+$ no. 5, 12, 17, 18, 31, 33 and 46; subgroup II = bA$^+$ no. 10, 16 and 43 (Table IV). The phage tails of subgroup I and II superficially resembled those of the *S. marcescens* bacteriophages SMB1 and SMB3 of Bradley (1967), respectively. The bacteriocins defining the two subgroups proved serologically unrelated as determined by neutralisation tests (Table III). The receptors of control indicator *S. marcescens* strain no. S 9*i* and of *S. marcescens* strain no. 08 for the three bacteriocins of subgroup II were shown to reside in the lipopoly-saccharide (LPS) fraction of the outer membrane (Traub and Kleber, 1974d). The receptors for subgroup I bacteriocins, with the exception of that of *S. marcescens* strain no. S 9*i* for bacteriocin no. 31, were not associ-ated with the LPS-moiety. As to the mode of action whereby these group A (phage tail) bacteriocins exert their lethal activity against susceptible cells of *S. marcescens*, preliminary experiments (Traub and Kleber, 1974c) indicated that the bacteriocins of both subgroups inhibit protein synthesis.

TABLE IV

Bacteriocinogenic (bA$^+$) and indicator (i-) strains of *S. marcescens* in current use

bA$^+$ isolate no.	Subgroup	Serotype*	i isolate no.	Serotype
5	I	rel. O6: H undeterm.	1i	rel. O15: H undeterm.
10	II	O14: H1	9i	O14: H12
12	I	O14: H10	21i	O14: H10
16†	II	O4: H1	16i†	O4: H1
17	I	O14: H2		
18	I	O5: H2		
31	I	O5: H3		
33	I	rel. O3: H4		
43	II	O3: H1		
46	I	O5: H4		

* Serotyping was kindly performed by Miss B. R. Davis, Enterobacteriology Unit, Center for Disease Control, Atlanta, Georgia, U.S.A.
† Isolates bA$^+$ 16 and 16i are one and the same strain.

This was measured in terms of inhibition of β-galactosidase activity (hydrolysis of ortho-nitrophenyl-β-D-galactopyranoside = ONPG). *S. marcescens* is known to ferment lactose slowly, and all strains are ONPG-positive (Edwards and Ewing, 1972). However, it was not determined whether this inhibitory effect was a consequence of possible inhibition of DNA or of ribonucleic acid (RNA) synthesis. In this context, the recent finding of Eichenlaub and Winkler (1974) is of relevance; a filamentous, phage tail-like, group A bacteriocin of *S. marcescens* strain HY inhibited synthesis of DNA, RNA, and protein in susceptible cells and caused degradation of DNA and, at higher concentrations, of RNA.

II. METHODS OF BACTERIOCIN TYPING

A. Based on susceptibility of isolates to selected group A bacteriocins

As stated in Section I.C, a simple method for typing clinical and environmental isolates of *S. marcescens*, based on the susceptibility or resistance/tolerance of the isolates to 10 selected group A (phage tail) bacteriocins, was developed (Traub *et al.*, 1971). Through trial and error, the following procedure evolved.

1. Induction of group A (phage tail) bacteriocins

The 10 bA$^+$ strains of *S. marcescens* (Table IV) were maintained on Brain Heart Infusion agar (BHIA, Difco) slants at 4°C, and in a mixture of one volume of Brain Heart Infusion broth (BHIB, Difco) plus one volume of heat-inactivated (56°C, 30 min) bovine serum at −65°C. The BHIA slants were transferred at monthly intervals. For induction with Mitomycin C (MC), the following procedure, which was modified slightly from the original procedure, has been used routinely since 1972: a loopful of growth of each of the bA$^+$ strains from BHIA slants is transferred to 2·5 ml of Tryptic Soy broth (TSB, Difco)* and incubated without agitation at 32°C for 5 h, at which time 1 ml of the broth culture is transferred to 8 ml of TSB in screw-capped (plus rubber lining) test-tubes (150 × 16 mm); these tubes are incubated without agitation for an additional hour, after which 1 ml of a 10 μg/ml aqueous, membrane filter-sterilised stock solution of MC (source of MC: Sigma Chemical Co., St. Louis, Missouri, U.S.A.) is added (final MC concentration = 1 μg/ml). Control tubes receive 9 ml of TSB plus 1 ml of 10 μg/ml of MC. After stationary incubation at 32°C overnight (16–18 h), 1 ml of chloroform is added to each tube, the tube contents shaken with 20 vigorous manual strokes and decanted into sterile glass (*not* plastic) centrifuge tubes. After centrifugation at 2500 × *g* at 4°C for 10 min, the supernatants are carefully removed so as not to disturb the bottom chloroform layer and transferred to 90 mm diameter plastic or glass Petri dishes which are left open and aerated for a minimum of 10 min. After aeration, the cell-lysates are stored in screw-capped test-tubes at 4°C (the biological activity of the group A phage tail bacteriocins remains essentially unchanged at 4°C for a period of at least 3 months).

2. Titration of crude cell-lysates

The group A bacteriocins are titrated against one of the 4 control indicator (*i*-) strains of *S. marcescens* as shown schematically in Table V. Serial dilutions of the cell-lysates are prepared in TSB (0·3 ml cell-lysate +2·7 ml TSB = 1:10; 1 ml of 1:10 + 1 ml TSB = 1:20, etc., including 1:1280). The *i*-strains are pre-grown in 2·5 ml TSB at 32°C overnight. The turbidity of the cultures of *i* strains is adjusted to that of McFarland barium sulphate standard no. 0·5 (corresponding to approximately 1·5 × 10^8 colony-forming-units/ml) in 0·154 M sterile saline, in a manner

---

* Optimal broth media for induction of group A bacteriocins are TSB, BHIB, but Mueller-Hinton broth is less good; nutrient broth and MacConkey broth have been found unsuitable for this purpose. Induction of representative bA$^+$ strains of both subgroups with 5, 2·5 and 1·0 μg/ml of MC yielded cell-lysates with identical bacteriocin titres.

TABLE V

Scheme of titration of 10 selected group A bacteriocins
against control indicator (*i*-) strains of *S. marcescens*

i-Strain employed	Bacteriocin of bA+ *S. marcescens* strain no.									
	5	10	12	16	17	18	31	33	43	46
1*i*	×		×		×					
9*i*	×		×			×	×		×	
21*i*								×		
16*i*										×

analogous to the Bauer–Kirby method of antibiotic susceptibility testing
(Bauer *et al.*, 1966). Next, 0·25 ml of these cell suspensions are transferred to 2·5 ml of TSB, respectively; these latter cell suspensions represent the *i*-cell inocula. Large (140 mm diameter) plates of MacConkey
agar with added Crystal violet (MAC–CV, Difco) are swab-inoculated in
three directions according to the Bauer–Kirby technique*. The inoculated
plates are divided into 9 sectors (1:0, 1:10 ... 1:1280). With the aid of a
Pasteur pipette (length 15 cm), one drop (0·05 ml) of each cell-lysate
dilution, starting with the highest dilution, is delivered to the appropriate
sector, including one drop of the undiluted cell-lysate. The drop inocula
are allowed to "air-dry", a process that requires roughly 10 to 15 min. The
inverted plates are incubated at 32°C overnight, after which they are
examined for the presence of bacteriocin-produced plaques, i.e., zones of
inhibition. Sectors are scored as positive if there are completely clear
plaques (diameters of 1·5 to 2·5 cm), or if the areas of inhibition contain
fewer than 50 bacteriocin-resistant/tolerant variant colonies against a clear
background. Those plaques that reveal greater than 50 variant colonies
and those sectors showing no inhibition or trace inhibition of growth only,
are interpreted as negative. The "titre" of a given cell-lysate is the highest
dilution that yields a positive response and is expressed as lethal units/
0·05 ml. Optimal cell-lysates for bacteriocin typing contain between 80 and
320 lethal units/0·05 ml.

3. *Bacteriocin typing of clinical isolates*

Clinical isolates of *S. marcescens* are identified by production of DNase,
fermentation of glucose with no or very little gas (less than 10%), hydrolysis
of ONPG, utilisation of citrate, motility, decarboxylation of lysine and
ornithine, and lack of fermentation of lactose, arabinose, and rhamnose
(Edwards and Ewing, 1972). Following identification, isolates are sub-

* It was found that MAC–CV or MAC agar without added Crystal violet
yielded larger bacteriocin inhibition zones than either Tryptic Soy (TSA) or
Mueller-Hinton agar.

cultured on standard MAC–CV (90 mm diameter) plates and incubated at 32°C overnight (check for viability and purity). Next morning, 3 colonies of each isolate are transferred to 2·5 ml of TSB and incubated at 32°C for 5 h* after which they are processed for inoculation of large MAC–CV plates as described in Section II.A.2 for the control i strains. The inoculated plates are divided into 11 sectors (designated 5, 10, 12, 16, 17, 18, 31, 33, 43 and 46, as well as a control sector). One drop of undiluted crude cell-lysate of each group A bacteriocin and of the control TSB–MC mixture (see Section II.A.1) is delivered to the appropriate sectors. The plates are incubated at 32°C overnight. Next morning, the sectors are examined and interpreted as stated in Section II.A.2. Results are recorded as susceptibility (+) or resistance/tolerance (−) of a given isolate against each of the 10 group A bacteriocins employed.

4. *Alternate method*

Independently, Farmer (1972b) developed a method for typing isolates of *S. marcescens* by their sensitivity to a standard set of 12 bacteriocins, designated P1 to P12. The 12 bacteriocinogenic *S. marcescens* strains, which have been selected from a total of 100 isolates, are induced in 5 ml volumes with 5 μg/ml of MC in Trypticase Soy broth (BBL) at 32°C for 5 h. This is achieved as follows: The isolates are pre-grown on TSA and a small amount of growth transferred to 4 ml of TSB. After incubation in a water bath at 32°C for 24 h, 0·7 ml of the growth is added to 3·3 ml of TSB. Following incubation at 32°C for 1 h, 1 ml of 25 μg/ml of MC (Sigma) is added, and the tubes held in a water bath at 32°C for 5 h, when 0·3 ml of chloroform is added, and the tube contents vigorously mixed for 10 sec using an orbital shaker. After the chloroform has settled, 2 ml are removed from the top layer and transferred to sterile tubes, which are aerated for 30 min.

The isolates to be typed are pre-grown overnight on BHIA at 37°C, after which a small amount of growth is removed and the cell concentration adjusted to that of a turbidity standard (*S. marcescens* strain I2 cell suspension in TSB; OD = 0·10 at 550 nm). From this suspension, 0·1 ml is added to 3 ml of "medium B" (0·24% Oxoid Ionagar or 0·7% Difco Bacto-Agar in distilled water), which has been melted and cooled to 50°C. The contents are mixed and overlaid onto BHIA. The overlay is allowed to solidify and air-dried at room temperature (RT) for 3·5 h.

Uniform drops of the 12 crude cell-lysates (P1 to P12) and a control (TSB+MC) are added simultaneously with an ACCU–DROP multi-

---

* It has been found that cells of *S. marcescens* either in the logarithmic or early stationary phase of growth yield identical bacteriocin type patterns and cell-lysate titres.

dropper (Sylvana Corp., Millburn, N.J.; Farmer, 1970a) or manually using tuberculin syringes. After the drop inocula have dried, the plates are incubated overnight at 37°C. Any inhibition greater than that noted on the control sector is scored as positive. According to Farmer, this end-point is more reproducible than any other.

B. Based on host range of group A bacteriocins elaborated by (non-typable) isolates against selected indicator strains

In late 1971, an outbreak of nosocomial cross-infection in an intensive care unit due to a non-typable strain of *S. marcescens* (Traub, 1972b) necessitated the development of an alternative procedure in order to determine strain relationships. This procedure was based on the determination of the host range of group A bacteriocins produced after MC induction of each non-typable isolate.

1. *Induction of group A bacteriocins*

The procedure is essentially as described in Section II.A.1.

2. *Determination of host range of group A bacteriocins*

For this purpose, the following 15 strains of *S. marcescens* are employed: the ten bA^+ isolates no. 5, 10, 12, 16, 17, 18, 31, 33, 43 and 46; the control *i*-strains no. 1*i*, 9*i* and 21*i*; isolate no. S 274 (this isolate has been more recently replaced with isolate no. S 354*i* which proved of identical bacteriocin- and sero-type, i.e., bacteriocin type 49, serotype O1:H5), and isolate no. S 326 n.t. The latter isolate, which was also serologically non-typable (B. R. Davis, personal communication), had been derived from the index patient of the outbreak referred to above.

The 15 *S. marcescens* isolates are pre-grown in TSB at 32°C overnight and processed for inoculation of large MAC–CV plates according to Section II.A.2. Each of the 15 MAC–CV plates is divided into sectors, depending on the number of non-typable isolates under study. Appropriate sectors receive 1 drop (0·05 ml) of undiluted crude cell-lysate, including 1 drop of the control TSB–MC mixture. The plates are incubated overnight at 32°C after which the sectors are evaluated and interpreted as stated in Section II.A.3.

3. *Alternate method*

Independently, Farmer (1972a) developed a similar bacteriocin typing method for *S. marcescens* based on the host range against a standard set of 9 indicator strains (designated I1 to I9).

The strains to be typed are induced with MC as in Section II.A.4.

The 9 indicator (*i*-) strains had been rendered Streptomycin-resistant, so that the chloroform treatment of crude cell-lysates could be eliminated. The few viable cells that survive induction by MC, fail to form colonies on "medium A" (18·5 g BHIB, Difco, plus 9·3 g Oxoid Ionagar, Colab, or plus 17·4 g Difco Bacto-Agar, in 500 ml distilled water), which is supplemented with 1000 µg/ml of Streptomycin sulphate, since the bacteriocinogenic strains are Streptomycin-sensitive.

The *i*-strains are processed as follows: plates of "medium A" are inoculated in a 1 cm diameter circle with a loopful from litmus milk working cultures. After incubation for 24 h at 37°C, some growth is removed with the aid of a cotton swab. The swab is twirled in 9 ml of TSB until the turbidity matches that of the turbidity standard (Section II.A.4) and 0·1 ml inoculated into 3 ml of "medium B" (Section II.A.4) which has been held at 50°C. The contents are mixed and overlaid on a plate of "medium A", and the plates are dried for up to 4 h at room temperature.

The bacteriocins (crude cell-lysates) are applied to each *i*-strain, including a TSB–MC control, either manually or with the aid of the ACCU–DROP applicator (Section II.A.4). The plates are incubated at 37°C for 18 to 24 h, after which they are examined as follows: any inhibition in the bacteriocin sector greater than the control is interpreted as positive; for example, + = completely clear; + (50) = clear zone with 50 colonies, etc. Furthermore, the bacteriocin responses of the 9 *i*-strains are converted into 3 groups of three-digit numbers, in accordance with the mnemonic method proposed by Farmer (1970b) for reporting bacteriophage and bacteriocin types.

III. RESULTS

A. Provisional typing schemes

To date, 63 provisional bacteriocin types, based on the susceptibility of clinical isolates of *S. marcescens* to 10 selected group A (phage tail) bacteriocins, have been identified, as shown in Table VI. Of these, bacteriocin types no. 1, 4, 9, 14, 15, 18 and 21 are encountered most frequently (Table VII). Of 826 isolates only 86 proved non-typable (10·4%), i.e., 89·6% of all clinical isolates could be typed with the simple procedure outlined in Section II.A.1–3. This procedure permitted epidemiological investigations of a number of outbreaks of nosocomial cross-infections (Traub and Raymond, 1971; Traub, 1972b; Traub and Kleber, 1974a; and unpublished observations).

The "reverse" typing procedure, based on the respective host ranges of group A bacteriocins elaborated by non-typable isolates of *S. marcescens*

TABLE VI

Bacteriocin types of *S. marcescens* (provisional scheme)*

Bacteriocin type no.	Bacteriocin sensitivity pattern: susceptibility and resistance/ tolerance to group A bacteriocin of *S. marcescens* isolate no.									
	5	10	12	16	17	18	31	33	43	46
1	+†	—	+	—	+	—	—	—	—	—
2	+	—	—	—	+	+	—	—	—	—
3	+	+	—	+	+	—	—	—	+	—
4	—	+	—	+	—	—	—	—	+	—
5	+	—	—	—	+	—	—	—	—	—
6	—	+	—	+	—	+	+	—	+	—
7	—	—	—	—	—	—	—	—	—	+
8	—	—	—	—	+	—	—	—	—	—
9	—	+	—	+	—	—	—	+	+	—
10	+	—	—	—	+	+	+	—	—	+
11	+	+	—	+	—	—	—	—	+	—
12	+	—	—	—	+	+	—	—	—	+
13	+	+	+	+	+	—	+	—	+	—
14	—	—	—	—	—	+	—	—	—	—
15	—	—	—	—	—	+	—	—	—	+
16	+	+	+	+	+	—	—	—	+	—
17	+	—	—	—	+	+	+	—	—	—
18	—	+	—	+	—	—	+	+	+	—
19	—	—	—	—	+	+	—	—	—	—
20	—	—	—	—	+	+	—	—	—	+
21	—	+	—	+	—	—	+	—	+	—
22	+	+	—	+	+	—	+	—	+	—
23	—	+	—	+	—	—	—	+	+	+
24	—	—	—	—	—	—	+	—	—	—
25	+	—	+	—	+	+	—	—	—	—
26	—	+	—	+	—	+	—	—	+	—
27	—	+	—	+	+	—	+	—	+	—
28	—	+	—	+	+	—	—	—	+	—
29	+	—	+	—	+	—	+	—	—	+
30	+	—	—	—	+	+	—	—	+	—
31	—	+	—	+	—	—	—	—	—	—
32	—	+	—	+	—	—	+	+	+	+
33	+	—	+	—	+	+	—	—	—	+
34	—	—	—	—	+	+	+	+	—	—
35	—	—	—	+	—	+	—	—	—	+
36	+	+	+	—	+	—	—	—	+	—
37	+	—	+	—	+	+	+	—	—	+
38	—	—	—	—	—	+	+	+	—	—
39	—	+	—	—	—	—	—	+	+	—
40	—	—	+	—	+	—	—	—	—	—
41	+	—	+	—	+	—	+	—	—	—

TABLE VI (*cont.*)

Bacteriocin type no.	Bacteriocin sensitivity pattern: susceptibility and resistance/ tolerance to group A bacteriocin of *S. marcescens* isolate no.									
	5	10	12	16	17	18	31	33	43	46
42	—	—	—	—	—	—	—	—	+	—
43	—	—	—	—	—	—	+	+	—	—
44	—	+	—	+	—	+	—	—	+	+
45	—	+	—	+	—	—	—	—	+	+
46	—	+	+	+	—	—	+	+	+	—
47	—	—	—	—	—	—	+	—	—	+
48	—	+	+	+	—	+	+	—	+	—
49	+	+	+	+	+	+	+	—	+	+
50	—	—	—	—	—	+	+	—	—	+
51	+	—	+	—	+	—	—	—	—	+
52	—	+	—	—	—	—	—	—	+	—
53	—	—	—	—	—	+	+	—	—	—
54	+	+	+	—	+	—	+	—	+	—
55	+	—	—	—	—	—	+	+	—	—
56	—	+	—	+	—	+	+	—	+	+
57	+	+	—	+	—	—	+	—	+	—
58	—	—	+	—	—	—	—	—	—	—
59	+	—	—	—	+	—	—	—	+	—
60	—	—	—	—	+	+	+	—	—	+
61	—	+	+	+	—	—	+	—	+	—
62	+	—	+	—	+	+	+	—	—	—
63	—	+	+	+	—	—	—	—	+	—

* Traub *et al.* (1971), Traub and Raymond (1971), Traub (1972b), Traub and Kleber (1974a), and unpublished observations (bacteriocin types no. 60–63).

† Symbols: +, sensitive; —, resistant (tolerant) to bacteriocin of respective producer strains.

against 15 selected indicator strains, has been reserved for those clinical situations, where a sudden, significant upsurge in the number of non-typable isolates necessitates detailed epidemiological investigations (Traub, 1972b). Consequently, our experience with this procedure is limited.

On the other hand, Farmer (1972b) was able to divide 93 strains of *S. marcescens* from various sources into 79 different bacteriocin types, based on his standard set of 12 group A bacteriocins. Typing by bacteriocin production (Farmer, 1972a) permitted identification of 72 different "host range types" among 129 strains of *S. marcescens* tested, of which only 11 (8·5%) proved non-typable. Farmer demonstrated the usefulness of both of his methods for several nosocomially significant strains of *S. marcescens* (Farmer, 1972a, b).

TABLE VII

Frequency distribution of bacteriocin types of
S. marcescens

Bacteriocin type no.	1969–1972 Winston–Salem	1971 CDC*	1972–1975 Erlangen	Total
1	28	10	6	44
2	2			2
3	9			9
4	31	9	11	51
5	4		1	5
6	2		2	4
7	5	2		7
8	1			1
9	42	10	32	84
10	4	1	1	6
11	1			1
12	3	1		4
13	12	2	1	15
14	27	10	26	63
15	1	17	29	47
16	18	4	5	27
17	5	3	1	9
18	38	6	159	203
19	2			2
20	2			2
21	16	6	11	33
22	7			7
23	2	1	3	6
24	5	2	1	8
25	3	1	1	5
26	8		5	13
27	1			1
28	3			3
29	2			2
30	1			1
31	1			1
32	1			1
33	2	1		3
34	1			1
35	1			1
36	1			1
37	3	2		5
38	1			1
39	1	1		2
40	1			1
41	3		2	5

TABLE VII (*cont.*)

Bacteriocin type no.	1969–1972 Winston-Salem	1971 CDC*	1972–1975 Erlangen	Total
42	1			1
43		1		1
44		2	12	14
45	1	2		3
46	1			1
47	1		1	2
48	1			1
49	2			2
50	1			1
51	2			2
52	1			1
53	1			1
54	1			1
55	1			1
56			10	10
57			1	1
58			2	2
59			1	1
60			1	1
61			4	4
62			1	1
63			1	1
Non-typable	63	7	16	86
Total	378	101	347	826

* The 101 isolates of *S. marcescens* were obtained from Miss B. R. Davis and Dr G. J. Hermann, Center for Disease Control, Enterobacteriology Unit, Atlanta, Georgia, U.S.A.

B. Correlation with serotyping/-grouping

In 1971, Miss B. R. Davis and Dr. G. J. Hermann of the Enterobacteriology Unit, Centre for Disease Control, Atlanta, Georgia, U.S.A., provided us with 101 serologically characterised isolates. Of these, 94 (93·1%) could be bacteriocin-typed (Traub, 1972a). In comparison, 80 isolates (79·2%) carried identifiable O- and H-antigens, whereas 91 of the isolates (90·1%) carried determinable O-antigens. It was noted, for example, that of 31 isolates of serogroup O14, one isolate was bacteriocin non-typable. The remainder could be subdivided into 10 different bacteriocin types (types no. 4, 9, 12, 14, 18, 21, 23, 39 and 43). Nine of 10 isolates comprising bacteriocin type 1 belonged to serotype O2: H1; two isolates to serotypes O4:H1 and O5:H2 belonged to bacteriocin types 7 and 45, respectively. The number of isolates however was too limited to allow any definitive conclusions to be drawn.

Similarly, Farmer (1972a) demonstrated (a) that *S. marcescens* isolates of identical serotype were not necessarily of identical bacteriocin type, as based on typing by the determination of bacteriocin host range of elaborated bacteriocins, and, (b) that several serologically non-typable isolates proved bacteriocin-typable. But, again, these observations were of a preliminary nature.

C. Pitfalls encountered

1. *Phenotypic variation of bacteriocinogenic and indicator strains*

In their recent review of bacteriocins, Mayr-Harting *et al.* (1972) admirably discussed the pitfalls of current bacteriocin typing procedures.

Several years ago, difficulties arose in our laboratory with respect to the production of sufficiently high-titred group A bacteriocin cell-lysate preparations (Traub and Kleber, 1973). Ten bA+ strains of both subgroups yielded different bacteriocin titres following induction with 1 μg/ml of MC. This problem was overcome by the induction of 10 or more single colonies of each bA+ strain and the subsequent substitution of the original bA+ strains by those phenotypic bA+ variants that had been found to yield the highest titres of group A bacteriocins.

On the other hand, the control indicator strains no. 1*i*, 9*i*, 21*i* and 16*i* showed quantitatively varying susceptibility to group A bacteriocins. This latter phenomenon has since been observed repeatedly, i.e., after the control *i*-strains have undergone 10 to 12 monthly passages on BHIA slants. Again, single colonies were examined, and the control *i*-strains were substituted by those *i*-strain variants, that had proved most susceptible to group A bacteriocins of subgroups I and II. It should be emphasised that all bA+ and *i*-strains of *S. marcescens* were originally derived from single colonies in the Clinical Microbiology Laboratory, North Carolina Baptist Hospital, Winston-Salem, North Carolina, U.S.A.

Very recently, we observed that colonial variants of *S. marcescens* strain no. 874–57, serotype O14:H12, a gift from Miss B. R. Davis, varied in susceptibility to two of our 10 selected group A bacteriocins. "Medium"-sized colonies were of bacteriocin type 18, as were "small"-sized colonies; "pinpoint"-sized colonies, however, were of bacteriocin type 4, i.e., these latter variants apparently lacked the receptors for bacteriocins nos. 31 and 33 as determined with the bacteriocin sensitivity typing procedure.

To compound matters further, potent group A bacteriocin cell lysates had previously been found to abolish the immunity of bA+ cells against autologous bacteriocins (Traub, 1972d).

2. *Phenotypic variation of clinical isolates from patients with chronic infections*

An additional pitfall of susceptibility to selected group A bacteriocins alluded to previously (Traub and Kleber, 1974a), is that of phenotypic variation of a nosocomially significant strain of *S. marcescens* of bacteriocin type 18 in patients with infections of long duration, i.e., in excess of 3 months. For example, patient H. K., a 16-year-old male, who developed tibial osteomyelitis after compound fracture, yielded a strain of *S. marcescens*, of bacteriocin type 18 and remained unchanged for the following 2·5 months (isolates no. SE 78, 86, and 102). During the subsequent 4-months, two isolates of bacteriocin type 9 (isolates nos. SE 114 and 140) were recovered. After 7 months, cells of bacteriocin type 4 (isolate no. SE 143–a) that were susceptible to Chloramphenicol, nalidixic acid, Trimethoprim, and Amikacin, and cells of bacteriocin type 18 (isolate no. SE 143–b), resistant to all antimicrobial drugs examined except Amikacin, were recovered from the same drainage specimen. Subcultures of isolate no. SE 143–a repeatedly yielded colonies of bacteriocin type 18. During the following 2·5-month-period, all subsequent isolates of *S. marcescens* (isolates no. SE 161, 166–a and 166–b, 168, 170, and 171) belonged to bacteriocin type 18.

From patient G.S., a 54-year-old female with a post-operatively infected hip endoprosthesis, the multiple-drug-resistant strain of *S. marcescens*, bacteriocin type 18, was repeatedly isolated from the fistular drainage during the first two weeks of infection (isolates nos. SE 154, 156, 157, 158, 159, and 160). A year later, the infectious process still yielded *S. marcescens*, which was of bacteriocin type 9 (isolates nos. SE 253, 270–a, and 284–b) and/or of bacteriocin type 18 (isolates no. SE 254, 270–b, 284–a, and 298). As before, the isolates were resistant against all chemotherapeutic agents except Amikacin.

Similarly, Bergan (1973) noted changes in phage type and serogroup among serial cultures of *Pseudomonas aeruginosa* from the same infectious lesion in patients. In addition, Bergan and Midtvedt (1975) recently demonstrated O-antigen and phage type changes, i.e., lysogenic conversion, in the non-lysogenic *P. aeruginosa* strain IC after lysogenisation with phage F 116, both *in vitro* and in mono-contaminated gnotobiotic rats. Whether or not this latter observation might apply to *S. marcescens*, including possible lysogenic conversion by defective prophages, is unknown at present.

D. Addendum: Phage typing

Recently, Hamilton and Brown (1972) isolated 34 *Serratia* bacteriophages from sewage water. The bacteriophages proved stable in nutrient broth at 4°C for 1 year. The host range of the 34 phages against a selected number of ATCC strains of *Enterobacteriaceae* was limited to *E. coli* K12, *Shigella dysenteriae*, *Shigella flexneri*, *Arizona hinshawii*, in addition to, *S. marescens*. These 34 phages were employed to type 185 of 204 isolates (90·6%) of *S. marcescens*. The typable isolates comprised 23 phage groups representing 71 distinct types. A correlation was noted between inositol fermentation and phage type, as well as between Carbenicillin susceptibility and phage type. However, there was no relationship between sources of the isolates and bacteriophage types. To date, this method had not been utilised for epidemiological investigations with respect to outbreaks of nosocomial infection due to *S. marcescens*.

IV. CONCLUSIONS

As is well known, all currently available, biological methods of bacteriocin typing, including those for *S. marcescens*, have not been standardised; consequently, these methods are fraught with a number of inherent variables, such as phenotypic variation of bacteriocinogenic and indicator strains as well as of repeatedly isolated specific types from chronically infected lesions, and lysogenic conversion by phages. In addition, outer envelope alterations, such as mutations affecting LPS– and proteinaceous receptors and the presence of surface K-antigens might possibly distort typing results. The majority of the cellular surface receptors for the 10 selected group A bacteriocins and those few receptors which seem to reside within the LPS– moiety of the outer membrane of certain strains of *S. marcescens* require further biochemical investigation. Very little is known about surface antigens (e.g., K– antigens) of *S. marcescens* (Edwards and Ewing, 1972). Finally, far too little is known about the mode of action of group A bacteriocins of *S. marcescens*, i.e., the precise biochemical target on which these bacteriocins act.

Ideally, one would wish to employ a combination of methods for the purpose of characterisation of the epidemiological behaviour of nosocomially significant strains of *S. marcescens*, preferably in a manner similar to the very painstaking investigations of Bergan (1973) with respect to *P. aeruginosa*. However, such combined typing procedures, i.e., sero-, phage- and bacteriocin-typing, would clearly overburden the routine bacteriology laboratory. Therefore, it is recommended that—for the time being—bacteriocin typing of clinical isolates of *S. marcescens* be performed only by central reference laboratories, possibly as a supplementary procedure to serotyping.

V. ACKNOWLEDGMENTS

It is a pleasure to acknowledge the expert technical assistance of Miss Ella A. Raymond, Winston-Salem, N.C., U.S.A., and of Miss Ingrid Kleber, Erlangen, Germany. Furthermore, I wish to thank Miss Betty R. Davis, Enterobacteriology Unit, Centre for Disease Control, Atlanta, Georgia, U.S.A., for kindly serotyping our bacteriocinogenic and indicator strains of *S. marcescens*, and for recently sending us the currently used O- and H-antigen reference strains of *S. marcescens*.

REFERENCES

Amati, P., and Ozeki, H. (1962). *VIII. Int. Congr. Microbiology*, A 5.1, p. 26.
Bauer, A. W., Kirby, W. M. M., Sherris, J. C., and Turck, M. (1966). *Am. J. Clinical Pathol.*, **45**, 493–496.
Bergan, T. (1973). *Acta path. microbiol. scand.*, Section B, **81**, 91–101.
Bergan, T., and Midtvedt, T. (1975). *Acta path. microbiol. scand.*, Section B, **83**, 1–9.
Bradley, D. E. (1967). *Bacteriol. Reviews*, **31**, 230–314.
Clayton, E., and von Graevenitz, A. (1966). *J. Am. Med. Assoc.*, **197**, 1059–1064.
Davis, J. T., Foltz, E., and Blakemore, W. S. (1970). *J. Am. Med. Assoc.*, **214**, 2190–2192.
Edwards, P. R., and Ewing, W. H. (1972). "Identification of *Enterobacteriaceae*", 3rd edn, pp. 308–317. Burgess Publishing Co., Minneapolis.
Eichenlaub, R., and Winkler, U. (1974). *J. gen. Microbiol.*, **83**, 83–94.
Ewing, W. H., Johnson, J. G., and Davis, B. R. (1962). "The occurrence of *Serratia marcescens* in nosocomial infections". Center for Disease Control, Atlanta, Georgia.
Farmer, J. J., III (1970a). *Appl. Microbiol.*, **20**, 517.
Farmer, J. J., III (1970b). *Lancet*, **2**, 96.
Farmer, J. J., III (1972a). *Appl. Microbiol.*, **23**, 218–225.
Farmer, J. J., III (1972b). *Appl. Microbiol.*, **23**, 226–231.
Foulds, J. (1971). *J. Bacteriol.*, **107**, 833–839.
Foulds, J. (1972). *J. Bacteriol.*, **110**, 1001–1009.
Foulds, J., and Shemin, D. (1969a). *J. Bacteriol.*, **99**, 655–660.
Foulds, J., and Shemin, D. (1969b). *J. Bacteriol.*, **99**, 661–666.
Fuller, A. T., and Horton, J. M. (1950). *J. gen. Microbiol.*, **4**, 417–433.
Hamilton, R. L., and Brown, W. J. (1972). *Appl. Microbiol.*, **24**, 899–906.
Hamon, Y., and Peron, Y. (1961). *Ann. Inst. Pasteur*, **100**, 818–821.
Mandel, M., and Mohn, F. A. (1962a). *Microbial Genetics Bull.*, **18**, 15.
Mandel, M., and Mohn, F. A. (1962b). *VIII. Int. Congr. Microbiol.*, A 5.3, p. 26.
Mayr-Harting, A., Hedges, A. J., and Berkeley, R. C. (1972). Methods for studying bacteriocins. *In* "Methods in Microbiology" (Eds. J. R. Norris, and D. W. Ribbons), vol. 7A, pp. 315–422. Academic Press, New York and London.
Pillich, J., Hradnecna, Z., and Kocur, M. (1964). *J. appl. Bacteriol.*, **27**, 65–68.

Prinsloo, H. E. (1966). *J. gen. Microbiol.*, **45**, 205–212.

Prinsloo, H. E., Mare, I. J., and Coetzee, J. N. (1965). *Nature, Lond.*, **206**, 1055.

Sedlak, J., Dlabac, V., and Motlikova, M. (1965). *J. Hyg. Epid. Microbiol. Immun.*, *Prague*, **9**, 45.

Sedlak, J. (1968). *Serratia. In "Enterobacteriaceae-*Infektionen" (Eds. J. Sedlak, and H. Rische), 2. Aufl., pp. 607–611. Edition Leipzig.

Traub, W. H. (1972a). *Appl. Microbiol.*, **23**, 979–981.

Traub, W. H. (1972b). *Appl. Microbiol.*, **23**, 982–985.

Traub, W. H. (1972c). *Zbl. Bakt. Hyg., I. Abt. Orig.*, **A222**, 232–244.

Traub, W. H. (1972d). *Experientia*, **28**, 986–988.

Traub, W. H., Acker, G., and Kleber, I. (1974). *Zbl. Bakt. Hyg., I. Abt. Orig.*, **A229**, 383–390.

Traub, W. H., and Kleber, I. (1973). *Zbl. Bakt. Hyg., I. Abt. Orig.*, **A225**, 296–304.

Traub, W. H., and Kleber, I. (1974a). *Zbl. Bakt. Hyg., I. Abt. Orig.*, **A229**, 372–382.

Traub, W. H., and Kleber, I. (1974b). *Experientia*, **30**, 494–495.

Traub, W. H., and Kleber, I. (1974c). *Zbl. Bakt. Hyg., I. Abt. Orig.*, **A229**, 482–489.

Traub, W. H., and Kleber, I. (1974d). *Zbl. Bakt. Hyg., I. Abt. Orig.*, **A229**, 490–502.

Traub, W. H., Kleber, I., and Schaber, I. (1973). *Nature, Lond.*, **245**, 144–145.

Traub, W. H., and Raymond, E. A. (1971). *Appl. Microbiol.*, **22**, 1058–1063.

Traub, W. H., Raymond, E. A., and Startsman, T. S. (1971). *Appl. Microbiol.*, **21**, 837–840.

Wilfert, J. N., Barrett, F. F., Ewing, W. H., Finland, M., and Kass, E. H. (1970). *Appl. Microbiol.*, **19**, 345–352.

Wilkowske, C. J., Washington, J. A., II, Martin, W. J., and Ritts, R. E., Jr (1970). *J. Amer. Med. Assoc.*, **214**, 2157–2162.

The References were compiled in November 1975.

CHAPTER VII

Phage Typing of *Proteus*

Tom Bergan

Department of Microbiology, Institute of Pharmacy, University of Oslo

I. TYPING METHODS FOR *PROTEUS*

In recent years, the possibility that *Proteus* species may cause nosocomial infections has been demonstrated (Chouteau *et al.*, 1974; McNamara *et al.*, 1967; Schmidt and Jeffries, 1974). Infection may be due to cross-infection. The presence of *Proteus* in the normal faecal and perineal flora makes endogenous infections a common occurrence, particularly with *P. mirabilis*, but cross-infections also occur. The indole positive proteae represent a difficult therapeutic problem on account of their multiple resistance to antibacterials. Antibiotic resistance transfer factors, transmissible by conjugation (R-factors) are common in the indole negative *P. mirabilis* which thereby may acquire a resistance resembling that of the other species (Mitsuhashi *et al.*, 1967; Tracy and Thomson, 1972).

To elucidate transmission and spreading, typing methods are necessary. Typing may be performed by serology (Belyavin, 1951; deLouvois, 1969; Penner *et al.*, 1976; Perch, 1948; Rustigian and Stuart, 1945) or by biochemical-cultural characterisation (Belyavin, 1951; Tracy and Thomson, 1972). The Dienes reaction may be utilised in *P. mirabilis* (Adler *et al.*, 1971, deLouvois, 1969; Skirrow, 1969; Story, 1954). In a number of cases, attempts have been made to use the antibiotic resistance pattern as identification marker (Adler *et al.*, 1971; Eisenberg *et al.*, 1958). *Proteus* produces bacteriophage-tail like, proteinaceous bacteriocins (Coetzee and Coetzee, 1968). Bacteriocin typing has been employed and was moderately successful in *P. mirabilis* where 60% were identified as bacteriocin producers and in *P. vulgaris* where only 10% were producers against an indicator set of 18 strains of *P. mirabilis* (Cradock-Watson, 1965). Vieu (1960) found 69% bacteriocinogenicity in *P. morganii* against indicators of that species. Coetzee and Coetzee (1968) found 60% producers among *P. vulgaris* as tested against a large number of *P. mirabilis* indicator strains. Bacteriocin typing sets have been developed (Al-Jumaili, 1975a,b; George, 1974; Senior, 1977).

Bacteriophage typing has been presented from a number of workers, and will be reviewed in this Chapter.

II. CHARACTERISTICS OF *PROTEUS* SPECIES

In order to ensure that the typing sets are applied on the same taxons at different centres, the major characteristics of the *Proteus* species will be reviewed. Although classification in the family *Enterobacteriaceae* is otherwise complicated by the considerable metabolic variation encountered among individual isolates, the situation is fairly simple for the genus *Proteus*.

The genus shares the general properties of the enterobacteria in being Gram-negative, asporogenous, rod-shaped, aerobic facultatively anaerobic bacteria.

Glucose is fermented (acid formation with or without a supply of oxygen), nitrate is reduced to nitrite and the indophenol oxidase test is negative. The bacteria grow readily on ordinary cultivating media.

The genus *Proteus* is distinguished from all other enterobacteria by having a strong phenylalanine deaminase reaction.† Other useful characteristics are a negative lactose reaction, motility (by peritrichous flagella-like all motile enterobacteria), and urease production. Urease, though, is negative in *P. inconstans*. Species differentiation follows the reactions of Table I. Here, the genetic variability of the various key properties are indicated as percentage positive reactions in the fashion elaborated by Ewing (1973) and Edwards and Ewing (1972).

† *Erwinia herbicola, Erw. cypripedii*, and *Erw. rhapontici* exhibit weak phenylalanine reactions (Buchanan and Gibbons, 1973). The *Enterobacter agglomerans* entity of Ewing and Fife (1972), synonymous with *Erw. herbicola* (Buchanan and Gibbons, 1974), is reported to have (a weak) phenylalanine reaction in 16% of cases (Ewing, 1973), but it is so biotypically heterogenous that its validity as a single species is in doubt (F. Grimont and P. Grimont, 1978, personal communication).

TABLE I

Differentiation of the genus *Proteus*†

	P. mirabilis	P. vulgaris	P. morganii	P. rettgeri	P. inconstans
Phenylalanin-deaminase	+(100)	+(100)	+(97)	+(100)	+(96)
Indole	+(1·9)	+(98)	+(100)	+(96)	+(99)
Citrate (Simmons)	+(59)	d(11)	−(0)	+(96)	+(97)
Ornithine-decarboxylase	+(98)	−(0)	+(96)	−(0)	−(0)
Urease	d(88)	+(95)	+(97)	+(100)	−(0)
Maltose	−(0·9)	+(96)	−‡	−‡	−‡
Mannitol	−‡	−‡	−(0)	d(89)	−(7)
Mannose	−(0)	−(0)	+(100)	+(100)	
Saccharose	d(19)	+(95)	−(1)	d(13)	−(8)
Gelatin (22°C)	+(92)	+(92)	−(0)	−(0)	−(0)
H₂S	+(95)	+(95)	−(0)	−(0)	−(0)
Mol% (G+C)	39·3	39·3	50·0	39·0	41·5

† Adapted from Ewing (1973) and Edwards and Ewing (1972). Numbers in parentheses indicate percentage of positive reactions. + =90% positive within two days; − =less than 10% positive within two days; d=different reactions; +,− or positive after three days or more.
‡ Buchanan and Gibbons (1974).

The terminology in Table I follows the eighth edition of *Bergey's Manual* (Buchanan and Gibbons, 1974). In some literature on phage typing, the species *P. mirabilis* and *P. vulgaris*, which produce H₂S, liquefy gelatin, and may exhibit swarming on the agar surface, have been referred to as *P. hauserii* (Chouteau et al., 1974; Coetzee, 1963; Pavlatou et al., 1965; Vieu, 1958, 1973). *P. inconstans* has been referred to as *Providencia stuartii* or *Prov. alcalifaciens* (Table II) which are now combined in one nomen-species in recognition of the small difference between the varieties.

TABLE II

Differentiation of biotypes of *Proteus inconstans*

	alcalifaciens	stuartii
Gas from glucose	d(85)	−(0)
Adonitol	+(94)	−(4)
Cellobiose	−(2)	d(10)
Inositol	−(0·6)	+(97)
Mannitol	−(2)	d(12)

Legend, see Table I.

III. PROPERTIES OF *PROTEUS* PHAGES

The existence of *Proteus* phages has been demonstrated by Dettling (1932), Fejgin (1924), Hartl (1956), d'Herelle (1926), Otto and Munter (1929) and Russ-Münzer and Kindermann (1933). Mostly, they studied phage–host interactions, plaque morphology, bacterial L-form development, and production of phage. Calcium is not needed in appreciable quantities for attachment of phage to bacterial receptors (Coetzee, 1958). The observation that attack by phages could interfere with bacterial motility (Brandis and Schwarzrock, 1956; Dettling, 1932), has implications of basic significance.

Proteus mirabilis surviving phage action may even temporarily or permanently lose the ability to swarm on an agar surface, even though no free phages may be detected, after multiple (15–60) passages (Brandis and Schwarzrock, 1956). *Proteus* bacteriophages may be isolated from sewage (Brandis and Schwarzrock, 1956; Coetzee, 1963), or from lysogenic strains (Coetzee, 1963; Fejgin, 1924; Vieu, 1960, 1963, 1973). All species of *Proteus* may be lysogenic. Some 50% appear to be lysogenic (Coetzee and Sacks, 1959). Vieu (1960) found 69% lysogenic among strains of *P. morganii*. He (1963) found a similar level of lysogenicity in *P. inconstans*, but only 3% among *P. rettgeri*.

The morphology of the phages has been studied by Ruska (1943) and Coetzee (1958). They have all exhibited the same morphology with a head and a tail. The head has a diameter of 60–110 μm, and the tail a length of 50–100 μm and a width of 15 μm. Heat stability varies. Whereas some are readily inactivated by 56°C for 30 min, others survive this treatment (Brandis and Schwarzrock, 1956; Coetzee, 1958).

Serologically, phages against *P. mirabilis* and *P. vulgaris* may be divided into seven groups (Coetzee, 1958). *Proteus* bacteriophages in several studies have lacked activity against bacteria of other enterobacterial genera (Brandis and Schwarzrock, 1956; Coetzee, 1958). Many of the phages are so restricted that they only attack their own host propagating strain. Two phages from *P. rettgeri* attacked *Escherichia coli* (Coetzee, 1963). Many phages from *P. mirabilis* attack *P. vulgaris* and vice versa (Coetzee, 1963). Phages from *P. mirabilis* and *P. vulgaris* attacked *P. morganii*, *P. rettgeri*, and *P. inconstans* (Coetzee, 1963).

IV. PHAGE TYPING

A. Basic studies

The fact that bacteriophages have different host ranges and, consequently, can be used for the differentiation of *Proteus*, is apparent from the works of Coetzee (1963), Cradock-Watson (1965), Taubeneck (1961), and Vieu (1960). Phages from *P. rettgeri* appear only to attack that species, whereas phages from *P. mirabilis*, *P. vulgaris*, and *P. morganii* may cross the species barriers of each other, but do not attack *P. rettgeri* or *P. inconstans*. Several alternative typing sets have been published, none of which is based on other work or compare different sets. The practice of *Proteus* phage typing is restricted to a few centres.

B. Methodology

The methods used have mostly been cursorily described. Schmidt and Jeffries (1974) were detailed and explicit in the presentation of their set. Since the methods would be applicable also to the other typing sets, and methodological detail is of utmost importance for satisfactory reproducibility, the technique of Schmidt and Jeffries will be described. Their procedures should also apply to the species *P. rettgeri* and *P. inconstans* even though their typing set was specifically directed towards *P. mirabilis*, *P. vulgaris*, and *P. morganii*.

Media. Plates for typing consist of:

Peptone (Difco)	0·5%
Beef Extract (Difco)	0·3%
Agar (Difco)	1·5%
CaCl₂	0·002M

Soft agar is the same, except that the agar concentration is 0·75%. NaCl is purposely omitted to avoid swarming of *P. mirabilis* and *P. vulgaris*. Four per cent agar or incubation at 42–44°C have also been tried to avoid swarming (Vieu, 1973; Vieu and Ducrest, 1961).

Phage propagation. For production of phage, a modification of the soft agar overlay method is used. The bacteria are 4 h log phase cultures ($3·5 \times 10^8$ colony forming units). Two drops of the bacterial culture and 0·1 ml of the appropriate phage dilution are added to 4 ml melted soft agar which is evenly spread on 9 cm plates. The phage dilution corresponds to the routine test dilution (RTD) (that dilution of phage which produces nearly confluent lysis). A modification consists of overlaying the soft agar with 5 ml nutrient broth. Incubation is carried out at room temperature for 4–6 h when the broth is removed and centrifuged. The supernatant after addition of a few drops of chloroform serves as phage stock.

France and Markham (1968) obtained good results when propagating by the traditional soft agar overlay method of Adams (1959) and the surface lawn method of Blair and Carr (1953). Harvesting was performed by freezing and thawing. Lytic spectra of the phages should be checked by the method of Blair and Williams (1961) before each new phage preparation is accepted.

Bacterial strain preparation. Strains to be typed have been prepared by different methods. Schmidt and Jeffries (1974) used nutrient broth cultures grown for 16 h at 35 °C and inoculated on to the typing plates by swabbing. France and Markham (1968) flooded plates with 3 h cultures incubated at 37 °C or a one-tenth dilution of a suspension prepared from an overnight slant culture (harvested into 10 ml broth).

Phage typing procedure. Preparation of the phage suspension is performed by first determining the proper dilution to be used, the routine test dilution (RTD). This is done by spotting calibrated drops (corresponding to the volume used in typing) of ten-fold dilutions of phage stock on to freshly inoculated plates of the host propagating strain (bacterial density sufficient to produce lawn growth). The dilution of phage which produces nearly confluent lysis corresponds to the RTD.

The RTD suspensions of the typing phages are applied drop-wise to well-dried plates of freshly inoculated (to produce lawn growth) plates of the strains to be typed. Care should be taken that the bacteria do not grow at room temperature before the phage suspension is applied. It is best to prepare at the most 100 plates at a time, alternatively the plates may be stored briefly at cool temperatures before applying the phage. For distribution of the phages, several alternatives are available—and technically suitable.

After spotting, the plates must be dried before incubation. Incubation takes place for 6–8 h at 35 °C (Schmidt and Jeffries, 1974) or 4–6 h at 37 °C (France and Markham, 1968). Vieu (1973) incubates at 37 °C, and reads after 6 and 18 h. Hickman and Farmer (1976) preferred to incubate 18–24 h before reading.

The reactions may be read and recorded as described by Blair and Williams (1961).

Strains which have not reacted with the phages at RTD, are retested at 1000 × RTD (Schmidt and Jeffries; France and Markham).

C. Alternative phage typing sets

1. *Typing set for* P. mirabilis, P. vulgaris *and* P. morganii

Set of Schmidt and Jeffries (1974). Schmidt and Jeffries (1974) have developed a set of 22 phages for the three species *P. mirabilis*, *P. vulgaris* and *P. morganii*. For *P. mirabilis* and *P. vulgaris* the 15 active phages were:

13/3a, 36a, 36/34, Fr2, 29/34a, 29/36b, 34a, 34/36, Fr5, 4b, 4a, 31/39, 39, 21b, and 21c. For *P. morganii* the active phages were: 16/21c, 16/34, 16c, 21b, 21c, 32a, and 32b.

The percentage of strains lysed by each phage appears in Table III. The phages 13/3a and Fr2 lysed isolates of both *P. mirabilis* and *P. vulgaris*. The phages 21b and 21c were the only ones which lysed both *P. morganii* and *P. mirabilis*.

The remainder lysed only *P. mirabilis* strains. The results have been interpreted by giving each phage pattern a distinct code (Tables III and IV). Thirty patterns have been published.

The typing set has successfully typed 86% of *P. mirabilis* and *P. vulgaris*, and 68% of *P. morganii* isolates. Overall, 84% of the isolates were lysed by at least one phage at RTD or 1000 × RTD. Among these, 13% were typed only at 1000 × RTD.

The reproducibility of the reactions has been satisfactory in that 50 strains yielded the same reactions upon retyping.

The phage suspensions have been stable at 4°C for months.

Set of France and Markham (1968). France and Markham (1968) developed a set of 20 phages: 1, 2, 4, 5, 6, 7, 8, 9, 11, 14, 15, 16, 18, 19, 20, 21, 22, 23, 24, 25. The lytic activity of each phage appears in Table V.

These typed 82% of the isolates. Six per cent were only typed at 1000 × RTD. Several phages lysed both *P. vulgaris* and *P. mirabilis*, whereas *P. morganii* was isolated in this regard.

The lytic patterns fell into ten main groups (Table VI).

Reproducibility of the phage patterns was elucidated by repeating the typing after six weeks on 30 strains. Changes were reported as not significant. In repeat specimens from 55 patients, only three gave a different phage pattern, but under such conditions, of course, superinfection might have occurred to explain the pattern change.

2. *Typing set for* P. mirabilis, P. vulgaris, P. morganii, *and* P. rettgeri

One phage typing set for all four classic *Proteus* species was developed by Pavlatou *et al.* (1965). It consisted of 15 phages designated by the numbers 1–15. Seven of them were derived from lysogenic bacteria (phages nos. 2, 3, 4, 5, 6, 14, and 15) and the rest from either sewage or river water (nos. 1, 7, 8, 9, 10, 11, 12, and 13). The phages were spotted as undiluted stock suspensions (Vieu, 1973).

This set typed 71% of the *P. mirabilis* and *P. vulgaris* strains which were distributed among 14 lysotypes (Table VII). The frequency of typable strains was less among *P. morganii* and *P. rettgeri* of which respectively 43 and 31% were typed.

10

TABLE III

Lysis patterns of *P. mirabilis* and *P. vulgaris* (162 isolates) in phage set of Schmidt and Jeffries (1968)

Group no.	Lysis by phage															No. in group	% in group
	13/3a	36a	36/34	Fr2	29/34a	29/36b	34a	34/36	Fr5	4b	4a	31/39	39	21b	21c		
1	+															56	34·6
5	+			+												11	6·8
8	+	+	+													10	6·1
9	+	+	+	+												2	1·2
7	+	+	+		+	+	+	+								24	14·8
6	+								+							1	0·25
4	+									+						1	0·6
3	+									+	+					1	0·6
2	+										+					2	1·2
10	+											+	+			9	5·5
11	+													+	+	2	1·2
15				+												1	0·6
17		+	+													2	1·2
16		+	+	+												2	1·2
20		+	+		+	+	+	+								7	4·3
18		+	+		+	+			+							1	0·6
13										+						1	0·6
12										+	+					1	0·6
14										+	+					6	3·6
19	+											+	+			23	14·2
21	+															1	0·25
No. strains lysed by phages	120	48	48	16	32	32	31	31	2	10	9	32	32	2	2		
% lysed	74	29	29	10	19	19	19	19	1	6	5	19	19	1	1		

TABLE IV

Lysis patterns of P. *morganii* (12 isolates) in phage set of Schmidt and Jeffries (1968)

Group no.	$\frac{16}{21c}$	$\frac{16}{34}$	16c	21b	21c	32a	32b	No. in group
				Lysis of phage				
1	+		+	+	+	+	+	1
2	+			+	+	+	+	2
3			+	+	+	+	+	1
4			+	+	+			1
5		+	+	+				1
6					+			1
7							+	3
8					+		+	1
9						+	+	1

TABLE V

**Lytic activity of phages in set of France and Markham (1968)
(on 170 bacterial strains)**

Phage	Lytic activity (% of strains)
1	5·3
2	38·8
4	63·5
5	2·9
6	34·7
7	70·7
8	0·59
9	29·4
11	68·9
14	69·5
15	5·9
16	2·4
18	12·9
19	82·4
20	74·8
21	7·1
22	2·9
23	78·3
24	75·9
25	77·1

TABLE VI

Phage patterns of bacteriophage typing groups and their frequency according to the typing set of France and Markham (1968)

Typing group	Phage pattern	Frequency (%)†
I	4/7/11/14/19/20/23/24/25	30
II	2/6/9/21	10
III	18/23/24	5
IV	1/15/16	1
V	19	5
VI	19/20/23/25	3
VII	2/6/15/22	2
VIII	1/7/11/14/20	1
I–II	2/4/6/7/9/11/14/19/20/21/23/24/25	20
I–II	4/7/11/14/18/19/20/23/24/25	3
Non-typable		20

† Pertains only to strains of *P. mirabilis* and *P. vulgaris*. When *P. rettgeri* strains included, 18% non-typable.

TABLE VII

Phage pattern of lysotypes among *P. mirabilis* and *P. vulgaris* with the set of Pavlatou et al. (1965)

Lysotype	Phages with mandatory reactions†		Frequency‡ (%)
	Lysis	No lysis	
I	1		8·8
II	2, 5, 6	1	12·5
III	3	1, 2, 7, 9, 10	2·6
IV	4	1, 2, 3, 7	2·6
V	5, 6, 8, 10, 11–15	4, 7, 9	0·5
VI	6, 10, 11	1–5	1·6
VII	7, 8	1–6	10·9
VIII	8, 12, 13	1–7	3·6
IX	9	1–8	9·3
X	10	1–9, 12, 15	1·0
XI	11	1–10	6·7
XII	12	1–11	9·3
XIII	13	1–12, 14	1·6
XIV	14, 15	1–13	0·5
Non-typable			28·5

† Other phages may show either lysis or no reaction.
‡ Based on reactions with 187 strains of *P. mirabilis* and six of *P. vulgaris*.

3. *Typing set for* P. mirabilis *and* P. vulgaris

Vieu (1958) and Vieu and Ducrest (1961) have developed a set of 12 serologically unrelated phages from sewage for the typing of *P. mirabilis* and *P. vulgaris*. The phages are designated by numbers 1–12. The phages were spotted as undiluted stock solutions. The patterns are interpreted as lysotypes according to Table VIII. This set typed 82% of the strains.

TABLE VIII

Lysotypes of *P. mirabilis* and *P. vulgaris* by set of Vieu (1958)

Lysotype	Phages of lytic reactions	Frequency (%)
I	1, 2, 3	13·3
II	4, 5	3·3
III	6, 7	5·6
IV	6, 8	1·1
V	6, 7, 8	37·8
VI	7, 8	3·3
VII	9, 10	6·7
VIII	10	2·2
IX	11	4·4
X	12	4·4
Non-typable		17·8

An extension of the set to 27 phages typed 85% of the strains differentiable into 14 types. In this set, lysotype 1 is characterised by reactions against nine serologically related phages.

A comparison of serology and lysotyping, showed that strains of individual serogroups could be differentiated by their susceptibility to bacteriophages (Vieu and Capponi, 1965).

4. *Typing set for* P. mirabilis

A phage typing set designed particularly for *P. mirabilis* has been developed by Hickman and Farmer (1976). The set consists of 23 bacteriophages designated R1–R23. The lytic activity of the phages appears in Table IX. The phages were isolated from sewage. The most suitable phages were selected from a total of 73 by computer processing. Also phages for other *Proteus* species were considered, but they had little activity and were not included in the final set.

The typing procedure is roughly like that presented in this Chapter, with the modification that 18–24 h elapsed before reading.

TABLE IX

Characteristics of typing phages in set of Hickman and Farmer (1976)

New phage designation	Old phage designation	*P. mirabilis* host strain	% of *P. mirabilis* strains lysed
R1	1002	1002	3·8
R2	1003	1003	49·7
R3	1006	1006	43·4
R4	1013	1013	4·4
R5	1017	1017	38·8
R6	1022	1022	36·3
R7	1029	1029	7·2
R8	11A	11A	15·7
R9	27	27	5·4
R10	29	29	9·8
R11	41	41	5·1
R12	48B	48B	13·8
R13	73	73	10·3
R14	75	75	35·5
R15	81D	81D	36·1
R16	82	82	8·4
R17	134	134	19·3
R18	135	135	23·2
R19	173	173	51·5
R20	193B	193B	17·0
R21	195	195	44·4
R22	196	196	28·3
R23	275	275	17·8

The lysis patterns have been converted to the mnemonic system of Farmer (1970) (Table X).

The set typed 94% (of 200 strains) which is the highest typability reported for *P. mirabilis* by any set developed so far. The set differentiated 113 lysotypes which is a larger number of types than obtained by the other sets, although only 24 types were represented by more than one isolate.

The set was compared in parallel by Hickman and Farmer with the typing set of Schmidt and Jeffries (1974). The latter set performed less satisfactorily in that only 20% of the strains were typable. The reason for this low level, however, appears to be that a different typing medium was used, trypticase soy agar for which the Hickman and Farmer set was designed. This medium inhibited the reactions of several phages in the Schmidt and Jeffries set, which was developed on another medium.

Other sets for *P. mirabilis* have been developed by Izdebska–Szymona *et al.* (1971). This typed 80% (of 305 isolates). Popovici and Ghioni (1962) typed 48% (of 100 strains).

TABLE X

Method of reporting bacteriophage patterns employed for set of Hickman and Farmer (1976)

Phage reactions†	Code
+ + +	1
+ + −	2
+ − +	3
− + +	4
+ − −	5
− + −	6
− − +	7
− − −	8

† The reactions are divided into triplets and converted to numbers from the table. If the number of reactions is not evenly divisible by three, the following symbols are used to code the remaining reactions: $+ + = A$; $+ − = B$; $− + = C$; $− − = D$; $+ = E$; $− = F$. Example: lysis pattern $+ − +(=3) + + +(=1) − + −$ $(=6) − + −(=6) − + −(=6) + + +(=1) − − −$ $(=8) − (=F)$ would be coded: 316 661 8F.

5. *Typing set for* P. inconstans

To solve a local nosocomial situation, Chouteau *et al.* (1974) developed a phage typing set for *P. inconstans*. The elaborated set consisted of 24 phages. These were used at RTD. Thirty-four lysotypes were distinguished. Thirteen per cent were untypable. The most frequent types and their frequency are listed below.

Lysotype	Frequency (%)
22	33
20	11
24	9
4	5
3 and 15	4
6, 23 and 28	3

V. PARTICULAR ASPECTS OF *PROTEUS* PHAGE TYPING

The phages derived from *P. mirabilis* and *P. vulgaris* may show small and hazy plaques on *P. morganii* (Schmidt and Jeffries, 1974).

No comparisons have been made between existing phage typing sets, and there is little evidence indicating preference of one over the others. The percentage of typable strains would be one criterion, but the figures derived from the locations where the sets were developed may not be comparable. Often, typing sets are developed on a limited number of strains from a circumscribed epidemiological situation. The typing phages may be selected to resolve differences in local, nosocomially associated, bacterial isolates. A comparable function on bacterial strains derived from widely different locations is not certain. On strains from other geographical areas than that of the original set, the typing system of Schmidt and Jeffries (1974), though, had the same portion of typable strains. Other useful criteria such as number of different lysotypes observed, distribution between types and length of pattern reactions are likewise difficult to assess. This is because the number of types is a function of both the variety of sources and the number of bacterial strains studied with each set. Most sets operate with one group comprising about one-third of the strains. Vieu (1958) has 46% in one group. Pavlatou et al. (1961) have a more satisfactory distribution. However, when the strains are isolated from one place within a short time span, sheer nosocomial association may influence the results in the very direction of a skewed frequency distribution.

Reproducibility appears good for the sets of Schmidt and Jeffries (1974) and Hickman and Farmer (1976). This criterion is more difficult to determine from some of the other published reports. The high level of lysogenicity (Vieu, 1973) implies that there may be instability of type reactions.

Nothing is known regarding the relationship between phage types and serological or bacteriocin typing. It may be mentioned, though, that typing by the Dienes phenomenon, antibiotic sensitivity, or biochemical properties yield unsatisfactory results (Burke et al., 1971; France and Markham, 1968).

The phage typing sets so far published have covered P. morganii and P. rettgeri less adequately than the others. This requires future attention. The reason is associated with the more common isolation of P. mirabilis and P. vulgaris. The set for P. inconstans, which appears fully satisfactory, was devised in response to a nosocomial situation although that species is usually quite rare.

VI. CLINICAL ACHIEVEMENTS OF PHAGE TYPING

By typing, one may distinguish between endogenous and exogenous infections, identify the likely sources and, in a group of patients, differentiate between cross-contamination and isolated cases. Burke et al. (1970) using phage typing in addition to other methods could detect that a nurse carried two strains of *P. mirabilis*. One of these was identical to isolates from six infected babies with meningitis or septicaemia.

France and Markham (1968) demonstrated the superiority of phage typing compared to other methods. In two wards, the isolates had the same biotype and had related flagellar antigens as determined by the Dienes phenomenon. By phage typing, it was resolved that infections in the two wards were due to cross-infections as opposed to autoinfection. In *P. inconstans*, lysotyping (Chouteau *et al.*, 1974) has been instrumental in determining the clinical importance of R-factors. When the same patients are followed over a period of time, it has been found that acquisition of resistance against various antibiotics may occur even though the phage type remains unchanged.

In the hospital environment, antibiotics represent possible selective advantages for multiresistant strains. Lysotyping has been employed to show that successive acquisition of *P. inconstans* is not necessarily associated with several types of multiple drug resistance. Several phage types may also exist concomitantly.

The nosocomial characteristics, spreading routes, pathogenicity, and opportunistic properties of *P. inconstans* bear a marked resemblance to those of *Pseudomonas aeruginosa* (Chouteau *et al.*, 1974; Hickman and Farmer, 1976).

No correlation has been determined in published works between phage types and bacterial properties, pathogenicity or occurrence at predilected sites.

REFERENCES

Adams, M. H. (1959). "Bacteriophages". Interscience Publishers, Inc. New York.
Adler, J. L., Burke, J. P., Martin, D. F., and Finland, M. (1971). *Ann. intern. Med.*, **75**, 517–530.
Al-Jumaili, I. J. (1975a). *J. clin. Path.*, **28**, 784–787.
Al-Jumaili, I. J. (1975b). *J. clin. Path.*, **28**, 788–792.
Belyavin, G. (1951). *J. gen. Microbiol.*, **5**, 197–207.
Blair, J. E., and Carr, M. (1953). *J. infect. Dis.*, **93**, 1–13.
Blair, J. E., and Williams, R. E. O. (1961). *Bull. Wld Hlth Org.*, **24**, 771–784.
Brandis, H., and Schwarzrock, A. (1956). *Zbl. Bakt. I. Abt. Orig.*, **165**, 266–269.
Buchanan, R. E., and Gibbons, N. E. (1974). "Bergey's Manual of Determinative Bacteriology", 8th edn. Williams and Wilkins Company, Baltimore, Md.
Burke, J. P., Adler, J. L., and Finland, M. (1971). *Antimicrobial Ag. Chemother.*, 1970, pp. 328–331.
Chouteau, J., Vieu, J. F., and Brault, G. (1974). *Med. Malad. infect.*, **4**, 575–578.
Coetzee, J. N. (1958). *S. Afr. J. Lab. clin. Med.*, **4**, 147–157.
Coetzee, J. N. (1963). *J. gen. Microbiol.*, **31**, 219–229.
Coetzee, J. N., and Sacks, T. G. (1959). *Nature*, **184**, p. 1340.
Coetzee, H. L., deKlerk, H. C., and Coetzee, J. N. (1968). *J. gen. Virol.*, **2**, 29–36.
Cradock-Watson, J. E. (1965). *Zbl. Bakt. I. Abt. Orig.*, **196**, 385–388.

Dettling, H. (1932). *Arch. Hyg.*, **108**, 359–364.

Edwards, P. R., and Ewing, W. H. (1972). "Identification of *Enterobacteriaceae*", 3rd edn. Burgess Publishing Company, Minneapolis, Minn.

Ewing, W. H., and Fife, M. A. (1972). *Int. J. syst. Bact.*, **22**, 4–11.

Eisenberg, G. M., Weiss, W., and Flippin, H. F. (1958). *J. clin. Path.*, **30**, 20–24.

Ewing, W. H. (1973). "Differentiation of *Enterobacteriaceae* by Biochemical Reactions", revised edn, U.S. Department of Health, Education, and Welfare Publication No. (CDC) 74–8270, Atlanta, Ga.

Farmer, J. J. (1970). *Lancet*, **2**, p. 96.

Fejgin, B. (1924). *C. R. Soc. Biol., Paris*. **90**, 1106–1108.

France, D. R., and Markham, N. P. (1968). *J. clin. Path.*, **21**, 97–102.

George, R. H. (1974). *J. clin. Path.*, **28**, 25–28.

Hartl, W. (1956). *Schweiz. Z. Path. Bakt.*, **19**, 26–50.

d'Herelle, F. (1926). "Le Bacteriophage et son Comportement". Masson et Cie, Paris. Cit. Coetzee (1958).

Hickman, F. W., and Farmer, J. J. (1976). *J. clin. Microbiol.*, **3**, 350–358.

Izdebska-Szymona, K., Monczak, E., and Lemczak, B. (1971). *Expl. Med. Microbiol.*, **23**, 18–22.

deLouvois, J. (1969). *J. clin. Path.*, **22**, 263–268.

McNamara, M. M., Hill, M. C., Balows, A., and Tucker, E. B. (1967). *Ann. intern. Med.*, **66**, 480–488.

Mitsuhashi, S., Hashimoto, H., Egawa, R., Tanaka, T., and Nagai, Y. (1967). *J. Bact.*, **93**, 1242–1245.

Otto, R. and Munter, H. (1929). *In* "Handuch der pathologischen [sic] Mikroorganismen", Vol. 1, p. 361. Gustav Jena, Coetzee.

Pavlatou, M., Hassikou-Kaklamani, E., and Zantioti, M. (1965). *Ann. Inst. Pasteur*, **108**, 402–407.

Penner, J. L., Hinton, N. A., and Hennessy, J. H. (1976). *J. clin. Microbiol.*, **3**, 385–389.

Perch, B. (1948). *Acta path. microbiol. scand.*, **25**, 703–714.

Popovici, M. and Ghioni, E. (1962). *Arch. Roum. Pathol. Exp. Microbiol.*, **21**, 307–314.

Ruska, H. (1943). *Erg. Hyg.*, **25**, 437–498.

Russ-Münzer, A. and Kindermann, V. (1933). *Zbl. Bakt. I. Abt. Orig.*, **129**, 401–404.

Rustigan, R. and Stuart, C. A. (1945). *J. Bact.*, **49**, 419–436.

Schmidt, W. C., and Jeffries, C. D. (1974). *Appl. Microbiol.*, **27**, 47–53.

Senior, B. W. (1977). *J. med. Microbiol.*, **10**, 7–71.

Skirrow, M. B. (1969). *J. med. Microbiol.* **2**, 471–477.

Story, P. (1945). *J. Path. Bact.*, **68**, 55–62.

Taubeneck, U. (1961). *Zbl. Bakt. I. Abt Orig.*, **185**, 416–418.

Tracy, O., and Thomson, E. J. (1972). *J. clin. Path.*, **25**, 69–72.

Vieu, J.-F. (1958). *Zbl. Bakt. I. Abt. Orig.*, **171**, 612–615.

Vieu, J.-F. (1960). *Ann. Inst. Pasteur*, **98**, 769–772.

Vieu, J.-F. (1963). *C. R. Acad. Sci. Paris*, **256**, 4317–4319.

Vieu, J.-F. (1973). Lysotypie und andere spezielle epidemiologische Laboratoriumsmethoden" (Ed. H. Rische), pp. 369–375. VEB Gustav Fischer Verlag, Jena.

Vieu, J.-F., and Capponi, M. (1965). *Ann. Inst. Pasteur*, **108**. 103–106.

Vieu, J.-F., and Ducrest. P. (1961). *Zbl. Bakt. I. Abt. Orig.*, **182**, 49–56.

CHAPTER VIII

Serological Typing Methods of Leptospires

H. Dikken

Chief, WHO/FAO Leptospirosis Reference Laboratory, Royal Tropical Institute, Amsterdam, The Netherlands

AND E. Kmety

Institute of Epidemiology, Medical Faculty of the Komensky University, Bratislava, Czechoslovakia

I. INTRODUCTION

Leptospirosis is one of the most wide-spread zoonoses in the world today. As investigations have intensified during the past few decades, it has been recognised that leptospirosis is of increasing importance, to both medicine and veterinary medicine, especially in the subtropics and tropics.

The exact determination of which serotype causes an outbreak of the disease is a basic requirement for the control of leptospirosis and is also indispensable in many other aspects of scientific work with this group of organisms.

Years ago it was recognised that serotyping represents the most practical basis for strain differentiation (Borg-Peterson, 1938; Gispen and Schüffner, 1939; Wolff, 1953). The principle of serological differentiation has been formulated in a report by a group of experts of the "Taxonomic Sub-committee on *Leptospira*" (1954). However, the serological methods used in typing procedures such as the agglutination and cross-agglutinin absorption test, have not been completely standardised. These tests are performed in specialised laboratories according to the same principles, but with local differences. This Chapter presents the recommended typing procedures developed over the years. As representative examples, the typing methods used in the Amsterdam and Bratislava laboratories will be described.

II. TAXONOMY

Leptospires are members of the *Treponemataceae* family. Noguchi (1917) was the first to recognise leptospires as a separate genus and to differentiate them on a morphological basis from the genera *Borrelia* and *Treponema*.

A. Genus

The description of the genus *Leptospira* Noguchi 1917 (I.C.N.B. S.T.Lep., 1971) was reconsidered and amended in 1973 and 1978 by the Sub-committee on the Taxonomy of *Leptospira* (I.C.N.B. S.T.Lep., 1974). The suggested definition has not yet been published, but the draft reads as follows:

Flexible helicoidal organisms which are usually 6–20 μm in length and about 0·1 μm in diameter. The coils have an amplitude of about 0·1–0·15 μm and a wavelength (pitch) of about 0·5 μm. In liquid milieux one or both ends are usually hooked or bent, but either one or both ends may become straightened temporarily. In some cultures the majority or all of the cells regularly have straight ends which are not seen to become hooked. The organism is not visible by ordinary bright-field microscopy; it is observed clearly by dark-field illumination and phase-contrast microscopy. The cells are not easily visualised after staining by analine dyes, but they can be demonstrated by silver-deposition techniques and by other special staining methods (e.g. Giemsa). The structure of *Leptospira* consists of a helicoidal protoplasmic cylinder delineated by a cytoplasmic membrane-peptidoglycan complex. An external membrane envelops the whole organism. There are two flagella ("axial filaments") each inserted subterminally at opposite ends of the protoplasmic cylinder. The flagella are located between the cytoplasmic membrane peptidoglycan complex and the outer enveloping membrane. Each flagellum is inserted by a basal body and is intertwined with the proplasmic cylinder; the free ends are directed towards the middle region of the organism where they do not usually overlap. The basal bodies of the flagella are similar to those of Gram-negative bacteria with the exception of those of *L. illinii* where they are similar to those of Gram-positive bacteria.

Nuclear material, ribosomes and mesosomes can be differentiated in the protoplasmic cylinder. The G + C content of the DNA of the strains tested ranges from 35·3 to 41 moles %.

Leptospires contain catalase, peroxidase, esterase and diaminopimelic acid and O-methylmannose in their cell walls.

In liquid milieux the characteristic movements appear to be alternating rotation around the long axis, and translation without polar differentiation; dividing cells often flex sharply and vigorously at the point of impending division; in semi-solid milieux flexion, boring and serpentine movements also occur.

Leptospires can pass through membrane filters of average pore diameter of 0·22 μm and readily migrate through 0·1% agar gel.

Can routinely be cultivated at about 30°C in media containing serum or serum-albumin and a pH 7·2–7·6. They are obligate aerobes. Do not grow at temperatures below 13°C. Exposure to 40°C is deleterious. When grown in semisolid tubed medium, a characteristic linear disc of growth is formed 1–3 cm below the surface. In plated soft agar medium (1% agar) they produce diffuse to discrete subsurface colonies. Generation time in media is c. 12–16 h. Successful cultivation usually depends on the availability of long chain fatty acids (14 carbons or more) as a readily utilisable carbon and energy source; ammonia in the form of inorganic salts, rather than amino acids, usually satisfies the nitrogen requirements. Purines, but not pyrimidines, are utilised. Requirements for thiamine (vitamin B_1) and cyanocobalamine (vitamin B_{12}) have been established.

The genus includes strains which are parasitic or free living. Some strains may be pathogenic for man or various other animal species.

B. Species

For many years, the genus *Leptospira* has been divided into two groups.

One group comprised the pathogenic strains, the other consisted of apparently free-living water or saprophytic leptospires. The latter are usually found in fresh surface waters. Various tests have been described to differentiate the two groups more precisely. On the basis of the findings with these reactions, the Subcommittee on the Taxonomy of *Leptospira* (Rep. Tax. Subc. Lep., 1963) considered the genus to consist of two species: *L. interrogans* (pathogenic) and *L. biflexa* (predominantly non-pathogenic).

Since then, various tests have been developed to improve differentiation (e.g. the ability to infect animals, the relative resistance to 8-azoguanine and copper sulphate, oxidase reaction, lipolytic activities, salt and temperature tolerances and others). However, it appeared doubtful whether the results obtained by these tests were sufficiently reliable for the circumscription of the two species (Kmety *et al.*, 1966). Therefore, a WHO expert group on leptospirosis research recommended that the genus *Leptospira* should be considered as monospecific, until the species within the genus could be differentiated with sufficient confidence (WHO, 1967). Then the term "complex", having no official standing as a taxon in the sense of a group, was proposed and accepted as a provisional designation for each of the two categorised leptospira groups.

In 1973 the Subcommittee on the Taxonomy of *Leptospira* (I.C.N.B. S.T.Lep., 1974) reconsidered this matter and came to the final conclusion that reliable tests to divide these groups seemed to have been developed and that, consequently, the two "complexes" should again be recognised as two separate species. But the species have not yet been described according to the latest requirements of the International Committee on Systematic Bacteriology.

C. Serogroups

Already, during the early years of research on *Leptospira*, it was recognised that *Leptospira* strains could be distinguished by their antigens. As more different strains were isolated, serological relationships were demonstrated between some of them. For practical reasons, strains belonging to separate, but closely related serotypes were grouped together into serogroups.

Serogroups have not as yet been defined accurately and have no official taxonomic status. However, for practical purposes, they will remain as a necessary component of the present classification system.

The pathogenic serotypes have been officially divided into 16† serogroups (W.H.O. Expert Group, 1967). On several occasions subdivision of a few of the serogroups into new serogroups has been suggested (Kmety, 1974; Wolff, 1962). More detailed antigenic studies (Kmety, 1967) showed the possibility of distinguishing subgroups, within some serogroups.

Cinco (1970) and Cinco *et al.* (1972) suggested that saprophytic strains could be divided into eight serogroups, later extended by another three

† It seems now that 18 serogroups have been unofficially recognised.

serogroups by Hlavata (1974). Only a few investigators have worked on saprophytic strains making it likely that new serogroups may be discovered when investigation on water leptospires has been extended.

D. Serotypes

Serotypes are the basic taxa of *Leptospira*. Each is represented by a reference strain, to which the description of the serotype is attached.

The presently valid definition of a serotype is formulated (WHO, 1967) as follows:

two strains are considered to belong to different serotypes if after cross-absorption with adequate amounts of heterologous antigen, 10% or more of the homologous titre regularly remains in at least one of the two antisera in repeated tests.

An alternative definition based on a study of antigenic composition has been suggested by Kmety (1967). Accordingly a serotype is characterised by a specific arrangement of its "major"† agglutinogens.

The official definition does not exactly characterise each serotype, but rather states the degree of serological difference necessary for the description of a strain to represent a new serotype. Within the same serotype less than 10% of the homologous antibodies should be present when absorbed by another strain belonging to that serotype. The limit of 10% residual antibodies in absorbed sera when compared with the homologous titre, was arbitrarily chosen and is essentially quantitative. However, this 10% limit has been quite valuable for practical purposes, even though the terms "adequate", "regularly" and "in repeated tests" used in this connection are not fully specified in the definition.

The alternative definition of Kmety (1967) is based on a more detailed study of the antigenic structure of the serotype reference strains, which has revealed "major" agglutinogens ("factors") and their specific arrangement within each serotype so far studied.

The basic method used to distinguish serotypes is the agglutinin-absorption test.

E. Varieties

In 1971, Borg-Peterson discovered the presence of a thermolabile antigen in strain Ictero I. Kmety (1972) later suggested distinguishing two agglutinogenic systems within leptospires. One was based on thermostable antigens, considered the fundamental one. A second was based on thermolabile antigens. The second system gives additional information and should be used to subdivide serotypes into two varieties; a thermolable (Vi-positive) and a thermostable (Vi-negative). At present such subdivision is only practised in the serotype icterohaemorrhagiae.

† Also called "main" agglutinogens.

F. Other biological properties of possible taxonomic value

Recent studies have revealed several biological groups of leptospires distinguishable by other biological properties. None of those other biological characteristics corresponds with the current scheme of classification based on agglutinogens. Using DNA base composition and DNA hybridisation, Haapala et al. (1969) described four distinct genetic groups of leptospires among selected parasitic and saprophytic strains. In later studies, Brendle (1974) described six groups on the basis of DNA annealing affinities. The strains within these groups were not necessarily antigenically related. The immunochemical characteristics of antigenic components have been extensively studied by Chang et al. (1970). Among others axistile antibodies were disclosed although their relevance for taxonomic purposes needs further investigation.

Also, the lipolytic activity of pathogenic and saprophytic strains has been investigated (Memoranda, 1972). The lipase activity of pathogenic strains was well correlated with an initial growth inhibition by 2,6-diamino-purine (DAP). On the basis of trioleinase activity, and sensitivity to purine analogues DAP and 8-azaguanine, Johnson and Harris (1968) differentiated three leptospira groups. These were associated with four genetic groups described by Haapala (1969).

Other easily discernible phenotype characters were explored for taxonomic purposes, such as growth in the presence of copper sulphate or sodium bicarbonate, growth at 13°C, oxidase, lecithinase and sphingomyelinase activity, survival in laboratory animals, etc. None of these techniques has so far been sufficiently developed and described to provide additional information to serotyping.

III. SEROTYPING PROCEDURES

A. History

During the first years after the discovery of pathogenic leptospires, Inada and his associates (1916) recognised the organisms by their typical microscopical appearance and their pathogenicity for laboratory animals. Martin et al. (1917) found that suspensions of living leptospires were agglutinated by sera of patients who had suffered from Weil's disease. This discovery was further developed by Schüffner (1927) and is still a basic diagnostic procedure in leptospirosis. It resulted in the early recognition that the agglutination test is most practical for the differentiation of the causative agent in the various leptospiral syndromes (Weil's disease, seven-days' fever, ricefield fever, etc.). The diagnostic absorption test was described by Ruys and Schüffner (1934).

Borg-Peterson (1938) made the important observation that two strains, indistinguishable by agglutination, could be clearly differentiated by absorption tests. This resulted in the discovery of two biotypes in the

Icterohaemorrhagiae group, complete biotype AB and incomplete biotype A. This work is the basis of the development of the present classification system of *Leptospira* classification. Gispen and Schüffner (1939) confirmed this observation by the use of more strains.

Wolff (1953) was the first to summarise the observations of those years and formulated, together with Broom (1954) the first officially recognised classification system of leptospires. They distinguished 20 serogroups with a total of 36 serotypes by means of agglutination and cross-agglutinin absorption reactions.

Subsequently, this first system was further amended and the theoretical background studied in more detail. Many new serotypes were described in different parts of the world.

In 1967, the last officially recognised list of serotypes was published (WHO, 1967) (see Table I). Since then the number of newly described

TABLE I

List of serotypes of leptospiras isolated from man and animals

Serogroup	Serotype
Icterohaemorrhagiae	icterohaemorrhagiae
	copenhageni
	mankarso
	naam
	nwogolo
	dakota
	sarmin
	birkini
	smithi
	ndambari
	ndahambukje
	budapest
	weaveri
Javanica	javanica
	poi
	sorex-jalna
	coxi
	sofia
Celledoni	celledoni
	whitcombi
Canicola	canicola
	bafani
	kamituga
	jonsis
	sumneri
	broomi
	bindjei
	schueffneri

TABLE I (*continued*)

Serogroup	Serotype
	benjamin
	malaya
Ballum	ballum
	castellonis
	arboreae
Pyrogenes	pyrogenes
	zanoni
	myocastoris
	abramis
	biggis
	hamptoni
	alexi
	robinsoni
	manilae
Cynopteri	cynopteri
	canalzonae
	butembo
Autumnalis	autumnalis
	rachmati
	fort-bragg
	sumatrana
	bulgarica
	bangkinang
	erinacei-auriti
	mooris
	sentot
	louisiana
	orleans
	djasiman
	gurungi
Australis	australis
	lora
	muenchen
	jalna
	bratislava
	fugis
	bangkok
	peruviana
	pina
	nicaragua
Pomona	pomona
	kennewicki
	monjakov
	mozdok

TABLE I (*continued*)

Serogroup	Serotype
	tropica
	proechimys
Grippotyphosa	grippotyphosa
	valbuzzi
Hebdomadis	hebdomadis
	nona
	kambale
	kremastos
	worsfoldi
	jules
	maru
	borincana
	kabura
	mini
	szwajizak
	georgia
	perameles
	hardjo
	recreo
	medanensis
	wolffi
	trinidad
	sejroe
	balcanica
	polonica
	saxkoebing
	nero
	haemolytica
	ricardi
Bataviae	bataviae
	paidjan
	djatzi
	kobbe
	balboa
	claytoni
	brasiliensis
Tarassovi	tarassovi
	bakeri
	atlantae
	guidae
	kisuba
	bravo
	atchafalaya
	chagres
	rama
	gatuni

TABLE I (*continued*)

Serogroup	Serotype
Panama	panama
Shermani	shermani
Semaranga	semaranga
	patoc
	sao-paulo
Andamana	andamana

serotypes has increased. Some years later, Borg-Peterson (1971) discovered a thermolabile antigen which necessitated a revision of the methodological basis for classification. Therefore, Kmety in 1972 and 1974 suggested that the whole classification system should be based on thermostable aggluti-nogens. Where a thermolabile antigen is present, varieties should be distinguished within the serotype.

For practical reasons, the official list of serotypes, which is more than ten years old now, and no longer up to date, has been extended by new serotypes, the serological status of which has been confirmed by various laboratories. There are still a number of strains under investigation, which will be included in the list after their correct status has been determined and confirmed. This list, which has *no official* status, is given in Appendix I, and should serve as a practical guide for the interpretation of typing results.

B. Determination of the serogroup

When leptospira strains are classified, first the serogroup to which a strain belongs has to be determined by the microscopic agglutination test using selected rabbit "group sera".

As representative "group serum" is chosen a rabbit immune serum which agglutinates optimally all serotypes of a given serogroup and which gives the least cross-agglutination with strains of other serogroups.

For instance, in the Canicola serogroup immune serum of strain Vleermuis 90 C is suggested as "group serum" because it reacts, in high titres, with all the other serotypes of the group, while cross-agglutination with serotypes of the Icterohaemorrhagiae group is low. This is very helpful with the differentiation of strains belonging to these two groups.

When strains from Europe are investigated, Canicola "group serum" against strain Hond Utrecht IV can be chosen also or alternatively, as the occurrence of serotypes belonging to the Schueffneri subgroup (Appendix I) has not been reported from Europe.

The determination of the serogroup status of the strain is usually not difficult, provided that the suggested battery of "group sera" is used (see

Table II). Local modifications may be epidemiologically warranted. In exceptional cases, when no clear results are obtained, the use of "subgroup sera" is suggested (see Appendix I). For instance, "group serum" Hardjoprajitno selected to represent the newly suggested "Sejröe serogroup", reacts only weakly with strains of serotype *haemolytica* and *ricardi*. In such a case, an additional investigation with immune sera against strains belonging to the Saxkoebing subgroup should be performed. As long as new serotypes are discovered, some of the suggested "group sera" might be replaced by antisera produced with more suitable strains. Therefore, and for other reasons explained above, the suggested table with the representative serogroup sera will continue to be revised.

As for the saprophytic serogroups, no "group sera" has so far been suggested. As the number of serotypes within the different serogroups is limited (eleven groups are represented only by one serotype, see Table III), the determination of the serogroup status of the strain will not cause

TABLE II

Suggested group sera

Serogroup	Serotype	Strain
Icterohaemorrhagiae	icterohaemorrhagiae	R.G.A.
Hebdomadis	hebdomadis	Hebdomadis
Autumnalis†	bangkinang	Bangkinang I
Bataviae	bataviae	van Tienen
Pyrogenes	robinsoni	Robinson
Grippotyphosa	valbuzzi	Valbuzzi
Canicola	schueffneri	Vleermuis 90 C
Pomona	pomona	Pomona
Australis	lora	Lora
Javanica	poi	Poi
Sejröe†	hardjo	Hardjoprajitno
Djasiman	djasiman	Djasiman
Cynopteri	cynopteri	3522 C
Tarassovi	hyos	Mitis Johnson
Mini	tabaquite	TVRL 3214
Ballum	ballum	Mus 127
Celledoni	celledoni	Celledoni
Louisiana	louisiana	LSU 1945
Panama	panama	CZ 214 K
Shermani	shermani	LT 821

† Alternative group serum:

Autumnalis	rachmati	Rachmat
Sejröe	medanensis	Hond H.C.
Pyrogenes	pyrogenes	Salinem
Canicola	canicola	Hond Ultrecht IV
Australis	bratislava	Jez-Bratislava
Mini	szwajizak	Szwajizak

difficulties, provided that the strain under investigation is not representing a new serogroup.

The practical procedure for determining the serogroup of a strain is as follows.

A well grown culture of the strain under investigation is first submitted to the microscopic agglutination test (MAT) with the 20 suggested "group

TABLE III

List of saprophytic serotypes of leptospiras†

Serogroup	Serotype	Reference strain
Semaranga	semaranga	Veldrat S 173
	patoc	Patoc 1
	sao paulo	Sao Paulo
	montevalerio	Monte Valerio
Andamana	andamana	CH 11
	bovedo	Bovedo
Cau	cau	Cau
Doberdo	doberdo	Doberdo 1
	rupino	RPE
	acquamaria	AM-20
	drahovce	V-2
	zoo	V-6
Percedol	percedol	Percedol
Basovizza	basovizza	Basovizza
	sangiusto	San Giusto
Aurisina	aurisina	Aurisina
	farneti	Farneti
Botanica	botanica	Botanica
	monfiascone	Monte Fiascone 2
Khoshamian	khoshamian	Khoshamian
Holland	holland	WaZ Holland
	roma	AM-6
	tredici	AM-13
	danubica	V-11
	timavo	8M
Tevere	tevere	AR-18
Orvenco	orvenco	44
Maritza	valderio	37
Udine	udine	48
Madunice	madunice	V-1
Dowina	dowina	V-10
Kyjovka	kyjovka	V-15
Pulpudeva	pulpudeva	Bulgaria 6

† Most serotypes are not officially recognised.

sera" used at low dilutions (1:200 or 1:300 provided that the antisera have titres of about 1:10,000 with the homologous strain). Then, the positively reacting sera are further titrated with the strain under investigation and with the homologous strain. Endpoint titres are also compared with those given in Appendix I. This usually allows the serogroup to be determined.

The titres given in Appendix I are only indicative, and a variation of two- or exceptionally four-fold difference may be allowed.

When the results are not quite comparable with the data given in the tables of Appendix I, or when the strain shows a certain relationship to a group subdivided into subgroups, additional agglutination tests with antisera against the reference strain belonging to one of the serotypes of the indicated subgroup(s) are recommended.

For instance, in the Sejröe group, an additional investigation with immune serum against M 84 and Mus 24, may be useful. In unusual cases, the results may not be comparable to those of any listed serotype. Such a strain may be a saprophytic strain or a new serotype and should either be submitted by the finder to the extensive investigations required, or should be sent to a reference laboratory.

C. Determination of the serotype status

1. *Principles*

The procedures used for the determination of the serotype status of a strain are derived from the definition of serotypes (see page 263).

According to the official definition which distinguishes serotypes by a certain degree of serological difference, every new strain must be compared with all reference strains within the serogroup to which it belongs. This is considered the classical method.

The alternative definition defines serotypes by the specific arrangement of "major" antigens. Using this method, the serotypes are classified by factor sera.

In general, these two different typing methods lead to the same results.

2. *Classical typing method*

A well grown culture of the strain under investigation is first submitted to MAT (see Appendix II) with antisera of all reference strains of the serogroup. An immune serum is prepared against the strain under investigation (see Appendix IV) and this antiserum is similarly investigated with all reference strains of the determined serogroup to which the strain belongs.

The next step is to carry out cross-agglutinin absorption tests (AAT) (see Appendix III) with all the reference strains which have some degree of serological relatedness to the unknown. Relatedness is considered to exist when the strain under investigation reacts with the immune sera of a reference strain, to at least a titre of 10% of that obtained with the homologous strains. If the strain under investigation gives a reaction with a (reference) serum less than 10% of the homologous titre, an absorption test with the homologous strain is redundant. When, during the comparative absorption tests, serological identity is proved with one of the selected reference strains, no further tests are necessary. Two strains are considered serologically identical if after absorption with the heterologous antigen less than 10% residual homologous antibodies remain in both sera (i.e. antiserum against unknown strain and reference strain).

In case residual antibodies remain in the absorbed sera at a titre near to the 10% limit, it is recommended to repeat the test with the same serum (or sera) or with newly made antisera, prepared by three rabbits. Each serum should be submitted to the cross-absorption test individually.

If the strain under study cannot be attached to a known serotype, it might be a new serotype. Sometimes the results of the agglutinin absorption test are equivocal. In either case it is advisable to send the strain to a reference laboratory for further investigation.

3. *Factor sera method*

The official classification of *Leptospira* was extended by analysis of antigenic factors (Kmety, 1967). This has resulted in a better understanding of the mosaic of leptospiral agglutinogens. A number of different agglutinogens are present. Some are dominating; they are responsible for the homologous titre and the high cross-agglutinations with closely related serotypes. The dominating antigens are called "major" antigens, others "minor" antigens.

(a) *Major antigens*. Major antigens are determined by means of the antibodies they elicit. Antibodies are considered to reflect the presence of a major antigen if after absorption of sera with strains lacking this main antigen, high titres are still obtained with the homologous strain and other strains sharing the same "major" antigens. By "high titre" antibodies in absorbed sera are understood antibodies which react to at least a five titre steps dilution higher than with the absorbing strain. Provision should be made for the absorption tests to be performed under standard conditions (Kmety *et al.*, 1970).

In the Ballum group for example, antiserum against strain Arborea reacts with strains Mus 127, Castellon 3 and the homologous strain to a titre of 1:12,800–1:25,600. After absorption of this serum with strain Mus 127, the corresponding antibodies are completely removed (<1:100, lowest dilution), but antibodies to the strains Castellon 3 and Arborea—titre 6400—still remain in the absorbed serum, which corresponds with a difference of at least seven titre steps. This reflects the presence of a common "major" antigen in the serotypes castellonis and arborea. Strains belonging to different serotypes but sharing a common major antigen do not always initiate the production of corresponding antibodies to the same level. For instance, immune serum Lora absorbed with Ballico or immune serum Jalna absorbed with Ballico and München 90 C contain antibodies reacting with strain Jež Bratislava in a titre of 1:6400 or 1:1600 respectively (corresponding to factor As-3). But in immune serum Jež Bratislava after absorption with München 90 C and LT 990 antibodies corresponding to this common major antigen are present only in a titre of 1:100 and 400 respectively (Kmety, 1974). The practical impact of this is that for the preparation of factor sera only sera anti-Lora or anti-Jalna are suitable. Consequently, sometimes only specially chosen serotypes are suitable for the preparation of factor sera.

(b) *Minor antigens.* As minor antigens are considered all the agglutinogens which are not major antigens. They are mostly responsible for cross-agglutination with reference strains of other serogroups. Being on the borderline of detection, with the methods presently available, they are not considered important for typing.

(c) *Antigenic schemes.* Each detected major or minor antigen is designated by a number. This means that every serotype can be represented by a numerical code (see Appendix V). The scheme has been extended by adding a lettered abbreviation for every serogroup name. For example:

  Hebdomadis group   Hb→Hb-1, Hb-2, Hb-3
  Grippotyphosa group   Gt→Gt-1, Gt-2, Gt-3
  Sejröe group   Sj→Sj-1, Sj-2, Sj-3
  etc.

  To distinguish major and minor antigens, the minor antigens were prefixed by an "o", e.g. Hb-o1, Hb-o2, Sj-o4, etc.
  Antigens present in two serogroups are designated by both abbreviations of those serogroups, e.g. Sj/Mn-o40 (minor antigen present in the Sejröe and Mini group).
  As an example, the numerical scheme of the Hebdomadis group may be considered (see Appendix V). Sixteen major antigens are recognised in this group. Every serotype contains three to six major antigens. The major antigens Hb-4 and Hb-5 seem to divide the group into the two subgroups Hebdomadis and Borincana. Only one "major" antigen Hb/Sj-12 appears to be expressed also in a serotype of the Sejröe serogroup. Although unimportant for serotyping purposes, a number of minor antigens complete this scheme. The other schemes of the serogroups so far studied are given in Appendix V.

(d) *Schemes of typing.* The arrangement of "major" antigens, typical for each serotype allows us to determine the serotype status of a strain by means of factor sera, provided that the status of the whole serogroup has been submitted to analysis of antigenic factors previously. For instance, in the Hebdomadis group strains are first screened by factor sera Hb-4 and Hb-5, to put them in the correct subgroup.

Strains of the Hebdomadis subgroup (Hb-4) are further investigated with factor sera Hb-2, -14, -15 and -16. The provisional serotype status is then determined according to the scheme.

Strains reacting with Hb-5 factor serum belong to the Borincana subgroup, and are subsequently tested with factor sera Hb-3, -6, -7, -9 and -10. The agglutination test with factor sera is preferably performed with a starting dilution of 1:200 or 1:400, to save sera and to avoid low titre cross-agglutinations.

Every investigation with factor sera should be executed with the necessary controls, i.e.

(a) homologous strain, used for the preparation of antisera (= positive control),

(b) absorbing strain(s) (= negative control).

Investigation of an unknown strain by factor sera *never proves* the definite serotype status of the strain. It is just a strong indication. Therefore, every typing by factor sera has to be *completed* by the *classical cross-agglutinin absorption test* with the indicated reference strain. Only this test definitely proves the serological identity of the compared strain. The cross-absorption test is described in Appendix 3.

If the results of the absorption test indicate serological differences between both strains (unknown strain and strain representing the suggested serotype), the strain is considered possibly to represent a new serotype and it is advised that it be sent to a reference laboratory for detailed investigations.

(e) *Preparation and specificity of factor sera.* The same factor serum can be prepared in different ways. For instance, a suitable factor serum As-3 may be prepared from immune serum Jalna by absorption with the strains Ballico and München 90 C. The same factor serum can also be obtained by absorption of Lora-antiserum with the Ballico reference strain (see Appendix V, serogroup Australis). The last method is more practical as only one antigen is needed for absorption. Appendix V gives recommendations for the preparation of factor sera.

Every factor serum contains a few agglutinins against usually "minor" antigens in addition to the main one. For instance, factor serum Hb-4 (see Appendix V) prepared from antiserum against Hebdomadis, absorbed with the reference strain Kremastos, contains antibodies against "major" antigen Hb-2 and "minor" antigens Hb-o2, Hb-o3 and Hb-o4, in addition to antibodies against "major" antigen Hb-4. Antibodies against "major" antigen Hb-2 do not influence the diagnostic value of the factor serum, as "major" antigen Hb-4 is only present together with "major" antigen Hb-2. Therefore, Hb-2 is always expressed together with Hb-4 and will not lead to false positive reactions.

Since "minor" antigen antibodies are only evoked in small quantities, their presence will not influence the results when the sera are used at the recommended dilution of 1 : 400.

Factor sera should always be checked for their quality by reference strains of serotypes which are determined by the serum (positive control). As a negative control, the absorbing strains are to be used. The serum is stored in small quantities at −20°C to avoid repeated freezing and thawing and possible decrease in titre.

4. *Combined method*

Both typing systems, the classical and factor sera methods, can be advantageously combined. The subgroup status of a strain under investigation may then be determined by use of factor sera, and the further final typing by the classical method.

This combination is especially useful, when the required factor sera are lacking, in circumstances when factor sera can only be prepared on a limited scale, or new serotypes have been described in the meantime which have not yet been submitted to factor analysis. In such cases, the strain under investigation should be compared with indicated reference strains by the classical cross-absorption agglutination test, to determine the serotype status.

The combined method has the advantage of being very practical, as it reduces the number of laborious cross-absorption tests required by the classical method and shortens the time to reach diagnosis. It also uses less factor serum and only a limited number of factor sera must be kept at the laboratory. This diminishes again the laborious time-consuming preparation of so many factor sera.

D. Determination of serotype varieties

Serotypes should be subdivided, in our opinion, into varieties based on the presence of a thermolabile antigen (see Section II.E). At this moment, only one labile variety has been proved, in the serotype *icterohaemorrhagiae*, and it is represented by the reference strain Ictero I. The reference strain R.G.A. belongs to the same serotype, but it only possesses the heat stable antigens.

Consequently, the presence of a thermolabile antigen should be determined for all strains which have been typed as serotype *icterohaemorrhagiae*. This can be proven by the cross-absorption technique using living

TABLE IV

Example of form for recording agglutination test results

Name Institute
Leptospira Unit

Antiserum of:
Absorbed with strain:
Agglutinated with strain:
Town: Date: Hour:

(or double dilution scheme)

Dilution	Agglutination		Living or loose Leptospira	
	1st	2nd	1st	2nd
10				
30				
100				
300				
1000				
3000				
10,000				
30,000				
100,000				
300,000				

Name Institute: Royal Tropical Institute
Leptospira Unit

Antiserum of: LR 110
Absorbed with strain: Kisuba
Agglutinated with strain: Perepelicin
Town: Amsterdam   Date: 14.4.77   Hour: 11 hours

Dilution	Agglutination		Living or loose Leptospira	
	1st	2nd	1st	2nd
10	+ + + + +		+	
30	+ + + + +		−	
100	+ + + + +		−	
300	+ + + +		+ + + +	
1000	+ +		+ + + +	
3000	−		+ + + +	
10,000	−		+ + + +	
30,000	−		+ + + +	
100,000	−		+ + + +	
300,000				

Explanation:

Agglutination

+ + + + = 100% agglutination→
+ + + = 75% agglutination→
+ + = 50% agglutination→
+ = 25% agglutination→
− = nil agglutination→

Living or loose leptospiral organisms

− = no loose
+ = 25% of control
+ + = 50% of control
+ + + = 75% of control
+ + + + = 100% of control

Interpretation of reaction:

Positive—50% or less non-agglutinating leptospires in comparison with control.
Negative—more than 50% non-agglutinating leptospires in comparison with control.

antigens and antisera prepared by the immunisation of rabbits with living Ictero I antigen and the strain to be typed respectively.

If, after cross-absorption, antibodies are completely removed from both sera, the typed strain represents the thermolabile (Vi-positive) variety of serotype *icterohaemorrhagiae*. If the strain to be typed cannot remove all antibodies from immune serum Ictero I, it belongs to the thermolabile (Vi-negative) variety of *L. icterohaemorrhagiae*.

The presence of the thermolabile agglutinogen can also be determined by a factor serum prepared by the absorption of antiserum against living Ictero I antigen with strain R.G.A. The absorbed serum contains now the thermolabile agglutinins only, as the thermolabile fraction is removed by the absorption with R.G.A. When utilising this factor serum, strains Ictero I and R.G.A. should be included as positive and negative controls.

IV. RECORDING OF TYPING RESULTS

Different methods of recording can be used, depending on the working conditions and the method used for typing. Therefore each laboratory normally has its own way of recording typing results. Some general recommendations will be made and an example will be given showing how results can be recorded. All data, e.g. history, day of receipt, typing results, etc., of reference strains, strains to be typed or strains waiting for or under investigation, should be recorded. Every strain may conveniently have its own set of three to four cards on which all data are recorded.

A. Protocol for recording agglutination tests

The result of the agglutination (absorption) test can be recorded on a special pre-printed form. To limit recording errors, it is advisable to register the agglutination results of only one complete set of one antiserum dilutions with one antigen on *one* form (see Table IV).

The typing results recorded on these forms are entered, on the same day, on serotyping cards as well. This working system ensures that only these forms are allowed into the laboratory; the cards and all other paperwork used for recording never enter the laboratory.

B. Recording of serogroup-typing

The results of these investigations are registered on the "serogroup-typing" card. Titres obtained with the strain under investigation and the homologous strain are expressed in reciprocals. There is a third column to express the reciprocal of the titre with the examined strain in percentages of the reciprocal of the titre with the homologous strain.

Part of such a card will be given as an example:

TABLE V

Example of card recording serogroup typing results

| Strain: LR 110 | | | | | |
| Antiserum: — | | | | | |

Date	Antiserum	Reciprocal titre of antiserum with homologous strain	Reciprocal titre with strain	%	Observations and other information, e.g. history origin, day of receipt, etc.
14.8.77	R.G.A.	10,000	3000	33	
	Celledoni		—	0	
	Vleermuis 90 C	30,000	300	1	
	Hond Utr. IV				
	Mus 127	30,000	—	0	
	Robinson/Salinem	10,000	—	0	

C. Recording of agglutination results within the serogroup

These results are entered on preprinted "reference strain" cards with the known reference strains divided according to their serological relatedness over the serogroups (see Appendix I and Table I). Every card covers at least one, but usually more serogroups. Two cards are selected for every serogroup. One card registers the agglutination results of the examined strain against the reference antisera in the defined serogroup. The second card records the reaction of antiserum of the examined strain against the reference strains within that group. Examples are given in Tables VI and VII.

D. Recording of results of absorption tests

For practical reasons these results can be entered on the back of the cards already referred to (see IV.B and IV.C). The results can be listed as suggested in Table VIII.

V. GENERAL COMMENTS

A. Methodology of typing

Typing methods always depend on the definition of the basic taxa. This means that in typing leptospires, at present, the 10% limit criterion as explained on page 263 is decisive. This rule leaves a 0–10% margin of agglutinogenic difference in antigenic structure for strains belonging to the same serotype. This margin seems considerable but takes into consideration all possible quantitative differences in the typing procedure. For example, the determination of the endpoint titre is often influenced by the agglutin-

TABLE VI

Recording results of agglutination tests within a serogroup

Strain: LR 110
Antiserum: —

Date	Antiserum	Reciprocal titre of antiserum with homologous strain	Reciprocal titre with strain	%	Observations
15.8.77	Icterohaemorrhagiae				
	R.G.A.	10,000	3000	33	
	Ictero I	30,000	10,000	33	
	M 20	10,000	10,000	100	
	Mankarso	30,000	3000	10	
	Naam	30,000	300	1	
	etc.				
	Ballum				
	Mus 127				
	Castellon 3				
	Arborea				
	etc.				

TABLE VII

Recording results of agglutination tests within a serogroup

Strain:
Antiserum: LR 110

Date	Strain	Reciprocal titre of antiserum with homologous strain	Reciprocal titre with strain	%	Observations
16.8.77	Icterohaemorrhagiae				
	R.G.A.	30,000	30,000	100	
	Ictero I	30,000	10,000	33	
	M 20	30,000	30,000	100	
	Mankarso	30,000	3000	10	
	Naam	30,000	1000	3	
	etc.				
	Ballum				
	Mus 127				
	Castellon 3				
	Arborea				
	etc.				

TABLE VIII

Recording results of absorption tests

Date of test	No. of absorption test	Titre of antiserum used	Antiserum used of strain	Absorbed with strain	Reciprocal titres of the absorbed sera expressed as a percentage of the reciprocal titre of the unabsorbed sera with the homologous antigen				
					M 20	LR 110	R.G.A.	Man-karso	Naam
22.8.77	2304	3000	M 20	LR 110	3	0			
22.8.77	2305	3000	LR 110	M 20	0	3			
22.8.77	2306	3000	LR 110	R.G.A.		10	0		
22.8.77	2307	3000	R.G.A.	LR 110		0	33		
23.8.77	2308	3000	Mankarso	LR 110		0		10	
23.8.77	2309	3000	LR 110	Mankarso		33		0	
23.8.77	2310	>3000	LR 110	Naam		33			0
23.8.77	2311	3000	Naam	LR 110					33

ability of the strain, a certain degree of subjectivity in reading the test, the density of the antigen and some other details such as temperature of incubation, incubation time, etc.

Moreover, the definition which is still accepted does not take into account the recent finding of thermolable antigen, and the preparation of antisera has not been completely described. In other words, neither the agglutination test nor the absorption test has been sufficiently standardised.

We therefore wish to give a description of serotyping methods in which the influence of the factors mentioned above, and others, is minimised as far as possible. The laboriousness of the typing methods and the scrupulous care needed in maintaining the reference strains of *Leptospira* restrict the typing of strains today mainly to specialised reference laboratories.

A better understanding of the presence of defined major antigenic components and their specific combinations may provide an alternative basis for the definition of a serotype. Based on this knowledge, the factor analysis method has been developed, and will soon be applicable to all serogroups. Provided that investigations with this method prove satisfactory in reference laboratories and sufficient factor sera can be supplied, it seems possible that classification could be accomplished in less specialised laboratories.

B. Type stability

Leptospires under normal laboratory maintenance conditions are antigenically sufficiently stable to justify serological typing. Strains which were isolated some 40 years ago do not show agglutinogen differences from similar strains, isolated at present. Comparative absorption tests in the Amsterdam laboratories, with antisera freeze-dried 20 years ago and selected strains, which had been subcultured *c.* 250 times, indicate that the agglutinogen pattern has not changed over these years.

From time to time, changes in antigenic structure have been reported but most of the time they were explained as labelling errors or mixed cultures. In recent years more authors have described an induced variability of antigenic structure of leptospiral strains under specific controlled laboratory conditions (Babudieri, 1971; Bakoss, 1974). These results have shown that the stability of the antigenic structure is in fact not absolute. Whether similar changes occur exceptionally under normal laboratory conditions is at present hypothetical. They cannot be excluded, but have not been proven.

C. Practical impact

Leptospirosis is a typical zoonosis. Many different kinds of animals serve as reservoirs of infection. The relatively high specificity of host–serotype relationships means that accurate typing of isolated strains is one of the basic methods for epidemiological investigations. It has been applied in

11

practice for many decades. It indicates the necessary epidemiological measures and helps in the treatment of outbreaks of disease as well as in their prevention.

New serotypes appearing in a certain area can only be detected by the use of typing methods. For instance, not long ago the serotypes *valbuzzi* and *muelleri* (Grippotyphosa serogroup) were detected for the first time in Europe (Lataste-Dorolle and Kmety, in press). Similar results have been reported on strains of the Australis and Sejröe serogroup. Exact knowledge of antigenic structure and the occurrence of particular strains circulating in a given area are fundamental factors in the choice of strains for use in routine serological diagnostic procedures. This is also important for selecting strains for a vaccine to be used in a specific area.

Typing has a considerable impact in scientific work. Antigenic variability studies are hardly possible without exact typing methods. In studies dealing with pathogenic properties of strains—production of haemolysis, lipases, etc., knowledge of the serotype status is also essential. For example, it is found that Pomona and Grippotyphosa strains especially produce a very active haemolysin which is of decisive importance in the pathogenesis of icterohaemoglobinuria in cattle.

D. Possible additional typing methods

Several papers have been published during the last decade on the genetics of *Leptospira*. Using DNA base composition and DNA:DNA hybridisation, several distinct genetic groups have been described (Haapala *et al.*, 1969). These genetic studies have not been advanced to a degree where they can form an alternative basis for classification.

Great efforts to define leptospiral antigens immunochemically have been made by Chang and Faine (1970), aimed at isolating chemically defined type specific antigens. Lately Japanese workers have described the identification of some strains as different serotypes by the precipitin-absorption test in gel (Yanagawa and Adachi, 1977). It is hoped that progress in this field may bring forward findings which could have impact on leptospira classification. Many new and interesting findings have been reported on the metabolism and other biochemical activities of leptospires. Various enzymes have been studied and attempts made to correlate their presence with taxonomic properties. However, the results obtained have seemed inconclusive (Memoranda, 1972). Research in this broad field is advancing and it is quite possible that future findings may influence the theoretical considerations on which the present classification is based.

Finally, it must be pointed out that every classification system is always undergoing continuous evolution. It reflects the present state of knowledge and is influenced by scientific progress. Therefore, the system presented, based on serological characteristics, should not be considered as a rigid and inflexible construction. We believe that it reflects the present advances achieved in this field and hope that it will be of some help for laboratory workers and will stimulate research workers who can apply the methods in practice.

REFERENCES

Babudieri, B. (1971). *Zbl. Bakt. I. Abt. Orig.*, **218**, 75–86.
Bakoss, P. (1974). *Folia Fac. Med. Comenianae Bratisl.*, **12** (Suppl.), 75–83.
Borg-Peterson, C. (1938). Int. Congr. Trop. Med. 3, Amsterdam, *Acta*, **1**, 396.
Borg-Peterson, C. (1971). *Trop. Geogr. Med.*, **23**, 282–285.
Brendle, J. J. (1974). *Folia Fac. Med. Comenianae Bratisl.*, **12** (Suppl.), 91–100.
Chang, A., and Faine, S. (1970). *Bull. Wld Hlth Org.*, **43**, 571–577.
Chang, A., Palit, A., Graves, S. R., and Faine, S. (1970). Abstracts of 10th Int. Congr. for Microbiology, Mexico, 77.
Cinco, M. (1970). *Trop. Geogr. Med.*, **22**, 237–244.
Cinco, M., and Faidutti, B. M. (1972). *Experientia*, **28**, 598–600.
Gispen, R., and Schüffner, W. (1939). *Zbl. Bakt. I. Abt. Orig.*, **144**, 427–434.
Haapala, D. K., Rogul, M., Evans, L. B., and Alexander, A. D. (1969). *J. Bact.*, **98**, 421–428.
Hlavata, Z. (1974). *Folia Fac. Med. Comenianae Bratisl.*, **12**, (Suppl.), 157–161.
I.C.N.B. Subcommittee on the Taxonomy of *Leptospira* (1971). *Int. J. Syst. Bact.*, **21**, 142–146.
I.C.N.B. Subcommittee on the Taxonomy of *Leptospira* (1974). *Int. J. Syst. Bact.*, **24**, 381–382.
Inada, R., Yutaki, I., Hoki, R., Kaneko, R., and Ito, H. (1916). *J. exp. Med.*, 377–403.
Johnson, R. C., and Harris, V. G. (1968). *Appl. Microbiol.*, **16**, 1584–1590.
Kmety, E., Plesko, I., and Bakoss, P. (1966). *Ann. Soc. Belge Med. Trop.*, **46**, 111–122.
Kmety, E. (1967). Thesis, Slovak Academy of Sciences, Bratislava.
Kmety, E., Galton, M. M., and Sulzer, C. R. (1970). *Bull. Wld Hlth Org.*, **42**, 733–738.
Kmety, E. (1972). *Zbl. Bakt. I. Abt. Orig.*, **221**, 343–351.
Kmety, E. (1974). *Folia Fac. Med. Comenianae Bratisl.*, **12**, (Suppl.), 101–110.
Martin, L., Petit, A., and Vaudremer, A. (1917). *C.R. Soc. Biol.*, Paris, **80**, 949–951.
Memoranda (1972). *Bull. Wld Hlth Org.*, **47**, 113–122.
Noguchi, H. (1917). *J. exp. Med.*, **25**, 755–789.
Postmus, S., and Schultsz, Th. W. (1932). *Nederl. Tijdschrift v. Geneeskunde* **76**, 252–261.
Report Meeting Taxonomic Subcommittee on *Leptospira* (1954). *Int. Bull. Bact. Nom. Taxon.*, **4**, 115–117.
Report Meeting Taxonomic Subcommittee on *Leptospira* (1963). *Int. Bull. Bact. Nom. Taxon.*, **13**, 161–165.
Ruys, A. Ch., and Schüffner, W. (1934). *Nederl. Tijdschrift v. Geneeskunde*, **78**, 3110–3114.
Schüffner, W., and Bohlander, H. (1939). *Zbl. Bakt. I. Abt. Orig.*, **144**, 434–439.
Schüffner, W. (1927), *Münch. Med. Wchr.*, **74**, 857.
Schüffner, W., and Mochtar, A. (1927). *Proc. R. Acad. Sci. Amsterdam*, **30**, 25–33.
Wolff, J. W. (1953). "Advances in the Control of Zoonoses", *WHO Monogr. Series*, **19**.
Wolff, J. W. (1954). "The Laboratory Diagnosis of Leptospirosis". Charles Thomas, Springfield, Illinois.
Wolff, J. W. (1962). "Proposed Subdivision of the Hebdomadis Serogroup". WHO Scientific Group on Research in Leptospirosis, Geneva.
Wolff, J. W., and Broom, J. C. (1954). *Doc. Med. Geogr. Trop.*, **6**, 78–95.
World Health Organisation (1967). *Tech. Rep. Series*, **380**, 7.
Yanagawa, R., and Adachi, Y. (1977). *Zbl. Bakt. I. Abt. Orig.*, **237**, 96–103.

Titres of agglutination reactions of sta[

Serogroup serotype		Strain	I. R.G.A.	II. Hebdomadis	III. Bangkinang I	IV. Van Tienen
I. Icterohaemorrhagiae (1915)						
(a) Icterohaemorrhagiae						
1. icterohaemorrhagiae Vi-positive variety	1915	R.G.A. Ictero I	25,000		200	
2. copenhageni	1935	M 20	25,000		800	
3. mankarso	1938	Mankarso	12,800		800	
4. naam	1936	Naam	12,800		200	
(b) Sarmin						
1. sarmin	1939	Sarmin	12,800			
2. ndahambukuje	1946	Ndahambukuje	12,800			
3. weaveri	1966	CZ 390 U	3,200			
4. waskurin		LT 63–68	3,200			
(c) Smithi						
1. birkini	1957	Birkin	12,800		800	
2. smithi	1957	Smith	12,800		400	
3. ndambari	1946	Ndambari	6,400		200	
4. mwogolo	1946	Mwogolo	3,200			
5. dakota	1962	Grand River	12,800		200	
6. gem		LT 16–67	400			
7. bog-vere		LT 60–69	1,600			
8. tonkini		LT 96–68	6,400			
II. Hebdomadis (1918)						
(a) Hebdomadis						
1. hebdomadis	1918	Hebdomadis		25,000		
2. kambale	1960	Kambale		12,800		
3. nona	1960	Nona		25,000		
4. maru	1966	CZ 285 B		12,800		
(b) Borincana						
1. borincana	1959	HS-622		12,800		
2. worsfoldi	1957	Worsfold		12,800		
3. jules	1958	Jules		12,800		
4. kabura	1958	Kabura		6,400		
5. kremastos	1958	Kremastos		12,800		
III. Autumnalis (1923)						
(a) Autumnalis						
1. rachmati	1923	Rachmat			25,000	
2. autumnalis	1925	Akiyami A			12,800	
3. bangkinang	1932	Bangkinang I			50,000	
4. erinacei-auriti	1951	Erin. auritus 670			50,000	
5. butembo	1959	Butembo			3,200	
(b) Fortbraggi						
1. fort-bragg	1957	Fort Bragg	400		50,000	
2. mooris	1954	Moores	200		25,000	
3. bulgarica	1958	Nikolaevo	200		50,000	
IV. Bataviae (1925)						
(a) Bataviae						
1. bataviae	1925	Van Tienen				25,000
2. paidjan	1953	Paidjan				12,800
3. djatzi	1955	HS 26				12,800
4. claytoni	?	LT 818				6,400
(b) Kobbe						
1. kobbe	1956	CZ 320 K				100
2. balboa	1956	LT 761				800
3. brasiliensis	?	LT 966				1,600
4. argentinensis	1968	LT 1019				12,800
V. Pyrogenes (1927)						
(a) Pyrogenes						
1. pyrogenes	1927	Salinem	800			
2. hamptoni	1957	Hampton				
3. alexi	1955	HS 616				
4. robinsoni	1954	Robinson				
5. manilae	1963	LT 398				
6. kwale		Julu				

	Reference serum													
VII. Vleermuis 90 C	VIII. Pomona	IX. Lora	X. Poi	XI. Hardjoprajitno	XII. Djasiman	XIII. 3522 C	XIV. Mitis Johnson	XV. TURL 3214	XVI. Mus 127	XVII. Celledoni	XVIII. LSU 1945	XIX. CZ 214 K	XX. LT 821
800			800								200		
800			1,600								100		
100			1,600										
200			400										
			1,600							100			
200		200	800								100		
200			200										
			1,600										
400	400		100								100		
400	100		200										
100			100										
			200								100		
100			400										
	200												
							200						
					200								
					400								
					200			400					
					200								
					400			100					
					400			400					
					200			200					
					800								
					400			200					
			200		200	200							
			200		100	100							
		200	200										
			800									200	
							100					200	
200					400	400							
	800	100	100		3,200	3,200							
	800		100										
							1,600						
							200						
							200						
							800						
							6,400						
							6,400						
							800						
										3,200			
			400						100				

Serogroop serotype	Strain	I. R.G.A.	II. Hebdomadis	III. Bangkinang I	IV. Van Tienen
V. Pyrogenes (1927) *continued*					
(b) Zanoni					
1. zanoni	1937 Zanoni	3,200			
2. abramis	1957 Abraham				
3. biggis	1957 Biggs				
4. myocastoris	1963 LSU 1551	3,200			
VI. Grippotyphosa (1928)					
(a) Grippotyphosa					
1. grippotyphosa	1928 Moskva V	200		100	
2. valbuzzi	1955 Valbuzzi				
3. canalzonae	1966 CZ 188 K				
4. muelleri	1955 RM-2				
(b) Vanderhoedeni					
1. vanderhoedeni	1969 Kipod 179			800	
VII. Canicola (1931)					
(a) Canicola					
1. canicola	1931 Utrecht IV	3,200			
2. bafani	1946 Bafani	3,200			
3. kamituga	1946 Kamituga	800			
4. jonsis	1957 Jones	200			
5. sumneri	1957 Sumner	6,400		200	
6. bindjei	1959 Bindjei				
7. broomi	1960 Patane	800			
8. galtoni	1967 LT 1014	1,600			
(b) Schueffneri					
1. schueffneri	1938 Vleermuis 90 C	800		200	
2. benjamini	1940 Benjamin	400			
3. malaya	1955 H 6	3,200			
VIII. Pomona (1937)					
1. pomona	1937 Pomona			400	
2. tropica	1956 CZ 299 U			800	
3. proechimys	? LT 796			800	
IX. Australis (1937)					
(a) Australis					
1. australis	1937 Ballico			200	
2. lora	1941 Lora			200	
3. fugis	1957 Fudge				
4. peruviana	1967 LT 941				
5. nicaragua	? LT 990	200		200	
6. ramisi	Musa				
(b) Jalna					
1. muencheni	1953 München C 90			200	
2. jalna	1960 Jalná			200	
3. bratislava	1960 Jež Bratislava	200		3,200	
X. Javanica (1938)					
1. javanica	1938 Veldrat Bataviae 46	400			
2. poi	1941 Poi	800			
3. sorex-jalna	1955 Sorex Jalná	800			
4. coxi	1957 Cox				
5. sofia	1961 Sofia 874	400			
6. ceylonica	1967 LT 1009	400		200	
7. menoni	Kerala				
XI. Sejröe (1938)					
(a) Sejröe					
1. sejroe	1938 M 84		400		
2. balcanica	1961 1627 Burgas		200		
3. polonica	1964 493 Poland		400		
4. istrica	Bratislava				
(b) Saxkoebing					
1. saxkoebing	1944 Mus 24		200		
2. haemolytica	1957 Marsh		400		
3. ricardi	1957 Richardson				
4. dikkeni	Mannuthi				
5. nyanza	Kibos				

	Reference serum													
VII. Vleermuis 90 C	VIII. Pomona	IX. Lora	X. Poi	XI. Hardjoprajitno	XII. Djasiman	XIII. 3522 C	XIV. Mitis Johnson	XV. TURL 3214	XVI. Mus 127	XVII. Celledoni	XVIII. LSU 1945	XIX. CZ 214 K	XX. LT 821
1,600													
100													
					12,800	400					1,600		
					12,800	200							
					1,600	800							
	100				800								
	200				200	100				100			
3,200									100				
6,400									400				
3,200									100				
3,200													
6,400									100				
6,400		100	100										
6,400													
12,800			400										
12,800													
6,400													
6,400		200											
	12,800					400					400		
	25,000					800							
	25,000				800	200							
					800								
		12,800											
		25,000											
		12,800											
		25,000											
		25,000											
		6,400											
		6,400											
		25,000											
	400	25,000				400						200	
				12,800						200			
				25,000									
				12,800									
				12,800							3,200		
				12,800							1,600		
				3,200									
				12,800				200					
				6,400									
				6,400									
				12,800									
				3,200									
				100									
				400									

Serogroup serotype		Strain	I. R.G.A.	II. Hebdomadis	III. Bangkinang I	IV. Van Tienen
XI. Sejröe (1938) *continued*						
(c) Wolffi						
1. medanensis	1948	Hond HC		3,200		
2. wolffi	1948	3705		100		
3. hardjo	1953	Hardjoprajitno				
4. recreo	1967	LT 957				
5. trinidad	1967	LT 1098		100		
6. gorgas	1967	LT 829				
7. roumanica		LM 294				
XII. Djasiman (1939)						
1. djasiman	1937	Djasiman				
2. sentot	1937	Sentot			800	
3. gurungi	1957	Gurung				
XIII. Cynopteri (1939)						
1. cynopteri	1939	3522 C				
XIV. Tarassovi (1941)						
1. tarassovi	1941	Mitis Johnson				400
2. bakeri	1957	LT 79				100
3. guidae	1957	RP 29				
4. atlantae	1957	LT 81				
5. kisuba	1956	Kisuba				100
6. bravo	1956	Bravo				
7. atchafalaja	1966	LSU 1013				
8. chagres	?	LT 924				400
9. rama	?	LT 955				400
10. gatuni	?	LT 839				
11. tunis	1969	P 2				
XV. Mini (1942)						
1. mini	1942	Sari		200		
2. szwajizak	1954	Szwajizak		400		
3. georgia	1960	LT 117		400		
4. perameles	1964	Bandicoot 343		100		
5. tabaquite	1972	TVRL 3214	100	200		
6. beye	1967	LT 844		800		
XVI. Ballum (1944)						
1. ballum	1944	Mus 127	800			
2. castellonis	1955	Castellon 3	800			
3. arboreae	1955	Arborea P-9719	400			
XVII. Celledoni (1954)						
1. celledoni	1954	Celledoni				
2. whitcombi	1957	Whitcomb				
3. anhoa		LT 90-68				
XVIII. Louisiana (1964)						
1. louisiana	1964	LSU 1945			200	
2. orleans	1964	LSU 2580			3,200	
3. lanka		LT 2567	400			
XIX. Panama (1966)						
1. panama	1966	CZ 214 K				
2. cristobali	?	LT 940				
3. pina	?	LT 932	100			
XX. Shermani (1966)						
1. shermani	1966	LT 821				

† This working table has no official status. The recorded titres are based on results of the Bratislava Leptospirosis Laboratory. The introduction of the serogroups Djasiman and Louisiana, the division of the groups into subgroups and the recorded reduced number of serotypes in the Pomona group are suggested by the same Laboratory.

‡ Serum titres with the homologous strains are 1:12,800 or 1:25,600.

	Reference serum												
VII. Vleermuis 90 C	VIII. Pomona	IX. Lora	X. Poi	XI. Hardjoprajitno	XII. Djasiman	XIII. 3522 C	XIV. Mitis Johnson	XV. TURL 3214	XVI. Mus 127	XVII. Celledoni	XVIII. LSU 1945	XIX. CZ 214 K	XX. LT 821
				3,200									
				3,200									
				12,800									
				6,400				100					
				6,400									
				6,400				100					
				3,200									
					12,800	100					400		
					1,600	1,600					400		
					12,800	100					1,600		
100						100	12,800				1,600		
							12,800						400
							25,000						1,600
							6,400						800
							12,800						400
							12,800						400
							6,400						12,800
							25,000						6,400
							6,400						3,200
							3,200						800
							12,800						1,600
				800				3,200					
				400				3,200					
				800				6,400					
				800				1,600					
				200				12,800					
				100				3,200					
									12,800				
									12,800				
									25,000				
									800				
			100										
			100							12,800			
			800							12,800			
										12,800			
			100		1,600	1,600				12,800			
			400		800	3,200				3,200			
					1,600	3,200				12,800			
		400										25,000	
		1,600										50,000	
		400										50,000	
											100	50,000	
													50,000

APPENDIX II

Agglutination test

The agglutination test can be performed with live or killed antigen. The killed antigen is generally prepared by adding neutralised formaldehyde at 0·05–0·5% (v/v) final concentration to live antigen. The use of formolised antigen has the advantages that the same antigen can be used for a longer period, and that contamination problems of the antigen are greatly reduced while it also helps to limit infections of laboratory personnel. Live antigen is more sensitive and often the titre is one step higher in comparison with formolised killed antigen. The use of live antigen also requires scrupulous working technique, which is indeed a basic requirement for a leptospirosis reference laboratory.

Live antigen is preferred in our laboratories. In general, a well growing fresh culture of six to seven days incubation at 30–32°C, containing a density of approximately 2×10^8 leptospira organisms/ml is used. In practice such cultures will be suitable for use for another week, when kept at room temperature.

The performance of the agglutination test in the Amsterdam and Bratislava laboratory is different. Therefore both tests will be described.

Split or interlocking scheme

In Amsterdam† the "split" or interlocking ten-fold dilution scheme, also referred to as the "Dutch" scheme is still in use. It was developed between 1927 and 1939 (Schüffner and Mochtar, 1927; Schüffner and Bohlander, 1939; Postmus and Schultsz, 1932) and is based on a droplet method, using Schüffner's pipettes and small porcelain plates (10·5–7·5 cm) holding four rows of six depressions or cups, each accommodating 0·5 ml of fluid. The final serum dilutions (including antigen) range from 1:10, 1:30, 1:100, 1:200, etc., up to 1:300,000. A detailed description of this method is given by Wolff (1954).

After the serum dilutions and antigen have been well mixed the suspensions are incubated for 2–4 h at 30°C. Thereafter the test is read by transferring some of the mixture of antiserum and antigen by way of a small platinum loop to a slide and examining it by darkfield microscopy at 160 (10 × 16)–200 (10 × 20) magnification.

Double dilution technique

The double dilution technique is used in the Bratislava laboratory starting with a first dilution of 1:100 (including antigen). Equal volumes of living antigen suspensions and diluted serum are mixed. After the mixture has been left to stand for about 2 h at room temperature (20–25°C), some of the mixture of antiserum and antigen is transferred by way of a pipette to a slide and the results are read by darkfield microscopy, using c. 100 × magnification, according to the 50% agglutination criterion. This method can be applied as a microlitre technique and is being automated.

Endpoint reaction

In both methods the endpoint is determined by the presence of 50% or just less than 50% free cells, compared with a control suspension with a 1 in 2 dilution in PBS. It is obvious that such judgement is a matter of subjective observation.

Automated dilution and reading techniques are being developed.

† Automated dilution and reading techniques are being developed.

APPENDIX III

Absorption test

The methods described below for the absorption tests used in Amsterdam and Bratislava differ on four essential points.

(a) *Formol treated—live antigen*

For laboratory safety a formol-treated absorbing antigen is used in the Dutch laboratory. The application of living antigen is given priority in the Czechoslovakian laboratory, as formalin might have some damaging effect on the antigen and antisera, and thus might have an adverse effect on the absorption test.

(b) *Titre of antisera*

The antisera used in the Amsterdam laboratory are diluted to a standard titre of 1:3000 prior to the absorption test. After absorption only weak reactions are allowed in the serum dilution 1:30 with the absorbing strain. However, in the Bratislava laboratory antisera with a titre of 1:12,800 or 1:25,600 are used and after absorption a remaining titre of 1:100 with the absorbing strain is preferred (Kmety, 1967) in order to prevent errors caused by non-specific absorption, when using too excessive amounts of antigen.

(c) *Mixing antigen–antisera*

In the past most laboratories mixed antiserum with antigen already present in the tube. This method is still practised in Holland with confidence. In 1970 Kmety *et al.* described the possible influence of the Danysz phenomenon, which led to a revision of the technique in adding a certain amount of antigen to a fixed volume of antiserum in three equal volumes at 10 min intervals.

(d) *Time and temperature necessary for absorption*

Under the test conditions prevalent in Bratislava a period usually of 90 min to overnight at room temperature is suggested to ensure the expected binding of antibodies to the corresponding antigen sites. In the Amsterdam laboratory the antiserum–antigen mixture is incubated overnight at 30°C, which has proved to give reliable results.

A detailed description of both absorption methods is now given starting with a description of a modified method first described by Wolff (1954).

The Dutch technique

1. A well-grown culture of leptospires is inoculated into four bottles each containing approximately 50 ml Modified Vervoort's medium or Ellinghausen's medium as modified by Johnson and Harris (1968). The bottles are incubated at approximately 30°C for one week.

2. After the incubation period the contents of three bottles containing well growing culture (min. 2×10^8 lept./ml) are mixed together and treated with formalin (0·5% final concentration of formalin). The formolised culture is allowed to stand for at least 30 min at room temperature (one bottle is kept for Section 11).

3. The 150 ml suspension of leptospires is divided into centrifuge tubes and centrifuged for 30 min at approximately $12,000 \times g$ (Sorvall SS3 high speed angle headed centrifuge).

4. The supernatant is discarded by careful decanting and the sediment assembled in one tube and centrifuged again at approximately $12,000 \times g$ for 30 min.

5. The supernatant fluid is discarded by decanting or careful pipetting.

6. The centrifuge tube with packed cells is placed upside down to dry on filter paper for 1–2 h.

7. A small amount of hyperimmune serum with a minimum titre of 1 : 10,000 chosen for the absorption test, is diluted to a *standard titre* of 1 : 3000 with the homologous strain. No more than about 0·25 ml of the diluted antisera is necessary for one test.

8. Of the diluted sera 0·1 ml is mixed with 0·9 ml phosphate buffered-formol-saline (0·5% formalin). This mixture is used to resuspend the sediment of leptospira cells. (The serum is diluted ten times.)

9. The mixture of cells and diluted serum is left to absorb for 16–24 h at 30°C and is thereafter centrifuged at $12,000 \times g$ for 30 min.

10. The supernatant fluid (absorbed sera) is carefully pipetted off and used for agglutination tests.

11. The absorbed serum is agglutinated with a living culture of the absorbing strain, to detect residual antibodies. If a strong titre is detected in the 1 : 30 dilution, further absorption is carried out with the 50 ml remaining culture (Section 3).

12. The agglutination of the absorbed sera is carried out in duplicate by two persons, using live and formolised antigen of the homologous strain. As the serum has been diluted ten-fold, the first dilution, according to the Dutch interlocking ten-fold dilution scheme, will start with the 1 : 30 final dilution. The duplicate tests should give the same titres. A one step lower titre is allowed for the reaction with formolised antigen compared with living antigen.

13. A control test with the diluted unabsorbed antisera (0·1 ml of the 0·15 ml remaining sera) and the homologous strain (living antigen and 0·5% formolised antigen) should be carried out in the same way as described under Section 12.

Bratislava technique

1. A 200–500 ml living antigen consisting of a 7–14 days old culture is used.

2. The antigen is centrifuged in a high speed angle headed centrifuge at $5000 \times g$ for 30 min.

3. The supernatant is discarded, and used to resuspend the packed cells to a density corresponding to McFarland Standard No. 10 (approximately 3–5 ml antigen). This corresponds to a concentration of about 40–80 times the original culture.

4. The antigen is added to the serum (not the serum to the antigen!) in three steps. If immune sera with a titre of c. 12,800 has to be absorbed, to one part (or 0·1 ml, or one drop) of serum, 24 parts (or 2·4 ml, or 24 drops) of antigen should be added in three approximately equal parts with an interval of c. 10 min each. The mixture is shaken each time after the antigen has been added. If immune sera with a lower titre have to be absorbed only 15 or 20 parts of antigen are added to one part of serum. The missing four to nine parts of antigen are now replaced by physiological saline to get a final serum dilution of 1 : 25. If the serum has to be absorbed with more antigens the scheme is modified as follows: to one part of serum, 24 parts of antigen 1 and 25 parts of antigen 2 are added. The final dilution is then 1 : 50. If smaller amounts of antigen 1 or 2 are used the remaining volume is replaced by saline to give a final dilution of 1 : 50.

5. The antigen–antiserum mixture should be allowed to stand at room temperature for 90 min or overnight.

6. The antigen is removed by centrifugation.

7. The absorbed antiserum is tested by the microscopic agglutination technique using living antigens.

8. The absorption test is considered to be optimally balanced if the absorbing strain still reacts in the first dilution ($\pm 1 : 100$) of the absorbed serum, since excessive amounts of antigen used for absorption may lead to non-specific reductions of the homologous titre.

Although the methods described above differ in several details, in most cases the typing results obtained are comparable.

Storage of absorbed sera

For short periods: one month at $+4\,^{\circ}\mathrm{C}$

For longer periods: at -70 or $-20\,^{\circ}\mathrm{C}$

APPENDIX IV

Preparation of rabbit hyperimmune serum

It is likely that in the near future typing of *Leptospira* strains will be based on rabbit immune serum containing antibodies only against thermostable agglutinogens. For the detection of heat-labile antibodies immune serum prepared with living cultures should be used. For immunisation, healthy rabbits weighing at least 2 kg but preferably 3–4 kg should be selected. They must be tested beforehand for the presence of leptospiral antibodies. It is essential that only rabbits with an entirely negative initial agglutination reaction are used.

Immunisation is carried out with a roughly seven days old well growing live or heat killed culture ($c.$ 2×10^8 leptospires/ml) injected intravenously in the ear vein. For classification of strains belonging to new serotypes the use of three different rabbit sera against such strains is recommended.

Immunisation scheme for the preparation of hyperimmune serum against thermostable antigens

A well-grown culture of leptospires is inactivated for 30 min at 56°C and used for immunisation.

1st dose, day 1, 4 ml inactivated culture,
2nd dose, day 6, 4 ml inactivated culture,
3rd dose, day 11, 4 ml living culture.

One week after the last inoculation a small blood sample is obtained from the ear vein and tested for titre against the homologous strain. If the titre is 1 : 10,000 (1 : 12,800) or higher, the animal is bled by heart puncture. When the titre is lower than 1 : 10,000 (1 : 12,800) a fourth injection of 4 ml living culture is given. About 10% of rabbits seem not to be able to produce sufficiently high antibody titres (1 : 10,000 (1 : 12,800) or higher). They are replaced by new rabbits.

Immunisation schemes for the preparation of hyperimmune serum against thermolabile antigen

A. The same immunisation schedule can be used as described above, provided that living cultures are used for the first two doses as well as for the third.

B. This scheme can replace the above mentioned schedule and is preferred when virulent strains are used for immunisation.

1st dose, day 1, 1 ml living culture,
2nd dose, day 6, 2 ml living culture,
3rd dose, day 11, 4 ml living culture,
4th dose, day 16, 6 ml living culture,
5th dose, day 21, 6 ml living culture.

One week after the last inoculation (day 28) a blood sample is taken from the ear vein and tested for titre against the homologous strain (see thermostable antigen).

Storage of serum

1. Preferably serum is dispensed in 2 ml amounts in sterile ampoules and freeze dried. The ampoules with freeze dried serum can be kept for years at +4°C.

2. The serum can be stored in small quantities (2 ml) at −70°C (or −20°C).

3. The serum can be treated with a preservative and stored at +4°C. The following preservatives can be used: merthiolate 0·02%, phenol 0·3% final concentrations, or the serum can be mixed with an equal volume of glycerin.

APPENDIX V

Schemes of antigenic factors within the serogroups, typing schemes and preparation of specific factor sera†

1. SEROGROUP PYROGENES

1. Scheme of antigenic structure†

Subgroup serotype	Reference strain	Antigens	
		Major	Minor
1. Pyrogenes			
pyrogenes	Salinem	Py-1, 2, 3, . . .	o1, o2, o3, o4, o5, o6, . . .
hamptoni	Hampton	−, 2, 4, . . .	o7, . . .
alexi	HS 616	−, 2, 5, . . .	o4, o6, o7, o8, . . .
robinsoni	Robinson	−, 2, −, . . .	o4, o5, o6, o9, o10, o11, . . .
kwalae‡	Julu	2	
2. Zanoni			
zanoni	Zanoni	Py-3, 7, 8, . . .	o2, o6, o10, o13, o14, . . .
abramis	Abraham	−, 7, 9, . . .	o2, o6, o15, o7, . . .
biggis	Biggs	−, 7, 8, 10, . . .	o1, o2, o6, o12, o15, o16, o7, o10, . . .
myocastoris	LSU 1551	3, 7, 8, 11, . . .	o3, o10, o16, o6, . . .
manilae‡	LT 398	7	

† Revised designation of antigens. Minor antigens common with serotypes of other serogroups are omitted.

‡ New serotypes not yet submitted to factor analysis.

2. Typing scheme

Subgroup serotype	Factor sera							
	Py-2	7	3	4	5	11	9	10
1. Pyrogenes								
pyrogenes	+	−	+	−	−			
hamptoni	+	−	−	+	−			
alexi	+	−	−	−	+			
robinsoni	+	−	−	−	−			
kwalae	+	−						
2. Zanoni								
zanoni	−	+	+			−		
abramis	−	+	−				+	−
biggis	−	+	−				−	+
myocastoris	−	+	+		+			
manilae	−	+						

† Of serogroups so far studied.

3. Preparation of factor sera

Factor serum	Immune serum	Absorbed with
Py-2, ...	Salinem	Zanoni + Biggs
Py-7, ...	Biggs	Salinem
Py-3, ...	Salinem	Biggs + Robinson
Py-4, ...	Hampton	HS 616
Py-5, ...	HS 616	Robinson
Py-9, ...	Abraham	Biggs
Py-10, ...	Biggs	Zanoni
Py-11, ...	LSU 1551	Zanoni

2. SEROGROUP GRIPPOTYPHOSA

1. Scheme of antigenic structure

Subgroup serotype	Reference strain	Antigens	
		Major	Minor
1. Grippotyphosa			
grippotyphosa	Moskva V	Gt-1, 2, ...	o1, o2, o3, o4, o7, o8, ...
valbuzzi	Valbuzzi	-1, 3, ...	o1, o2, o3, o5, ...
muelleri	RM 2	-1, 4, ...	o1, o4, ...
canalzonae	CZ 188 K	-1, -, ...	o1, o3, o6, o7, o9, ...
2. Vanderhoedeni			
vanderhoedeni	Kipod 179	Gt-5, ...	o1, o5, o8, o9, ...

2. Typing scheme

Subgroup serotype	Factor sera				
	Gt-1	2	3	4	5
1. Grippotyphosa					
grippotyphosa	+	+	−	−	−
valbuzzi	+	−	+	−	−
muelleri	+	−	−	+	−
cananalzonae	+	−	−	−	−
2. Vanderhoedeni					
vanderhoedeni	−				+

3. Preparation of factor sera

Factor serum	Immune serum	Absorbed with
Gt-1, ...	Moskva V	Kipod 179
Gt-2, ...	Moskva V†	Valbuzzi
Gt-3, ...	Valbuzzi	Moskva V + CZ 188 K + Kipod
Gt-4, ...	RM 2	Moskva V
Gt-5, ...	Kipod 179	Moskva V + Valbuzzi

† Better factor sera may be obtained from antiserum Nzirandukula.

3. SEROGROUP HEBDOMADIS

1. Scheme of antigenic structure

Subgroup serotype	Reference strain	Antigens Major	Antigens Minor
1. Hebdomadis			
hebdomadis	Hebdomadis	Hb-1, 4, 2, 3, ...	Hb-o1, o2, o3, o4, Hb/Sj-o30, o31, Hb/Mn-o32, o33, o35, ...
kambale	Kabale	-1, 4, -, -, 14, ...	Hb-o3, o8, o11, o12, o6, ...
nona	Nona	-1, 4, 2, 3, 15, ...	Hb-o9, o13, Hb/Sj-o37, ...
maru	CZ 285	-1, 4, -, -, 16, ...	Hb-o2, o5, o7, o12, o13, ...
2. Borincana			
borincana	HS-622	Hb-1, 5, -, 6, 7, ...	Hb-o1, o6, o7, o5, Hb/Sj-o34, ...
worsfoldi	Worsfold	-1, 5, 3, -, -, 8, 9, 10, ...	Hb-o8, Hb/Mn-o35, o36, ...
jules	Jules	-1, 5, 3, -, -, 8, 9, ...	Hb-o8, o9, o10, Hb/Mn-o36, o35, ...
kabura	Kabura	-1, 5, -, 6, -, -, -, Hb/Sj-12, ...	Hb-o1, o5, o11, Hb/Sj-o34, o38, ...
kremastos	Kremastos	-1, 5, 3, -, -, 8, -, -, -, 13, ...	Hb-o1, Hb/Sj-o31, o30, Hb/Mn-o33, o36, o32, o35, ...

2. Typing scheme

Subgroup serotype	Factor sera										
	Hb-4	5	2	3	6	7	9	10	14	15	16
1. Hebdomadis											
hebdomadis	+	−	+						−	−	−
kambale	+	−	−						+	−	−
nona	+	−	+						−	+	−
maru	+	−	−						−	−	+
2. Borincana											
borincana	−	+		−	+	+					
worsfoldi	−	+		+	−	~	+	+			
jules	−	+		+	−	−	+	−			
kabura	−	+		−	+	−	−	−			
kremastos	−	+		+	−	−	−	−			

3. Preparation of factor sera

Factor serum	Immune serum	Absorbed with
Hb-4, . . .	Hebdomadis	Kremastos
-5, . . .	Kremastos	Hebdomadis
-2, . . .	Hebdomadis	Kremastos, Kambale and CZ-285
-6, . . .	HS 622	Jules and CZ-285
-8, . . .	Worsfold	Hebdomadis and Kabura
-9, . . .	Worsfold	Kremastos
-10, . . .	Worsfold	Jules
-11, . . .	Kabura	HS 622 and Kambale
-14, . . .	Kambale	Kabura, Nona and CZ-285
-15, . . .	Nona	Hebdomadis and Kabura
-16, . . .	CZ-285	Hebdomadis, Kambale and Nona

4. SEROGROUP CANICOLA

1. Scheme of antigenic structure†

Subgroup serotype	Reference strain	Antigens	
		Major	Minor
1. Canicola			
canicola	Utrecht IV	Ca-1, 2, 3, ...	o1, o8, o14, o20, o23, o34, o40, o43, ...
bafani	Bafani	-1, -, 4, 5, 6, ...	o1, o2, o4, o7, o9, o14, o30, o33, o40, o43, ...
kamituga	Kamituga	-1, -, 7, ...	o1, o12, o15, o20, o3, o30, o41, ...
jonsis	Jones	-1, -, 5, 8, ...	o1, o4, o10, o13, o17, o19, o21, o22, o23, o32, o37, o40, o43, ...
sumneri	Sumner	-1, 2, 6, 9, ...	o1, o2, o4, o5, o9, o13, o16, o24, o3, o7, o21, o31, o35, o40, o42, o44, ...
bindjei	Bindjei	-1, 2, 8, 9, ...	o1, o4, o8, o10, o13, o19, o25, o26, o18, o29, o37, o39, o40, o42, o43, ...
broomi	Patane	-1, 2, 3, 8, 11, ...	o1, o2, o4, o10, o26, o27, o29, o30, o31, o33, o43, ...
2. Schueffneri			
schueffneri	Vleermuis 90 C	Ca-12, 13, ...	o1, o2, o9, o15, o27, o34, o35, o36, o37, o38, o39, ...
benjamini	Benjamin	-12, -, 14, ...	o1, o28, o34, o35, o36, o40, o41, o42, o43, o44, o37, ...
malaya	H 6	-12, -, 14, 15, ...	o1, o2, o11, o14, o24, o34, o35, o38, o08, o29, o37, o43, ...
galtoni‡	LT 1014	-12, ...	

† Revised designation of antigens. Minor antigens common with serotypes of other serogroups are omitted.
‡ New serotype not yet submitted to factor analysis.

2. Typing scheme

Subgroup serotype	Factor sera							
	3	8	9	6	7	12	14	15
1. Canicola								
canicola	+	−				−		
bafani	−	−	−	+	−	−		
kamituga	−	−	−	−	+	−		
jonsis	−	+	−			−		
sumneri	−	−	+			−		
bindjei	−	+	+			−		
broomi	+	+				−		
2. Schueffneri†								
schueffneri						+	−	−
benjamini						+	+	−
malaya						+	+	+

† Strains typed to be different from all the three serotypes also have to be compared with serotype galtoni.

3. Preparation of factor sera

Factor serum	Immune serum	Absorbed with
Ca-3, ...	Utrecht IV	Sumner + Bindjei
Ca-8, ...	Bindjei	Utrecht IV + Sumner
Ca-9, ...	Bindjei	Utrecht IV + Jones + Bafani + Patane
Ca-6, ...	Sumner	Utrecht IV + Kamituga + Bindjei
Ca-7, ...	Kamituga	Utrecht IV + Bafani + Jones + Sumner
Ca-12, ...	Benjamin	Utrecht IV + Kamituga + Sumner
Ca-14, ...	Benjamin	Vleermuis 90 C + Kamituga + Sumner
Ca-15, ...	H 6	Vleermuis 90 C + Benjamin + Bafani + Sumner

5. SEROGROUP AUSTRALIS

1. Scheme of antigenic structure†

Subgroup serotype	Reference strain	Antigens	
		Major	Minor
1. Australis			
australis	Ballico	As-1, 2, . . .	o1, o2, o3, o4, o5, o6, . . .
lora	Lora	-1, 2, 3, . . .	o2, o6, o7, o8, . . .
fugis	Fudge	-, -, 2, 4, 5, . . .	o9, o13, . . .
peruviana	LT 941	-1, 2, 6, . . .	o3, o9, o10, o11, . . .
nicaragua	LT 990	-1, 2, 7, . . .	o3, o10, o12, o13, . . .
ramisi‡	Musa		
2. Jalna			
muencheni	München C 90	As-1, -, 8, 9, 10, 11, . . .	o5, o8, . . .
jalna	Jalna	-1, 3, 8, 11, 12, 13, . . .	o6, o8, . . .
bratislava	Jež Bratislava	-, -, 3, 8, 5, 9, 12, 14, . . .	o11, o12, . . .

† Revised designation of antigens. Minor antigens common with serotypes of other serogroups are omitted.

‡ New serotype not yet submitted to analysis.

2. Typing scheme

Subgroup serotype	Factor sera							
	As-2	8	3	4	6	7	3	14
1. Australis								
australis	+	−	−	−	−	−		
lora	+	−	+	−	−	−		
fugis	+	−	−	+	−	−		
peruviana	+	−	−	−	+	−		
nicaragua	+	−	−	−	−	+		
2. Jalna								
muencheni	−	+					−	−
jalna	−	+					+	−
bratislava	−	+					+	+

3. Preparation of factor sera

Factor serum	Immune serum	Absorbed with
As-2, . . .	Lora	Jalna
As-8, . . .	Jalna	Lora
As-3, . . .	Lora	Ballico
As-4, . . .	Fudge	Ballico + Jež Bratislava
As-6, . . .	LT 941	Ballico + LT 990 + Jež Bratislava
As-7, . . .	LT 990	LT 941 + Jež Bratislava
As-14, . . .	Jež Bratislava	München C 90 + Jalna

6. SEROGROUP SEJROE

1. Scheme of antigenic structure†

Subgroup serotype	Reference strain	Antigens	
		Major	Minor
1. Sejroe			
sejroe	M 84	Sj-1, 2, 3, –, 5, 6, 7, ...	o1, o2, o3, o4, o5, o8, o9, o12, o19, o24, ...
balcanica	1627 Burgas	–, –, 3, –, 5, 6, –, ...	o1, o2, o3, o4, o5, ...
polonica	493 Poland	–, 2, –, 5, –, 7, ...	o1, o2, o6, o7, o8, o9, o12, o18, o19, o24, ...
istrica	Bratislava	–1, 2, 3, 4, 5, 6, 7, ...	o1, o6, o2, o3, o4, o5, o8, o9, o12, o19, o24, ...
2. Saxkoebing			
saxkoebing	Mus 24	Sj-7, 21, 22, Hb/Sj-12, ...	o26, o21, ...
haemolytica	Marsh	–, 21, –, 23, 24, ...	o20, o21, o22, o23, o24, o25, ...
ricardi	Richardson	–, 21, –, –, –, 25, ...	o20, o21, o23, o24, o25, o26, ...
dikkeni‡	Mannuthi	– 21	
nyanza‡	Kibos	– 21	
3. Wolffi			
wolffi	Hond HC	Sj-5, 6, 8, 9, 10, ...	o2, o3, o4, o5, o10, o11, o13, o14, ...
medanensis	3705	–5, –, 8, –, 10, 11, ...	o2, o3, o4, o5, o8, o10, o14, o15, ...
hardjo	Hardjoprajitno	–5, –, 8, 13, ...	o1, o2, o3, o4, o5, o8, o9, o10, o14, o15, o16, o17, o18, ...
recreo	LT 957	–5, –, 8, 14, 15, ...	o2, o3, o4, o8, o13, o14, o15, o17, o19, o25, o26, ...
trinidad	LT 1098	–5, –, 8, 16, ...	o2, o3, o5, o8, o11, o14, o15, o16, ...
gorgas	LT 829	–5, –, 8, 14, 17, 18, 23, ...	o2, o3, o8, o10, o14, o15, o20, ...
roumanica	LM 294	–, –, –, 17, 19, ...	o2, o14, o20, o21, o22, ...

† Antigens common with serotypes of other serogroups are omitted.
‡ New serotypes not yet submitted to factor analysis.

2. Typing scheme†

Subgroup serotype	Factor sera														
	Sj-2, 3	21	2	3	4	22	24	25	10	9	16	13	14, 15	18	19
1. Sejroe															
sejroe	+	−	+	+	−										
balcanica	+	−	−	+	−										
polonica	+	−	+	−	−										
istrica	+	−	+	+	+										
2. Saxkoebing															
saxkoebing	−	+				+	−	−							
haemolytica	−	+				−	+	−							
ricardi	−	+				−	−	+							
3. Wolffi															
medanensis	−	−							+	+	−				
wolffi	−	−							+	−	−				
hardjo	−	−							−	−	−	+			
recreo	−	−							−	−	−	−	+	−	−
trinidad	−	−							+	−	+				
gorgas	−	−							−	−	−		+	+	−
roumanica	−	−							−	−	−		−	−	+

† Strains found to be different from the three Saxkoebing subgroup serotypes should be also compared with the strains Mannuthi and Kibos.

3. Preparation of factor sera

Factor serum	Immune serum	Absorbed with
Sj-2, 3, ...	M 84	Hond HC or 3705
Sj-21, ...	Mus 24	394 Poland
Sj-2, ...	493 Poland	Hardjo + Mus 24
Sj-3, ...	1627 Burgas	Hond HC + Hardjoprajitno
Sj-4, ...	Bratislava	M 84 + 493 Poland
Sj-22, ...	Mus 24	Richardson + 493 Poland
Sj-24, ...	Marsh	Richardson + LT 829
Sj-25, ...	Richardson	Mus 24 + Marsh
Sj-10, ...	Hond HC	1627 Burgas + LT 957 + LT 827
Sj-9, ...	Hond HC	M 84 + 3705 + LT 1098
Sj-16, ...	LT 1098	Hardjoprajitno + 3705
Sj-13, ...	Hardjoprajitno	3705 + LT 829 + 493 Poland
Sj-14, 15, ...	LT 957	M 84 + Hardjoprajitno
Sj-18, ...	LT 829	LT 957 + LM 294
Sj-19, ...	LM 294	LT 829 + Mus 24

7. SEROGROUP MINI

1. Scheme of antigenic structure†

Serotype	Reference strain	Antigens	
		Major	Minor
mini	Sari	Mn-1, 2, 3, 4, ...	o1, o2, o3, ...
szwajizak	Szwajizak	-1, -, 3, 4, ...	o1, o2, o3, o4, ...
georgia	LT 117	-1, 4, 5, 6, ...	o1, o2, o4, ...
perameles	Bandicoot 343	-, 7, 8, ...	o1, o5, ...
tabaquite	TVRL 3214	-1, 5, 7, 9, ...	o1, o4, o5, ...
beye	LT 844	-1, 3, 4, 10, ...	o1, o4, o6, ...

† Minor antigens common with serotypes of other serogroups are omitted.

2. Typing scheme

Serotype	Factor sera				
	Mn-4	2	5	10	7
mini	+	+	−	−	−
szwajizak	+	−	−	−	−
georgia	+	−	+	−	−
perameles	−	−	−	−	+
tabaquite	−	−	+	−	+
beye	+	−	−	+	−

3. Preparation of factor sera

Factor serum	Immune serum	Absorbed with
Mn-4, . . .	LT 117	TVRL 3214
Mn-2, . . .	Sari	Szwajizak
Mn-5, . . .	LT 117	Szwajizak + LT 844
Mn-7, . . .	Bandicoot 343	LT 117
Mn-10, . . .	LT 844	Szwajizak + LT 117

8. SEROGROUP BALLUM

1. Scheme of antigenic structure†

Serotype	Reference strain	Antigens	
		Major	Minor
ballum	Mus 127	Bl-1, –, –, . . .	o1, o2, o3, . . .
catellonis	Castellon 3	-1, 2, 3, . . .	o3, . . .
arborea	Arborea	-1, –, 3, . . .	o2, o4, . . .

† Revised designation of antigens. Minor antigens common with serotypes of other serogroups are omitted.

2. Typing scheme

Serotype	Factor sera	
	Bl-2	3
ballum	–	–
castellonis	+	+
arborea	–	+

3. Preparation of factor sera

Factor serum	Immune serum	Absorbed with
Bl-2, . . .	Castellon 3	Arborea + Mus 127
Bl-3, . . .	Arborea	Mus 127

9. SEROGROUP CELLEDONI

1. Scheme of antigenic structure†

Serotype	Reference strain	Antigens	
		Major	Minor
celledoni	Celledoni	Ce-1, –, Jv/Ce-8, . . .	o1, . . .
whitcombi	Whitcomb	Ce-1, 2, Jv/Ce-8, . . .	
anhoa‡	LT 90–68	Ce-1, –, 3, . . .	

† Revised designation of antigens. Minor antigens common with serotypes of other serogroups are omitted.

‡ New serotype.

2. Typing scheme

Serotype	Factor sera	
	Ce-2	3
celledoni	–	–
whitcombi	+	–
anhoa	–	+

3. Preparation of factor sera

Factor serum	Immune serum	Absorbed with
Ce-2, . . .	Whitcomb	Celledoni
Ce-3, . . .	LT 90–68	Celledoni

10. SEROGROUP JAVANICA

1. Scheme of antigenic structure†

Serotype	Reference strain	Antigens	
		Major	Minor
javanica	Veldrat Bataviae 46	Jv-1, 2, 3, . . .	o1, o2, o3, o4, o5, . . .
poi	Poi	-1, 2, 3, 4, 5, . . .	o2, o6, o7, o8, . . .
sorex-jalna	Sorex Jalná	-1, –, 3, 4, 5, 6, . . .	o3, o7, o9, o10, . . .
coxi	Cox	-1, 2, 7, Jv/Ce-8, . . .	o1, o6, o7, o10, o11, . . .
sofia	Sofia 874	-1, –, 3, 5, 6, . . .	o3, o7, o11, o12, . . .
ceylonica	LT 1009‡		
dikkeni	Bandicoot XIV‡		

† Revised designation of antigens. Minor antigens common with serotypes of other serogroups are omitted.

‡ New serotypes not yet submitted to analysis.

2. Typing scheme

Serotype	Factor sera		
	Jv-4	2	3
javanica	−	+	+
poi	+	+	+
sorex-jalna	+	−	+
coxi	−	+	−
sofia	−	−	+

3. Preparation of factor sera

Factor serum	Immune serum	Absorbed with
Jv-4, . . .	Sorex Jalná	Sofia 874
Jv-2, . . .	Poi	Sorex Jalná
Jv-3, . . .	Poi or Sorex Jalná	Cox

Subject Index

A

Aerobacter, 226
Arizona, 53
A. hinshawii, 240

B

Bacterium rubidaeum, 66
Borrelia, 260 *et seq.*

C

Citrobacter, 6, 71, 186
  serotyping of, 51 *et seq.*
C. diversus, 51
C. freundii, 33, 51
C. intermedius, 51
C. koseri, 41, 51
Clostridium perfringens, 224
colicin typing, epidemiological value of, 84
Corynebacterium diptheriae, 224

E

Edwardsiella, 6
  serotyping of, 50 *et seq.*
E. tarda, 50
Enterobacter, 6, 11, 186
  serotyping of, 64 *et seq.*
E. aerogenes, 64
E. agglomerans, 244
E. cloacae, 64
E. liquefaciens, 66
Enterobacteria, bacteriocin typing of, 85
Enterobacteriaceae, 186, 194, 223, 240
  agglutination of, 12
  antigenic schemes for, 40
  antigens, cross-reactions of, 71
  antisera for, 35
    absorbed, 38
    H, 37
    O, 36
    OK, 37

polyvalent, 38
production of, 35
bacteriocin typing of, 79–85
biochemical characters of, 6
capsular polysaccharides of, 10, 11
chemistry of, 5
"common" antigens of, 29
fimbrial antigens of, 35
H antigens of, 27, 34
  H$^+$ to H$^-$ variation, 34
  phase variation, 34
haemagglutination of, indirect, 16
immunofluorescence of, 17
K antigens of, 2–77
  determination of, 25
  plasmid determined, 33
  polysaccharide, 32
  terminology, 25
lipopolysaccharide (LPS), 5
M antigen of, 35
morphology of, 4
O antigens of, 20, 31
  form variation of, 31
  phage induced changes in, 31
  smooth (S) to rough (R) variation, 31
  S–T–R variation, 31
O specific polysaccharide, structure of, 10
serological techniques for, 12
serotypes, definition of, 39
serotyping of, 1–77
  gel precipitation for, 17
subdivision of, 3
surface antigens of, other, 27
surface structures of, 4
Erwinia, serotyping of, 70 *et seq.*
E. cypripedii, 244
E. herbicola, 244
E. rhapontici, 244
Erwinieae, 4
Escherichia, 6, 8, 10, 11, 26, 27, 40, 41
  A antigens of, 23
E. coli, 5, 8, 11, 13, 15, 16, 28, 31, 32, 39, 40, 71, 80, 82, 85, 88, 186, 226, 246